**W. John Lee** is professor of petroleum engineering at Texas A&M U. After receiving a PhD degree from Georgia Inst. of Technology in 1963, he worked as a senior research specialist with Exxon Production Research Co. until 1968. He was associate professor of petroleum engineering at Mississippi State U. from 1968 to 1971 and technical adviser with Exxon Co. U.S.A. from 1971 to 1977. Lee has been with Texas A&M since 1977. He has been faculty adviser to the SPE student chapter during several school years.

## Acknowledgments

Many individuals have contributed to the completion of this textbook. The contributions of the following individuals have been particularly helpful: William D. McCain, Jr. who urged me to write the book and who, having survived a similar experience, urged me to persist; members of the SPE Textbook Committee (particularly Weldon Winsauer, Robert C. Earlougher Jr., and James T. Smith), who provided thorough and helpful reviews of early drafts; Georgeann Bilich, who effectively dealt with manuscript problems no author has any right to inflict on a technical editor, and who did so in her characteristic gracious way; and Erin Stewart, SPE Associate Editor, who successfully put it all together despite (at times) an almost total lack of cooperation from anyone. To all of you and the many others who helped—thank you!

**Publication of this first SPE Textbook was funded in part by the**

**SPE**

**SPE Foundation**

# Well Testing

**John Lee**
*Professor of Petroleum Engineering*
Texas A&M University

**First Printing**

**Society of Petroleum Engineers of AIME**
**New York 1982 Dallas**

**ISBN 0-89520-317-0**

## SPE Textbook Series

The Textbook Series of the Society of Petroleum Engineers of AIME was established in 1972 by action of the SPE Board of Directors. The Series is intended to ensure availability of high-quality textbooks for use in undergraduate courses in areas clearly identified as being within the petroleum engineering field. The work is directed by the Society's Textbook Committee, one of more than 40 Society-wide standing committees, through a member designated as Textbook Coordinator. The Textbook Coordinator and the Textbook Committee provide technical evaluation of the book. Below is a listing of those who have been most closely involved in the final preparation of this book. Many others contributed as Textbook Committee members or others involved with the book.

## Textbook Coordinators

Weldon Winsauer, Exxon Co. U.S.A., Houston
James T. Smith, Texas Tech. U., Lubbock

## Textbook Committee (1981)

Weldon Winsauer, chairman, Exxon Co. U.S.A., Houston
James T. Smith, Texas Tech. U., Lubbock
John L. Moore, Consolidated Natural Gas Services, Pittsburgh
H.M. Staggs, ARCO Inc., Dallas
Medhat Kamal, Amoco Production Co., Tulsa
Jack Evers, U. of Wyoming, Laramie

# Contents

# Introduction

This textbook explains how to use well pressures and flow rates to evaluate the formation surrounding a tested well. Basic to this discussion is an understanding of the theory of fluid flow in porous media and of pressure-volume-temperature (PVT) relations for fluid systems of practical interest. This book contains a review of these fundamental concepts, largely in summary form.

One major purpose of well testing is to determine the ability of a formation to produce reservoir fluids. Further, it is important to determine the underlying reason for a well's productivity. A properly designed, executed, and analyzed well test usually can provide information about formation permeability, extent of wellbore damage or stimulation, reservoir pressure, and (perhaps) reservoir boundaries and heterogeneities.

The basic test method is to create a pressure drawdown in the wellbore; this causes formation fluids to enter the wellbore. If we measure the flow rate and the pressure in the wellbore during production or the pressure during a shut-in period following production, we usually will have sufficient information to characterize the tested well.

This book begins with a discussion of basic equations that describe the unsteady-state flow of fluids in porous media. It then moves into discussions of pressure buildup tests; pressure drawdown tests; other flow tests; type-curve analysis; gas well tests; interference and pulse tests; and drillstem and wireline formation tests. Fundamental principles are emphasized in this discussion, and little effort is made to bring the intended audience – undergraduate petroleum engineering students – to the frontiers of the subject. This role is filled much better by other publications, such as the Society of Petroleum Engineers' monographs on well testing[1,2] and Alberta Energy Resources and Conservation Board's gas well testing manual.[3]

Basic equations and examples use engineering units. However, to smooth the expected transition to the Intl. System of Units (SI) in the petroleum industry, Appendix F discusses this unit system and restates major equations in SI units. In addition, answers to examples worked out in the text are given in SI units in Appendix F.

## References


1. Matthews, C.S. and Russell, D.G.: *Pressure Buildup and Flow Tests in Wells,* Monograph Series, SPE, Dallas (1967) **1.**
2. Earlougher, R.C. Jr.: *Advances in Well Test Analysis,* Monograph Series, SPE, Dallas (1977) **5.**
3. *Theory and Practice of the Testing of Gas Wells,* third edition, Pub. ECRB-75-34, Energy Resources and Conservation Board, Calgary, Alta. (1975).

# Chapter 1
# Fluid Flow in Porous Media

## 1.1 Introduction

In this initial chapter on fluid flow in porous media, we begin with a discussion of the differential equations that are used most often to model unsteady-state flow. Simple statements of these equations are provided in the text; the more tedious mathematical details are given in Appendix A for the instructor or student who wishes to develop greater understanding. The equations are followed by a discussion of some of the most useful solutions to these equations, with emphasis on the exponential-integral solution describing radial, unsteady-state flow. An appended discussion (Appendix B) of dimensionless variables may be useful to some readers at this point.

The chapter concludes with a discussion of the radius-of-investigation concept and of the principle of superposition. Superposition, illustrated in multiwell infinite reservoirs, is used to simulate simple reservoir boundaries and to simulate variable rate production histories. An approximate alternative to superposition, Horner's "pseudoproduction time," completes this discussion.

## 1.2 The Ideal Reservoir Model

To develop analysis and design techniques for well testing, we first must make several simplifying assumptions about the well and reservoir that we are modeling. We naturally make no more simplifying assumptions than are absolutely necessary to obtain simple, useful solutions to equations describing our situation – but we obviously can make no fewer assumptions. These assumptions are introduced as needed, to combine (1) the law of conservation of mass, (2) Darcy's law, and (3) equations of state to achieve our objectives. This work is only outlined in this chapter; detail is provided in Appendix A and the References.

Consider radial flow toward a well in a circular reservoir. If we combine the law of conservation of mass and Darcy's law for the isothermal flow of fluids of small and constant compressibility (a highly satisfactory model for single-phase flow of reservoir oil), we obtain a partial differential equation that simplifies to

$$\frac{\partial^2 p}{\partial r^2} + \frac{1}{r}\frac{\partial p}{\partial r} = \frac{\phi \mu c}{0.000264\, k}\frac{\partial p}{\partial t}, \quad \text{.......... (1.1)}$$

if we assume that compressibility, $c$, is small and independent of pressure; permeability, $k$, is constant and isotropic; viscosity, $\mu$, is independent of pressure; porosity, $\phi$, is constant; and that certain terms in the basic differential equation (involving pressure gradients squared) are negligible. This equation is called the *diffusivity equation*; the term $0.000264k/\phi\mu c$ is called the *hydraulic diffusivity* and frequently is given the symbol $\eta$.

Eq. 1.1 is written in terms of field units. Pressure, $p$, is in pounds per square inch (psi); distance, $r$, is in feet; porosity, $\phi$, is a fraction; viscosity, $\mu$, is in centipoise; compressibility, $c$, is in volume per volume per psi $[c=(1/\rho)(d\rho/dp)]$; permeability, $k$, is in millidarcies; time, $t$, is in hours; and hydraulic diffusivity, $\eta$, has units of square feet per hour.

A similar equation can be developed for the radial flow of a nonideal gas:

$$\frac{1}{r}\frac{\partial}{\partial r}\left(\frac{p}{\mu z}\, r\, \frac{\partial p}{\partial r}\right) = \frac{\phi}{0.000264\, k}\frac{\partial}{\partial t}\left(\frac{p}{z}\right), \quad \text{.... (1.2)}$$

where $z$ is the gas-law deviation factor.

For simultaneous flow of oil, gas, and water,

$$\frac{1}{r}\frac{\partial}{\partial r}\left(r\frac{\partial p}{\partial r}\right) = \frac{\phi c_t}{0.000264\, \lambda_t}\frac{\partial p}{\partial t}, \quad \text{.......... (1.3)}$$

where $c_t$ is the total system compressibility,

$$c_t = S_o c_o + S_w c_w + S_g c_g + c_f, \quad \text{........... (1.4)}$$

and the total mobility $\lambda_t$ is the sum of the mobilities of the individual phases:

$$\lambda_t = \left(\frac{k_o}{\mu_o} + \frac{k_g}{\mu_g} + \frac{k_w}{\mu_w}\right). \quad \text{................ (1.5)}$$

In Eq. 1.4, $S_o$ refers to oil-phase saturation, $c_o$ to oil-phase compressibility, $S_w$ and $c_w$ to water phase, $S_g$ and $c_g$ to gas phase; and $c_f$ is the formation compressibility. In Eq. 1.5, $k_o$ is the effective permeability to oil in the presence of the other phases, and $\mu_o$ is the oil viscosity; $k_g$ and $\mu_g$ refer to the gas phase; and $k_w$ and $\mu_w$ refer to the water phase. Because the formation is considered compressible (i.e., pore volume decreases as pressure decreases), porosity is not a constant in Eq. 1.3 as it was assumed to be in Eqs. 1.1 and 1.2.

## 1.3 Solutions to Diffusivity Equation

This section deals with useful solutions to the diffusivity equation (Section 1.2) describing the flow of a slightly compressible liquid in a porous medium. We also have some comments on solutions to Eqs. 1.2 and 1.3.

There are four solutions to Eq. 1.1 that are particularly useful in well testing: the solution for a bounded cylindrical reservoir; the solution for an infinite reservoir with a well considered to be a line source with zero wellbore radius; the pseudosteady-state solution; and the solution that includes wellbore storage for a well in an infinite reservoir. Before we discuss these solutions, however, we should summarize the assumptions that were necessary to develop Eq. 1.1: homogeneous and isotropic porous medium of uniform thickness; pressure-independent rock and fluid properties; small pressure gradients; radial flow; applicability of Darcy's law (sometimes called laminar flow); and negligible gravity forces. We will introduce further assumptions to obtain solutions to Eq. 1.1.

**Bounded Cylindrical Reservoir**

Solution of Eq. 1.1 requires that we specify two boundary conditions and an initial condition. A realistic and practical solution is obtained if we assume that (1) a well produces at constant rate, $qB$, into the wellbore ($q$ refers to flow rate in STB/D at surface conditions, and $B$ is the formation volume factor in RB/STB); (2) the well, with wellbore radius $r_w$, is centered in a cylindrical reservoir of radius $r_e$, and that there is no flow across this outer boundary; and (3) before production begins, the reservoir is at uniform pressure, $p_i$. The most useful form of the desired solution relates flowing pressure, $p_{wf}$, at the sandface to time and to reservoir rock and fluid properties. The solution is[1]

$$p_{wf}=p_i-141.2\frac{qB\mu}{kh}\left\{\frac{2t_D}{r_{eD}^2}+\ln r_{eD}-\frac{3}{4}+2\sum_{n=1}^{\infty}\frac{e^{-\alpha_n^2 t_D}J_1^2(\alpha_n r_{eD})}{\alpha_n^2\,[J_1^2(\alpha_n r_{eD})-J_1^2(\alpha_n)]}\right\}, \quad \ldots (1.6)$$

where, for efficiency and convenience, we have introduced the dimensionless variables

$$r_{eD}=r_e/r_w$$

and

$$t_D=0.000264\, kt/\phi\mu c_t r_w^2,$$

where the $\alpha_n$ are the roots of

$$J_1(\alpha_n r_{eD})\,Y_1(\alpha_n)-J_1(\alpha_n)\,Y_1(\alpha_n r_{eD})=0;$$

and where $J_1$ and $Y_1$ are Bessel functions. (Total compressibility, $c_t$, is used in all equations in this chapter because even formations that produce a single-phase oil contain an immobile water phase and have formation compressibility.)

The reader unfamiliar with Bessel functions should not be alarmed at this equation. It will not be necessary to use Eq. 1.6 in its complete form to calculate numerical values of $p_{wf}$; instead, we will use limiting forms of the solution in most computations. The most important fact about Eq. 1.6 is that, under the assumptions made in its development, it is an *exact solution* to Eq. 1.1. It sometimes is called the van Everdingen-Hurst constant-terminal-rate solution.[2] Appendix C discusses this solution more completely. Because it is exact, it serves as a standard with which we may compare more useful (but more approximate) solutions. One such approximate solution follows.

**Infinite Cylindrical Reservoir With Line-Source Well**

Assume that (1) a well produces at a constant rate, $qB$; (2) the well has zero radius; (3) the reservoir is at uniform pressure, $p_i$, before production begins; and (4) the well drains an infinite area (i.e., that $p \rightarrow p_i$ as $r \rightarrow \infty$). Under those conditions, the solution to Eq. 1.1 is

$$p=p_i+70.6\frac{qB\mu}{kh}\,Ei\left(\frac{-948\,\phi\mu c_t r^2}{kt}\right), \quad \ldots\ldots (1.7)$$

where the new symbols are $p$, the pressure (psi) at distance $r$ (feet) from the well at time $t$ (hours), and

$$Ei(-x)=-\int_x^{\infty}\frac{e^{-u}}{u}du,$$

the $Ei$ function or exponential integral.

Before we examine the properties and implications of Eq. 1.7, we must answer a logical question: Since Eq. 1.6 is an exact solution and Since Eq. 1.7 clearly is based on idealized boundary conditions, when (if ever) are pressures calculated at radius $r_w$ from Eq. 1.7 satisfactory approximations to pressures calculated from Eq. 1.6? Analysis of these solutions shows[3] that the $Ei$-function solution is an accurate approximation to the more exact solution for time $3.79\times10^5\ \phi\mu c_t r_w^2/k<t<948\ \phi\mu c_t r_e^2/k$. For times less than $3.79\times10^5\ \phi\mu c_t r_w^2/k$, the assumption of zero well size (i.e., assuming the well to be a line source or sink) limits the accuracy of the equation; at times greater than $948\ \phi\mu c_t r_e^2/k$, the reservoir's boundaries begin to affect the pressure distribution in the reservoir, so that the reservoir is no longer infinite acting.

A further simplification of the solution to the flow equation is possible: for $x<0.02$, $Ei(-x)$ can be approximated with an error less than 0.6% by

$$Ei(-x)=\ln(1.781x). \quad \ldots\ldots\ldots\ldots\ldots\ldots (1.8)$$

To evaluate the $Ei$ function, we can use Table 1.1 for $0.02<x\leq10.9$. For $x\leq0.02$, we use Eq. 1.8; and for $x>10.9$, $Ei(-x)$ can be considered zero for applications in well testing.

In practice, we find that most wells have reduced permeability (damage) near the wellbore resulting from drilling or completion operations. Many other wells are stimulated by acidization or hydraulic fracturing. Eq. 1.7 fails to model such wells properly; its derivation holds the explicit assumption of uniform permeability throughout the drainage area of the well up to the wellbore. Hawkins[4] pointed out that if the damaged or stimulated zone is considered equivalent to an altered zone of uniform permeability ($k_s$) and outer radius ($r_s$), the *additional* pressure drop across this zone ($\Delta p_s$) can be modeled by the steady-state radial flow equation (see Fig. 1.1). Thus,

$$\Delta p_s = 141.2 \frac{qB\mu}{k_s h} \ln(r_s/r_w)$$

$$- 141.2 \frac{qB\mu}{kh} \ln(r_s/r_w)$$

$$= 141.2 \frac{qB\mu}{kh} \left(\frac{k}{k_s} - 1\right) \ln(r_s/r_w). \quad \ldots . (1.9)$$

**TABLE 1.1* – VALUES OF THE EXPONENTIAL INTEGRAL, – *Ei* (– *x*)**

– *Ei* ( – *x*), 0.000 < 0.209, interval – 0.001

| *x* | 0 | 1 | 2 | 3 | 4 | 5 | 6 | 7 | 8 | 9 |
|---|---|---|---|---|---|---|---|---|---|---|
| 0.00 | + ∞ | 6.332 | 5.639 | 5.235 | 4.948 | 4.726 | 4.545 | 4.392 | 4.259 | 4.142 |
| 0.01 | 4.038 | 3.944 | 3.858 | 3.779 | 3.705 | 3.637 | 3.574 | 3.514 | 3.458 | 3.405 |
| 0.02 | 3.355 | 3.307 | 3.261 | 3.218 | 3.176 | 3.137 | 3.098 | 3.062 | 3.026 | 2.992 |
| 0.03 | 2.959 | 2.927 | 2.897 | 2.867 | 2.838 | 2.810 | 2.783 | 2.756 | 2.731 | 2.706 |
| 0.04 | 2.681 | 2.658 | 2.634 | 2.612 | 2.590 | 2.568 | 2.547 | 2.527 | 2.507 | 2.487 |
| 0.05 | 2.468 | 2.449 | 2.431 | 2.413 | 2.395 | 2.377 | 2.360 | 2.344 | 2.327 | 2.311 |
| 0.06 | 2.295 | 2.279 | 2.264 | 2.249 | 2.235 | 2.220 | 2.206 | 2.192 | 2.178 | 2.164 |
| 0.07 | 2.151 | 2.138 | 2.125 | 2.112 | 2.099 | 2.087 | 2.074 | 2.062 | 2.050 | 2.039 |
| 0.08 | 2.027 | 2.015 | 2.004 | 1.993 | 1.982 | 1.971 | 1.960 | 1.950 | 1.939 | 1.929 |
| 0.09 | 1.919 | 1.909 | 1.899 | 1.889 | 1.879 | 1.869 | 1.860 | 1.850 | 1.841 | 1.832 |
| 0.10 | 1.823 | 1.814 | 1.805 | 1.796 | 1.788 | 1.779 | 1.770 | 1.762 | 1.754 | 1.745 |
| 0.11 | 1.737 | 1.729 | 1.721 | 1.713 | 1.705 | 1.697 | 1.689 | 1.682 | 1.674 | 1.667 |
| 0.12 | 1.660 | 1.652 | 1.645 | 1.638 | 1.631 | 1.623 | 1.616 | 1.609 | 1.603 | 1.596 |
| 0.13 | 1.589 | 1.582 | 1.576 | 1.569 | 1.562 | 1.556 | 1.549 | 1.543 | 1.537 | 1.530 |
| 0.14 | 1.524 | 1.518 | 1.512 | 1.506 | 1.500 | 1.494 | 1.488 | 1.482 | 1.476 | 1.470 |
| 0.15 | 1.464 | 1.459 | 1.453 | 1.447 | 1.442 | 1.436 | 1.431 | 1.425 | 1.420 | 1.415 |
| 0.16 | 1.409 | 1.404 | 1.399 | 1.393 | 1.388 | 1.383 | 1.378 | 1.373 | 1.368 | 1.363 |
| 0.17 | 1.358 | 1.353 | 1.348 | 1.343 | 1.338 | 1.333 | 1.329 | 1.324 | 1.319 | 1.314 |
| 0.18 | 1.310 | 1.305 | 1.301 | 1.296 | 1.291 | 1.287 | 1.282 | 1.278 | 1.274 | 1.269 |
| 0.19 | 1.265 | 1.261 | 1.256 | 1.252 | 1.248 | 1.243 | 1.239 | 1.235 | 1.231 | 1.227 |
| 0.20 | 1.223 | 1.219 | 1.215 | 1.210 | 1.206 | 1.202 | 1.198 | 1.195 | 1.191 | 1.187 |

– *Ei* ( – *x*), 0.00 < *x* < 2.09, interval = 0.01

| | | | | | | | | | | |
|---|---|---|---|---|---|---|---|---|---|---|
| 0.0 | + ∞ | 4.038 | 3.335 | 2.959 | 2.681 | 2.468 | 2.295 | 2.151 | 2.027 | 1.919 |
| 0.1 | 1.823 | 1.737 | 1.660 | 1.589 | 1.524 | 1.464 | 1.409 | 1.358 | 1.309 | 1.265 |
| 0.2 | 1.223 | 1.183 | 1.145 | 1.110 | 1.076 | 1.044 | 1.014 | 0.985 | 0.957 | 0.931 |
| 0.3 | 0.906 | 0.882 | 0.858 | 0.836 | 0.815 | 0.794 | 0.774 | 0.755 | 0.737 | 0.719 |
| 0.4 | 0.702 | 0.686 | 0.670 | 0.655 | 0.640 | 0.625 | 0.611 | 0.598 | 0.585 | 0.572 |
| 0.5 | 0.560 | 0.548 | 0.536 | 0.525 | 0.514 | 0.503 | 0.493 | 0.483 | 0.473 | 0.464 |
| 0.6 | 0.454 | 0.445 | 0.437 | 0.428 | 0.420 | 0.412 | 0.404 | 0.396 | 0.388 | 0.381 |
| 0.7 | 0.374 | 0.367 | 0.360 | 0.353 | 0.347 | 0.340 | 0.334 | 0.328 | 0.322 | 0.316 |
| 0.8 | 0.311 | 0.305 | 0.300 | 0.295 | 0.289 | 0.284 | 0.279 | 0.274 | 0.269 | 0.265 |
| 0.9 | 0.260 | 0.256 | 0.251 | 0.247 | 0.243 | 0.239 | 0.235 | 0.231 | 0.227 | 0.223 |
| 1.0 | 0.219 | 0.216 | 0.212 | 0.209 | 0.205 | 0.202 | 0.198 | 0.195 | 0.192 | 0.189 |
| 1.1 | 0.186 | 0.183 | 0.180 | 0.177 | 0.174 | 0.172 | 0.169 | 0.166 | 0.164 | 0.161 |
| 1.2 | 0.158 | 0.156 | 0.153 | 0.151 | 0.149 | 0.146 | 0.144 | 0.142 | 0.140 | 0.138 |
| 1.3 | 0.135 | 0.133 | 0.131 | 0.129 | 0.127 | 0.125 | 0.124 | 0.122 | 0.120 | 0.118 |
| 1.4 | 0.116 | 0.114 | 0.113 | 0.111 | 0.109 | 0.108 | 0.106 | 0.105 | 0.103 | 0.102 |
| 1.5 | 0.1000 | 0.0985 | 0.0971 | 0.0957 | 0.0943 | 0.0929 | 0.0915 | 0.0902 | 0.0889 | 0.0876 |
| 1.6 | 0.0863 | 0.0851 | 0.0838 | 0.0826 | 0.0814 | 0.0802 | 0.0791 | 0.0780 | 0.0768 | 0.0757 |
| 1.7 | 0.0747 | 0.0736 | 0.0725 | 0.0715 | 0.0705 | 0.0695 | 0.0685 | 0.0675 | 0.0666 | 0.0656 |
| 1.8 | 0.0647 | 0.0638 | 0.0629 | 0.0620 | 0.0612 | 0.0603 | 0.0595 | 0.0586 | 0.0578 | 0.0570 |
| 1.9 | 0.0562 | 0.0554 | 0.0546 | 0.0539 | 0.0531 | 0.0524 | 0.0517 | 0.0510 | 0.0503 | 0.0496 |
| 2.0 | 0.0489 | 0.0482 | 0.0476 | 0.0469 | 0.0463 | 0.0456 | 0.0450 | 0.0444 | 0.0438 | 0.0432 |

2.0 < *x* < 10.9, interval = 0.1

| *x* | 0 | 1 | 2 | 3 | 4 | 5 | 6 | 7 | 8 | 9 |
|---|---|---|---|---|---|---|---|---|---|---|
| 2 | $4.89\times10^{-2}$ | $4.26\times10^{-2}$ | $3.72\times10^{-2}$ | $3.25\times10^{-2}$ | $2.84\times10^{-2}$ | $2.49\times10^{-2}$ | $2.19\times10^{-2}$ | $1.92\times10^{-2}$ | $1.69\times10^{-2}$ | $1.48\times10^{-2}$ |
| 3 | $1.30\times10^{-2}$ | $1.15\times10^{-2}$ | $1.01\times10^{-2}$ | $8.94\times10^{-3}$ | $7.89\times10^{-3}$ | $6.87\times10^{-3}$ | $6.16\times10^{-3}$ | $5.45\times10^{-3}$ | $4.82\times10^{-3}$ | $4.27\times10^{-2}$ |
| 4 | $3.78\times10^{-3}$ | $3.35\times10^{-3}$ | $2.97\times10^{-3}$ | $2.64\times10^{-3}$ | $2.34\times10^{-3}$ | $2.07\times10^{-3}$ | $1.84\times10^{-3}$ | $1.64\times10^{-3}$ | $1.45\times10^{-3}$ | $1.29\times10^{-3}$ |
| 5 | $1.15\times10^{-3}$ | $1.02\times10^{-3}$ | $9.08\times10^{-4}$ | $8.09\times10^{-4}$ | $7.19\times10^{-4}$ | $6.41\times10^{-4}$ | $5.71\times10^{-4}$ | $5.09\times10^{-4}$ | $4.53\times10^{-4}$ | $4.04\times10^{-4}$ |
| 6 | $3.60\times10^{-4}$ | $3.21\times10^{-4}$ | $2.86\times10^{-4}$ | $2.55\times10^{-4}$ | $2.28\times10^{-4}$ | $2.03\times10^{-4}$ | $1.82\times10^{-4}$ | $1.62\times10^{-4}$ | $1.45\times10^{-4}$ | $1.29\times10^{-4}$ |
| 7 | $1.15\times10^{-4}$ | $1.03\times10^{-4}$ | $9.22\times10^{-5}$ | $8.24\times10^{-5}$ | $7.36\times10^{-5}$ | $6.58\times10^{-5}$ | $5.89\times10^{-5}$ | $5.26\times10^{-5}$ | $4.71\times10^{-5}$ | $4.21\times10^{-5}$ |
| 8 | $3.77\times10^{-5}$ | $3.37\times10^{-5}$ | $3.02\times10^{-5}$ | $2.70\times10^{-5}$ | $2.42\times10^{-5}$ | $2.16\times10^{-5}$ | $1.94\times10^{-5}$ | $1.73\times10^{-5}$ | $1.55\times10^{-5}$ | $1.39\times10^{-5}$ |
| 9 | $1.24\times10^{-5}$ | $1.11\times10^{-5}$ | $9.99\times10^{-6}$ | $8.95\times10^{-6}$ | $8.02\times10^{-6}$ | $7.18\times10^{-6}$ | $6.44\times10^{-6}$ | $5.77\times10^{-6}$ | $5.17\times10^{-6}$ | $4.64\times10^{-6}$ |
| 10 | $4.15\times10^{-6}$ | $3.73\times10^{-6}$ | $3.34\times10^{-6}$ | $3.00\times10^{-6}$ | $2.68\times10^{-6}$ | $2.41\times10^{-6}$ | $2.16\times10^{-6}$ | $1.94\times10^{-6}$ | $1.74\times10^{-6}$ | $1.56\times10^{-6}$ |

*Adapted from Nisle, R.G.: "How To Use The Exponential Integral," *Pet. Eng.* (Aug. 1956) B171-173.

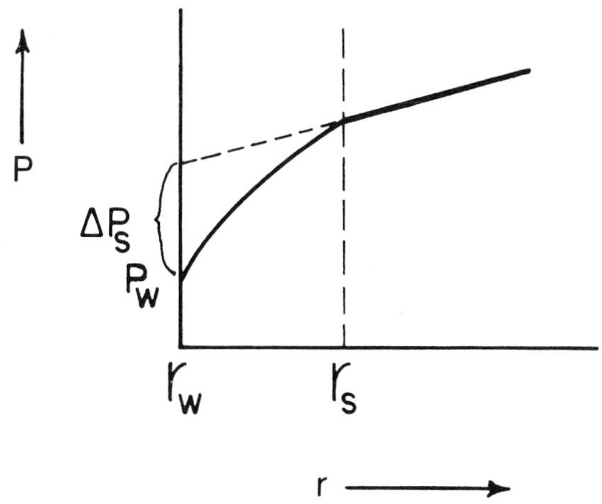


Fig. 1.1—Schematic of pressure distribution near wellbore.

Eq. 1.9 simply states that the pressure drop in the altered zone is inversely proportional to $k_s$ rather than to $k$ and that a correction to the pressure drop in this region (which assumed the same permeability, $k$, as in the rest of the reservoir) must be made. Combining Eqs. 1.7 and 1.9, we find that the total pressure drop at the wellbore is

$$p_i - p_{wf} = -70.6\frac{qB\mu}{kh}\,Ei\left(-\frac{948\,\phi\mu c_t r_w^2}{kt}\right) + \Delta p_s$$

$$= -70.6\frac{qB\mu}{kh}\left[Ei\left(-\frac{948\,\phi\mu c_t r_w^2}{kt}\right) - 2\left(\frac{k}{k_s}-1\right)\ln\left(\frac{r_s}{r_w}\right)\right].$$

For $r=r_w$, the argument of the $Ei$ function is sufficiently small after a short time that we can use the logarithmic approximation; thus, the drawdown is

$$p_i - p_{wf} = -70.6\frac{qB\mu}{kh}\left[\ln\left(\frac{1{,}688\,\phi\mu c_t r_w^2}{kt}\right) - 2\left(\frac{k}{k_s}-1\right)\ln\left(\frac{r_s}{r_w}\right)\right].$$

It is convenient to define a skin factor, $s$, in terms of the properties of the equivalent altered zone:

$$s = \left(\frac{k}{k_s}-1\right)\ln\left(\frac{r_s}{r_w}\right). \quad \text{.................. (1.10)}$$

Thus, the drawdown is

$$p_i - p_{wf} = -70.6\frac{qB\mu}{kh}\left[\ln\left(\frac{1{,}688\,\phi\mu c_t r_w^2}{kt}\right) - 2s\right]. \quad \text{..................... (1.11)}$$

Eq. 1.10 provides some insight into the physical significance of the sign of the skin factor. If a well is damaged ($k_s < k$), $s$ will be positive, and the greater the contrast between $k_s$ and $k$ and the deeper into the formation the damage extends, the larger the numerical value of $s$. There is no upper limit for $s$. Some newly drilled wells will not flow at all before stimulation; for these wells, $k_s \simeq 0$ and $s \rightarrow \infty$. If a well is stimulated ($k_s > k$), $s$ will be negative, and the deeper the stimulation, the greater the numerical value of $s$. Rarely does a stimulated well have a skin factor less than $-7$ or $-8$, and such skin factors arise only for wells with deeply penetrating, highly conductive hydraulic fractures. We should note finally that, if a well is neither damaged nor stimulated ($k = k_s$), $s = 0$. We caution the reader that Eq. 1.10 is best applied qualitatively; actual wells rarely can be characterized *exactly* by such a simplified model.

Before leaving the discussion of skin factor, we should point out that an altered zone near a particular well affects only the pressure near that well—i.e., the pressure in the unaltered formation away from the well is not affected by the existence of the altered zone. Said another way, we use Eq. 1.11 to calculate pressures at the sandface of a well with an altered zone, but we use Eq. 1.7 to calculate pressures beyond the altered zone in the formation surrounding the well. We have presented *no* simple equations that can be used to calculate pressures for radius, $r$, such that $r_w < r < r_s$, but this will offer no difficulties in well test analysis.

---

### *Example 1.1—Calculation of Pressures Beyond the Wellbore Using the Ei-Function Solution*

**Problem.** A well and reservoir have the following characteristics: The well is producing only oil; it is producing at a constant rate of 20 STB/D. Data describing the well and formation are

$\mu$ = 0.72 cp,
$k$ = 0.1 md,
$c_t$ = $1.5\times10^{-5}$ psi$^{-1}$
$p_i$ = 3,000 psi,
$r_e$ = 3,000 ft,
$r_w$ = 0.5 ft,
$B_o$ = 1.475 RB/STB,
$h$ = 150 ft,
$\phi$ = 0.23, and
$s$ = 0.

Calculate the reservoir pressure at a radius of 1 ft after 3 hours of production; then, calculate the pressure at radii of 10 and 100 ft after 3 hours of production.

**Solution.** The $Ei$ function is not an accurate solution to flow equations until $t > 3.79\times10^5\,\phi\mu c_t r_w^2/k$. Here,

$$\frac{3.79\times10^5\,\phi\mu c_t r_w^2}{k} = \left[(3.79\times10^5)(0.23)(0.72)\cdot(1.5\times10^{-5})(0.5)^2\right]/(0.1)$$

$$= 2.35 < t = 3 \text{ hours}.$$

Thus, we can use Eq. 1.7 with satisfactory accuracy *if*

the reservoir is still infinite acting at this time. The reservoir will act as an infinite reservoir until $t > 948\ \phi\mu c r_e^2/k$.

Here,

$$\frac{948\ \phi\mu c_t r_e^2}{k} = \left[(948)(0.23)(0.72) \cdot (1.5\times10^{-5})(3{,}000)^2\right]/0.3 = 211{,}900 \text{ hours.}$$

Thus, for times less than 211,900 hours, we can use Eq. 1.7. At a radius of 1 ft,

$$p = p_i + 70.6\frac{qB\mu}{kh}\,Ei\left(\frac{-948\ \phi\mu c_t r^2}{kt}\right)$$

$$= 3{,}000 + \frac{(70.6)(20)(1.475)(0.72)}{(0.1)(150)} \cdot Ei\left[\frac{-(948)(0.23)(0.72)(1.5\times10^{-5})(1)^2}{(0.1)(3)}\right]$$

$$= 3{,}000 + (100)Ei(-0.007849)$$

$$= 3{,}000 + 100\ln\left[(1.781)(0.007849)\right]$$

$$= 3{,}000 + (100)(-4.27)$$

$$= 2{,}573 \text{ psi.}$$

At a radius of 10 ft,

$$p = 3{,}000 + 100 \cdot Ei\left[\frac{-(948)(0.23)(0.72)(1.5\times10^{-5})(10)^2}{(0.1)(3)}\right]$$

$$= 3{,}000 + 100\,Ei(-0.7849)$$

$$= 3{,}000 + (100)(-0.318)$$

$$= 2{,}968 \text{ psi.}$$

In this calculation, we find the value of the *Ei* function from Table 1.1. Note, as indicated in the table, that it is a negative quantity.

At a radius of 100 ft,

$$p = 3{,}000 + 100 \cdot Ei\left[\frac{-(948)(0.23)(0.72)(1.5\times10^{-5})(100)^2}{(0.1)(3)}\right]$$

$$= 3{,}000 + 100\,Ei(-78.49)$$

$$= 3{,}000 \text{ psi.}$$

Here we note that for an argument of 78.49, the *Ei* function is essentially zero.

**Pseudosteady-State Solution.** We now discuss the next solution to the radial diffusivity equation that we will use extensively in this introduction to well test analysis. Actually, this solution (the pseudosteady-state solution) is not new. It is simply a limiting form of Eq. 1.6, which describes pressure behavior with time for a well centered in a cylindrical reservoir of radius $r_e$. The limiting form of interest is that which is valid for large times, so that the summation involving exponentials and Bessel functions is negligible; after this time ($t > 948\ \phi\mu c_t r_e^2/k$),

$$p_{wf} = p_i - 141.2\frac{qB\mu}{kh}\left(\frac{2t_D}{r_{eD}^2} + \ln r_{eD} - \frac{3}{4}\right),$$

or

$$p_{wf} = p_i - 141.2\frac{qB\mu}{kh}\left[\frac{0.000527kt}{\phi\mu c_t r_e^2} + \ln\left(\frac{r_e}{r_w}\right) - \frac{3}{4}\right]. \quad \text{(1.12)}$$

Note that during this time period we find, by differentiating Eq. 1.12,

$$\frac{\partial p_{wf}}{\partial t} = -\frac{0.0744\,qB}{\phi c_t h r_e^2}.$$

Since the liquid-filled pore volume of the reservoir, $V_p$ (cubic feet), is

$$V_p = \pi r_e^2 h\phi,$$

then

$$\frac{\partial p_{wf}}{\partial t} = -\frac{0.234\,qB}{c_t V_p}. \quad \text{(1.13)}$$

Thus, during this time period, the rate of pressure decline is inversely proportional to the liquid-filled pore volume $V_p$. This result leads to a form of well testing sometimes called reservoir limits testing, which seeks to determine reservoir size from the rate of pressure decline in a wellbore with time.

Another form of Eq. 1.12 is useful for some applications. It involves replacing original reservoir pressure, $p_i$, with average pressure, $\bar{p}$, within the drainage volume of the well.

The volumetric average pressure within the drainage volume of the well can be found from material balance. The pressure decrease $(p_i - \bar{p})$ resulting from removal of $qB$ RB/D of fluid for $t$ hours [a total volume removed of $5.615\,qB(t/24)$ cu ft] is

$$p_i - \bar{p} = \frac{\Delta V}{c_t V} = \frac{5.615\,qB(t/24)}{c_t(\pi r_e^2 h\phi)}$$

$$= \frac{0.0744\,qBt}{\phi c_t h r_e^2}. \quad \text{(1.14)}$$

Substituting in Eq. 1.12,

$$p_{wf} = \bar{p} + \frac{0.0744\,qBt}{\phi c_t h r_e^2} - \frac{0.0744\,qBt}{\phi c_t h r_e^2} - 141.2\frac{qB\mu}{kh}\left[\ln\left(\frac{r_e}{r_w}\right) - \frac{3}{4}\right],$$

or

$$\bar{p} - p_{wf} = 141.2\frac{qB\mu}{kh}\left[\ln\left(\frac{r_e}{r_w}\right) - \frac{3}{4}\right]. \quad \text{(1.15)}$$

Eqs. 1.12 and 1.15 become more useful in practice if they include a skin factor to account for the fact that most wells are either damaged or stimulated. For example, in Eq. 1.15,

$$\bar{p}-p_{wf}=141.2\frac{qB\mu}{kh}\left[\ln\left(\frac{r_e}{r_w}\right)-\frac{3}{4}\right]+(\Delta p)_s,$$

$$\bar{p}-p_{wf}=141.2\frac{qB\mu}{kh}\left[\ln\left(\frac{r_e}{r_w}\right)-\frac{3}{4}+s\right], \quad \ldots\ldots (1.16)$$

and

$$p_i-p_{wf}=141.2\frac{qB\mu}{kh}\left[\frac{0.000527\,kt}{\phi\mu c_t r_e^2}+\ln\left(\frac{r_e}{r_w}\right)-\frac{3}{4}+s\right]. \quad \ldots\ldots\ldots\ldots (1.17)$$

Further, we can define an average permeability, $k_J$, such that

$$\bar{p}-p_{wf}=141.2\frac{qB\mu}{k_J h}\left[\ln\left(\frac{r_e}{r_w}\right)-\frac{3}{4}\right]$$

$$=141.2\frac{qB\mu}{kh}\left[\ln\left(\frac{r_e}{r_w}\right)-\frac{3}{4}+s\right],$$

from which,

$$k_J=k\left[\ln\left(\frac{r_e}{r_w}\right)-\frac{3}{4}\right]\Big/\left[\ln\left(\frac{r_e}{r_w}\right)-\frac{3}{4}+s\right]. \quad \ldots (1.18)$$

This average permeability, $k_J$, proves to have considerable value in well test analysis, as we shall see later. Note that for a damaged well, the average permeability $k_J$ is lower than the true, bulk formation permeability $k$; in fact, these quantities are equal only when the skin factor $s$ is zero. Since we sometimes estimate the permeability of a well from productivity-index (PI) measurements, and since the productivity index $J$ (STB/D/psi), of an oil well is defined as

$$J\equiv\frac{q}{\bar{p}-p_{wf}}=\frac{k_J h}{141.2B\mu\left[\ln\left(\frac{r_e}{r_w}\right)-\frac{3}{4}\right]}, \quad \ldots (1.19)$$

this method *does not* necessarily provide a good estimate of formation permeability, $k$. Thus, there is a need for a more complete means of characterizing a producing well than exclusive use of PI information.

---

## *Example 1.2 – Analysis of Well From PI Test*

**Problem.** A well produces 100 STB/D oil at a measured flowing bottomhole pressure (BHP) of 1,500 psi. A recent pressure survey showed that average reservoir pressure is 2,000 psi. Logs indicate a net sand thickness of 10 ft. The well drains an area with drainage radius, $r_e$, of 1,000 ft; the borehole radius is 0.25 ft. Fluid samples indicate that, at current reservoir pressure, oil viscosity is 0.5 cp and formation volume factor is 1.5 RB/STB.

1. Estimate the productivity index for the tested well.
2. Estimate formation permeability from these data.
3. Core data from the well indicate an effective permeability to oil of 50 md. Does this imply that the well is either damaged or stimulated? What is the apparent skin factor?

**Solution.**

1. To estimate productivity index, we use Eq. 1.19:

$$J=\frac{q}{\bar{p}-p_{wf}}=\frac{100}{(2{,}000-1{,}500)}$$

$$=0.2\text{ STB/psi-D}.$$

2. We do not have sufficient information to estimate formation permeability; we can calculate average permeability, $k_J$, only, which is not necessarily a good approximation of formation permeability, $k$. From Eq. 1.19,

$$k_J=\frac{141.2\,JB\mu\left[\ln\left(\frac{r_e}{r_w}\right)-\frac{3}{4}\right]}{h}$$

$$=\frac{(141.2)(0.2)(1.5)(0.5)\left[\ln\left(\frac{1{,}000}{0.25}\right)-0.75\right]}{10}$$

$$=16\text{ md}.$$

3. Core data frequently provide a better estimate of formation permeability than do permeabilities derived from the productivity index, particularly for a well that is badly damaged. Since cores indicate a permeability of 50 md, we conclude that this well is damaged. Eq. 1.18 provides a method for estimating the skin factor $s$:

$$s=\left(\frac{k}{k_J}-1\right)\left[\ln\left(\frac{r_e}{r_w}\right)-\frac{3}{4}\right]$$

$$=\left(\frac{50}{16}-1\right)\left[\ln\left(\frac{1{,}000}{0.25}\right)-0.75\right]$$

$$=16.$$

### Flow Equations for Generalized Reservoir Geometry

Eq. 1.16 is limited to a well centered in a circular drainage area. A similar equation[5] models pseudosteady-state flow in more general reservoir shapes:

$$\bar{p}-p_{wf}=141.2\frac{qB\mu}{kh}\left[\frac{1}{2}\,ln\left(\frac{10.06\,A}{C_A r_w^2}\right)-\frac{3}{4}+s\right], \quad \ldots\ldots\ldots\ldots (1.20)$$

where

$A$ = drainage area, sq ft, and
$C_A$ = shape factor for specific drainage-area shape and well location, dimensionless.

Values of $C_A$ are given in Table 1.2; further explanation of the source of these $C_A$ values is given in

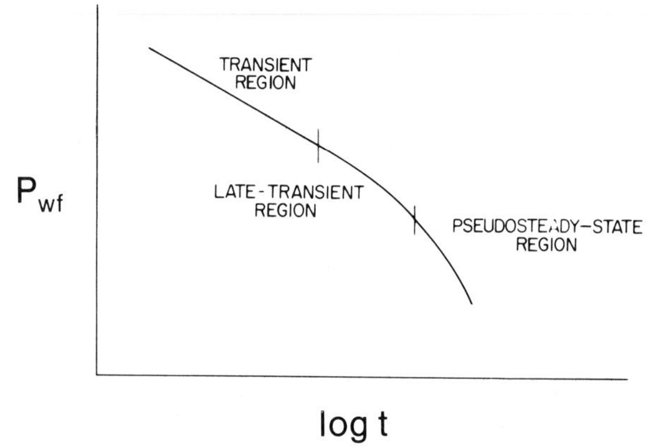


**Fig. 1.2**—Flow regions on semilogarithmic paper.

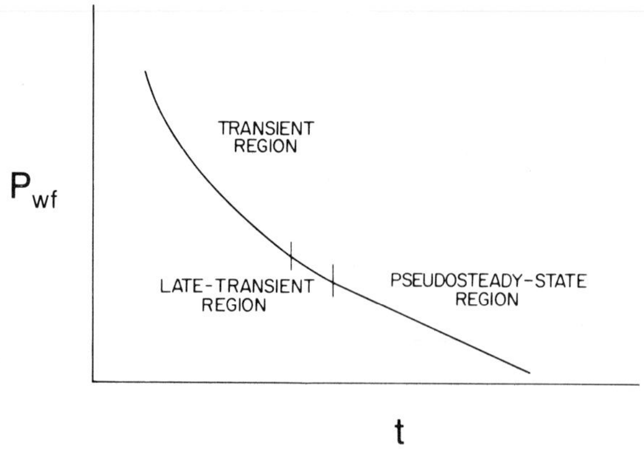


**Fig. 1.3**—Flow regions on Cartesian-coordinate graph.

Chap. 2.

Productivity index, $J$, can be expressed for general drainage-area geometry as

$$J=\frac{q}{\bar{p}-p_{wf}}=\frac{0.00708\,kh}{B\mu\left[\frac{1}{2}\ln\left(\frac{10.06\,A}{C_A r_w^2}\right)-\frac{3}{4}+s\right]}. \qquad (1.21)$$

Other numerical constants tabulated in Table 1.2 allow us to calculate (1) the maximum elapsed time during which a reservoir is infinite acting (so that the *Ei*-function solution can be used); (2) the time required for the pseudosteady-state solution to predict pressure drawdown within 1% accuracy; and (3) time required for the pseudosteady-state solution to be exact.

For a given reservoir geometry, the maximum time a reservoir is infinite acting can be determined using the entry in the column "Use Infinite-System Solution With Less Than 1% Error for $t_{DA}<$." Since $t_{DA}=0.000264\;kt/\phi\mu c_t A$, this means that the time in hours is calculated from

$$t<\frac{\phi\mu c_t A t_{DA}}{0.000264\,k}.$$

Time required for the pseudosteady-state equation to be accurate within 1% can be found from the entry in the column headed "Less Than 1% Error for $t_{DA}>$" and the relationship

$$t>\frac{\phi\mu c_t A t_{DA}}{0.000264\,k}.$$

Finally, time required for the pseudosteady-state equation to be exact is found from the entry in the column "Exact for $t_{DA}>$."

At this point, it is helpful to depict graphically the flow regimes that occur in different time ranges. Figs. 1.2 and 1.3 show BHP, $p_{wf}$, in a well flowing at constant rate, plotted as a function of time on both logarithmic and linear scales.

In the transient region, the reservoir is infinite acting, and is modeled by Eq. 1.11, which implies that $p_{wf}$ is a linear function of log $t$. In the pseudosteady-state region, the reservoir is modeled by Eq. 1.20 in the general case or Eqs. 1.15 and 1.12 for the special case of a well centered in a cylindrical reservoir. Eq. 1.12 shows the linear relationship between $p_{wf}$ and $t$ during pseudosteady-state. This linear relationship also exists in generalized reservoir geometries.

At times between the end of the transient region and the beginning of the pseudosteady-state region, this is a transition region, sometimes called the late-transient region, as in Figs. 1.2 and 1.3. No simple equation is available to predict the relationship between BHP and time in this region. This region is small (or, for practical purposes nonexistent) for a well centered in a circular, square, or hexagonal drainage area, as Table 1.2 indicates. However, for a well off-center in its drainage area, the late-transient region can span a significant time region, as Table 1.2 also indicates.

Note that the determination of when the transient region ends or when the pseudosteady-state region begins is somewhat subjective. For example, the limits on applicability of Eqs. 1.7 and 1.12 (stated in the text earlier) are not exactly the same as given in Table 1.2—but the difference is slight. Other authors[1] consider the deviation from Eq. 1.7 to be sufficient for $t>379\;\phi\mu c_t r_e^2/k$ that a late-transient region exists even for a well centered in a cylindrical reservoir between this lower limit and an upper limit of $1{,}136\;\phi\mu c_t r_e^2/k$. These apparently contradictory opinions are nothing more than different judgments about when the slightly approximate solutions, Eqs. 1.7 and 1.12, can be considered to be identical to the exact solution, Eq. 1.6.

These concepts are illustrated in Example 1.3.

---

### *Example 1.3—Flow Analysis in Generalized Reservoir Geometry*

**Problem.** 1. For each of the following reservoir geometries, calculate the time in hours for which (a) the reservoir is infinite acting; (b) the pseudosteady state

**TABLE 1.2 – SHAPE FACTORS FOR VARIOUS SINGLE-WELL DRAINAGE AREAS[10]**

| In Bounded Reservoirs | $C_A$ | $\ln C_A$ | $0.5 \ln\left(\frac{2.2458}{C_A}\right)$ | Exact for $t_{DA} >$ | Less Than 1% Error for $t_{DA} >$ | Use Infinite System Solution With Less Than 1% Error for $t_{DA} <$ |
|---|---|---|---|---|---|---|
|  | 31.62 | 3.4538 | 1.3224 | 0.1 | 0.06 | 0.10 |
|  | 31.6 | 3.4532 | – 1.3220 | 0.1 | 0.06 | 0.10 |
|  | 27.6 | 3.3178 | – 1.2544 | 0.2 | 0.07 | 0.09 |
|  | 27.1 | 3.2995 | – 1.2452 | 0.2 | 0.07 | 0.09 |
| 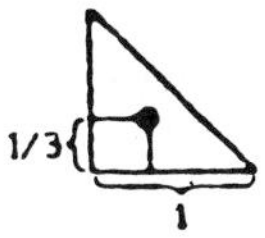 | 21.9 | 3.0865 | – 1.1387 | 0.4 | 0.12 | 0.08 |
|  | 0.098 | – 2.3227 | 1.5659 | 0.9 | 0.60 | 0.015 |
|  | 30.8828 | 3.4302 | – 1.3106 | 0.1 | 0.05 | 0.09 |
| 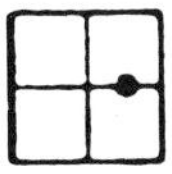 | 12.9851 | 2.5638 | – 0.8774 | 0.7 | 0.25 | 0.03 |
| 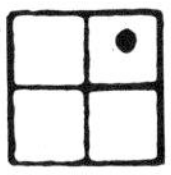 | 4.5132 | 1.5070 | – 0.3490 | 0.6 | 0.30 | 0.025 |
|  | 3.3351 | 1.2045 | – 0.1977 | 0.7 | 0.25 | 0.01 |
| 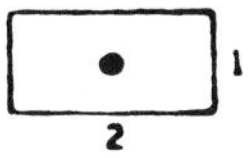 | 21.8369 | 3.0836 | – 1.1373 | 0.3 | 0.15 | 0.025 |
| 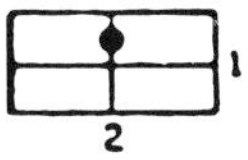 | 10.8374 | 2.3830 | – 0.7870 | 0.4 | 0.15 | 0.025 |
| 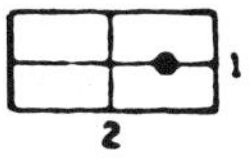 | 4.5141 | 1.5072 | – 0.3491 | 1.5 | 0.50 | 0.06 |
| 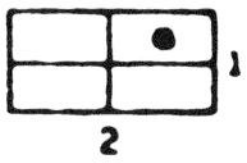 | 2.0769 | 0.7309 | 0.0391 | 1.7 | 0.50 | 0.02 |
| 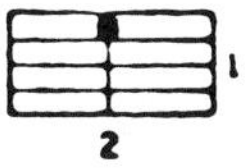 | 3.1573 | 1.1497 | – 0.1703 | 0.4 | 0.15 | 0.005 |

**TABLE 1.2 – SHAPE FACTORS FOR VARIOUS SINGLE-WELL DRAINAGE AREAS[10]**

| In Bounded Reservoirs | $C_A$ | $\ln C_A$ | $0.5 \ln\left(\frac{2.2458}{C_A}\right)$ | Exact for $t_{DA} >$ | Less Than 1% Error for $t_{DA} >$ | Use Infinite System Solution With Less Than 1% Error for $t_{DA} <$ |
|---|---|---|---|---|---|---|
| 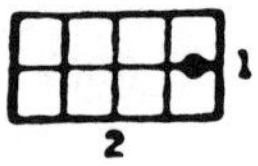 | 0.5813 | −0.5425 | 0.6758 | 2.0 | 0.60 | 0.02 |
| 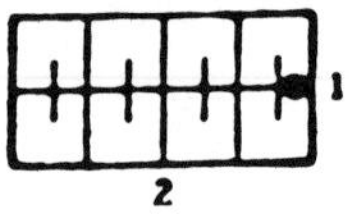 | 0.1109 | −2.1991 | 1.5041 | 3.0 | 0.60 | 0.005 |
| 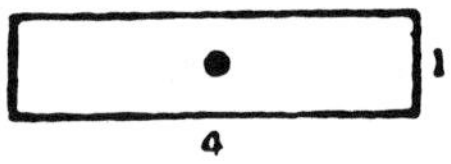 | 5.3790 | 1.6825 | −0.4367 | 0.8 | 0.30 | 0.01 |
| 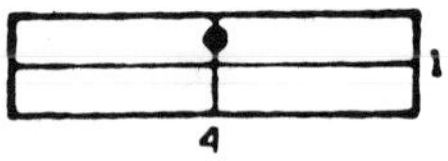 | 2.6896 | 0.9894 | −0.0902 | 0.8 | 0.30 | 0.01 |
| 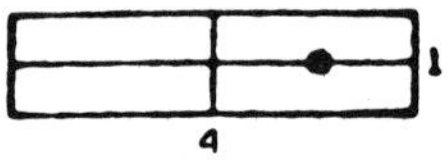 | 0.2318 | −1.4619 | 1.1355 | 4.0 | 2.00 | 0.03 |
| 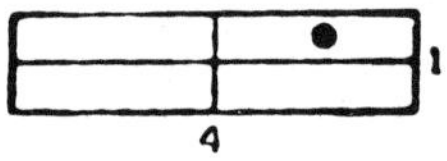 | 0.1155 | −2.1585 | 1.4838 | 4.0 | 2.00 | 0.01 |
| 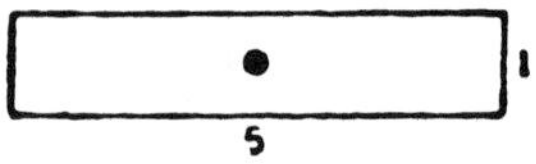 | 2.3606 | 0.8589 | −0.0249 | 1.0 | 0.40 | 0.025 |
| In vertically fractured reservoirs: use $(r_e/L_f)^2$ in place of $A/r_w^2$ for fractured systems | | | | | | |
| 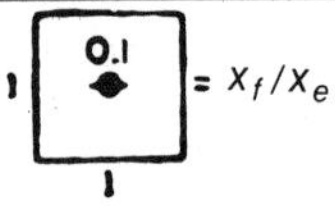 | 2.6541 | 0.9761 | −0.0835 | 0.175 | 0.08 | cannot use |
| 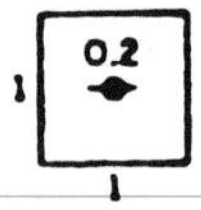 | 2.0348 | 0.7104 | 0.0493 | 0.175 | 0.09 | cannot use |
| 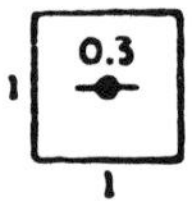 | 1.9886 | 0.6924 | 0.0583 | 0.175 | 0.09 | cannot use |
|  | 1.6620 | 0.5080 | 0.1505 | 0.175 | 0.09 | cannot use |
|  | 1.3127 | 0.2721 | 0.2685 | 0.175 | 0.09 | cannot use |
| 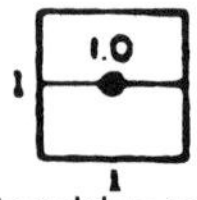 | 0.7887 | −0.2374 | 0.5232 | 0.175 | 0.09 | cannot use |
| In water-drive reservoirs | | | | | | |
|  | 19.1 | 2.95 | −1.07 | – | – | – |
| In reservoirs of unknown production character | | | | | | |
|  | 25.0 | 3.22 | −1.20 | – | – | – |

is exact; and (c) the pseudosteady-state equation is accurate to within 1%. (1) Well centered in circular drainage area, (2) well centered in square drainage area, and (3) well centered in one quadrant of square drainage area.

In each case,

$$\begin{aligned} A &= 17.42\times 10^6 \text{ sq ft (40 acres)},\\ \phi &= 0.2,\\ \mu &= 1 \text{ cp},\\ c_t &= 1\times 10^{-5} \text{ psi}^{-1}, \text{ and}\\ k &= 100 \text{ md}. \end{aligned}$$

2. For each of the wells in Part 1, estimate PI and stabilized production rate with $\bar{p}-p_{wf}=500$ psi, if

$$\begin{aligned} h &= 10 \text{ ft},\\ s &= 3.0,\\ r_w &= 0.3 \text{ ft, and}\\ B &= 1.2 \text{ RB/STB}. \end{aligned}$$

3. For the well centered in one of the quadrants of a square, write equations relating constant flow rate and wellbore pressure drops at elapsed times of 30, 200, and 400 hours.

**Solution.** 1. We first calculate the group $\phi\mu c_t A/0.000264\,k$.

$$\frac{\phi\mu c_t A}{0.000264\,k}=\frac{(0.2)(1)(1\times 10^{-5})(17.42\times 10^6)}{(0.000264)(100)}$$
$$=1{,}320.$$

We then prepare the following table (values from Table 1.2):

| | Infinite Solution | | Pseudosteady-State (Approximate) | | Pseudosteady-State (Exact) | |
|---|---|---|---|---|---|---|
| Geometry | $t_{DA}$ | $t$ (hours) | $t_{DA}$ | $t$ (hours) | $t_{DA}$ | $t$ (hours) |
| Circular | 0.1 | 132 | 0.06 | 79.2 | 0.1 | 132 |
| Square-centered | 0.09 | 119 | 0.05 | 66.0 | 0.1 | 132 |
| Square-quadrant | 0.025 | 33 | 0.3 | 396 | 0.6 | 792 |

For the off-center well, the reduction in time for which the infinite-system solution is accurate and the increase in time required to achieve pseudosteady-state flow are noteworthy.

2. To calculate PI and stabilized production rate, we use the equations

$$J=\frac{0.00708\,kh}{B\mu\left[\frac{1}{2}ln\left(\frac{10.06\,A}{C_A r_w^2}\right)-\frac{3}{4}+s\right]}$$

$$=\frac{(0.00708)(100)(10)}{(1.2)(1)\left[\frac{1}{2}ln\left(\frac{(10.06)(17.42\times 10^6)}{C_A(0.3)^2}\right)-\frac{3}{4}+3.0\right]},$$

$$J=5.9/(12.94-\frac{1}{2}lnC_A),$$

and

$$q=J(\bar{p}-p_{wf})=500\,J.$$

Thus, we can prepare the following table.

| Geometry | $C_A$ | $J$ (STB/D-psi) | $q$ (STB/D) |
|---|---|---|---|
| Circular | 31.62 | 0.526 | 263 |
| Square-centered | 30.88 | 0.526 | 263 |
| Square-quadrant | 4.513 | 0.484 | 242 |

3. (a) For $t=30$ hours, the reservoir is infinite acting, and

$$p_i-p_{wf}=-70.6\frac{qB\mu}{kh}\left[\ln\left(\frac{1{,}688\,\phi\mu c_t r_w^2}{kt}\right)-2s\right].$$

(b) For $t=200$ hours, the reservoir is no longer infinite acting and the pseudosteady-state equation is not yet accurate. Accordingly, no simple equation can be written.

(c) For $t=400$ hours, the pseudosteady-state equation is accurate, and

$$\bar{p}-p_{wf}=141.2\frac{qB\mu}{kh}\left[\frac{1}{2}ln\left(\frac{10.06\,A}{C_A r_w^2}\right)-\frac{3}{4}+s\right].$$

### Radial Flow in Infinite Reservoir With Wellbore Storage

The next solution to the radial diffusivity equation includes a phenomenon causing variable flow rates after production begins. This phenomenon is called wellbore, as shown in Fig. 1.4, and assume there is of wellbore storage before we can deal with the solution to the flow equations that includes its effect.

Consider a shut-in oil well in a reservoir with uniform and unchanging pressure. Reservoir pressure will support a column of liquid to some equilibrium height in the wellbore. If we open a valve at the surface and initiate flow, the first oil produced will be that stored in the wellbore, and the initial flow rate from the formation to the well will be zero. With increasing flow time, at constant surface producing rate, the downhole flow rate will approach the surface rate, and the amount of liquid stored in the wellbore will approach a constant value.

We now develop a mathematical relationship between sandface (formation) and surface flow rates. Consider a well with a liquid/gas interface in the wellbore, as shown in Table 1.4, and assume there is some mechanism (a pump or gas lift) to deliver liquid to the surface. Let the surface rate $q$ be variable in the general case.

From a mass balance in the wellbore, the rate of liquid in is $q_{sf}B$ in RB/D; the rate of liquid out is $q_B$ in RB/D; and the rate of liquid accumulation in the well is

$$\frac{d}{dt}\left(\frac{24\,V_{wb}}{5.615}\right)=\frac{24\,A_{wb}}{5.615}\frac{dz}{dt}.$$

Then, assuming constant wellbore area, $A_{wb}$, and constant oil formation volume factor, $B$, the same at the sandface and at the surface, the balance becomes

$$\frac{24}{5.615}A_{wb}\frac{dz}{dt}=(q_{sf}-q)B. \qquad (1.22)$$

For a well with surface pressure $p_t$,

$$p_w=p_t+\frac{\rho z}{144}\frac{g}{g_c}, \qquad (1.23)$$

Where $\rho$ is the density of the liquid in the wellbore (lbm/cu ft) and $g/g_c=$ lbf/lbm. Then,

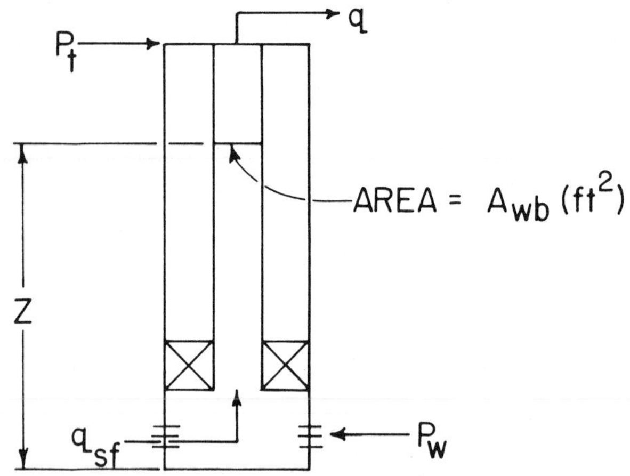


Fig. 1.4—Schematic of wellbore with moving liquid/gas interface.

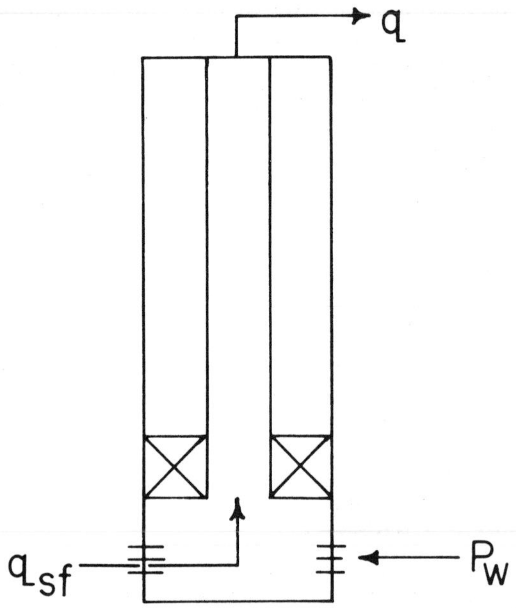


Fig. 1.5—Schematic of wellbore containing single-phase liquid or gas.

$$\frac{d(p_w-p_t)}{dt}=\frac{\rho}{144}\frac{g}{g_c}\frac{dz}{dt}. \quad \ldots\ldots\ldots\ldots (1.24)$$

Thus,

$$\frac{(24)(144)}{5.615\rho}\frac{g_c}{g}A_{wb}\frac{d(p_w-p_t)}{dt}$$

$$=(q_{sf}-q)B. \quad \ldots\ldots\ldots\ldots (1.25)$$

Define a wellbore storage constant, $C_s$:

$$C_s=\frac{144\,A_{wb}}{5.615\rho}\frac{g_c}{g}. \quad \ldots\ldots\ldots\ldots (1.26)$$

Then,

$$q_{sf}=q+\frac{24\,C_s}{B}\frac{d(p_w-p_t)}{dt}. \quad \ldots\ldots\ldots\ldots (1.27)$$

For zero or unchanging surface pressure, $p_t$ (a major and not necessarily valid assumption),

$$q_{sf}=q+\frac{24\,C_s}{B}\frac{dp_w}{dt}. \quad \ldots\ldots\ldots\ldots (1.28)$$

To understand the solution to flow problems that include wellbore storage, it is necessary to introduce dimensionless variables, similar to those discussed in Appendix B. Let $q_i$ be the surface rate at $t=0$ and introduce the definitions of dimensionless time and dimensionless pressure:

$$p_D=\frac{0.00708\,kh(p_i-p_w)}{q_iB\mu}, \quad \ldots\ldots\ldots\ldots (1.29)$$

$$t_D=\frac{0.000264\,kt}{\phi\mu c_t r_w^2}. \quad \ldots\ldots\ldots\ldots (1.30)$$

Substituting,

$$\frac{dp_w}{dt}=-\frac{q_iB\mu}{0.00708\,kh}\times\frac{0.000264\,k}{\phi\mu c_t r_w^2}\frac{dp_D}{dt_D}$$

$$=-\frac{0.0373q_iB}{\phi\mu c_t h r_w^2}\frac{dp_D}{dt_D}. \quad \ldots\ldots\ldots\ldots (1.31)$$

Thus,

$$q_{sf}=q-\frac{0.894\,q_iC_s}{\phi c_t h r_w^2}\frac{dp_D}{dt_D}, \quad \ldots\ldots\ldots\ldots (1.32)$$

If we define a dimensionless wellbore storage constant, $C_{sD}$, as

$$C_{sD}=0.894\,C_s/\phi c_t h r_w^2, \quad \ldots\ldots\ldots\ldots (1.33)$$

then

$$q_{sf}=q_i\left[\frac{q}{q_i}-C_{sD}\frac{dp_D}{dt_D}\right]. \quad \ldots\ldots\ldots\ldots (1.34)$$

For constant-rate production $[q(t)=q_i]$, Eq. 1.34 becomes

$$\frac{q_{sf}}{q}=1-C_{sD}\frac{dp_D}{dt_D}. \quad \ldots\ldots\ldots\ldots (1.35)$$

Eq. 1.35 is the inner boundary condition for the problem of constant-rate flow of a slightly compressible liquid with wellbore storage. Note that, for small $C_{sD}$ *or* for small $dp_D/dt_D$, $q_{sf}/q \cong 1$ (i.e., the effect of wellbore storage or sandface rate will be negligible).

As a second example, consider a wellbore (Fig. 1.5) that contains a single-phase fluid (liquid or gas) and that is produced at some surface rate, $q$. If we let $V_{wb}$ be the volume of wellbore open to formation (barrels) and $c_{wb}$ be the compressibility of the fluid in the wellbore (evaluated at *wellbore conditions*), the mass-balance components are (1) rate of fluid in = $q_{sf}B$, (2) rate of fluid out = $qB$, and (3) rate of fluid accumulation in wellbore = $24V_{wb}c_{wb} \times (dp_w/dt)$. The balance becomes

$$(q_{sf}-q)B=24\, V_{wb}c_{wb}\frac{dp_w}{dt}, \quad \ldots \ldots (1.36)$$

or

$$q_{sf}=q+\frac{24\, V_{wb}c_w}{B}\frac{dp_w}{dt}. \quad \ldots \ldots (1.37)$$

In this case, let $C_s=c_{wb}V_{wb}$. Then,

$$q_{sf}=q+\frac{24\, V_{wb}c_{wb}}{B}\frac{dp_w}{dt}. \quad \ldots \ldots (1.37)$$

Eq. 1.38 is identical to Eq. 1.28; the wellbore storage constant $C_s$ simply has a different definition. Note, however, that we are forced to make a significant simplifying assumption: when we apply Eq. 1.38 to a gas well, $c_{wb}$ is the compressibility of the gas in the wellbore, and it is a strong function of pressure (as an approximation, $c_{wb}=1/p_{wb}$). Thus, the wellbore storage "constant" for a gas well may be far from constant.

Since Eqs. 1.38 and 1.28 are identical, Eqs. 1.34 and 1.35 are also valid for a wellbore containing a single-phase fluid.

The radial diffusivity equation, with the wellbore storage equation (Eq. 1.35) as an inner boundary condition, infinite drainage radius, uniform initial formation pressure, and formation damage or stimulation (characterized by skin factor, $s$), has been solved both analytically and numerically in Refs. 6 and 7. The analytical solution is presented in Fig. 1.6. From this figure, values of $p_D$ (and thus $p_w$) can be determined for a well in a formation with given values of $t_D$, $C_{sD}$, and $s$.

Two properties of this log-log graph require special mention at this point.

### Presence of Unit-Slope Line

At earliest times for a given value of $C_{sD}$, and for most values of $s$, a "unit-slope line" (i.e., line with 45° slope) is present on the graph. This line appears and remains as long as *all* production comes from the wellbore and *none* comes from the formation. Eq. 1.35 leads us to expect this line. For $q_{sf}/q=0$, the equation becomes

$$1-C_{sD}\frac{dp_D}{dt_D}=0,$$

or

$$dt_D=C_{sD}dp_D. \quad \ldots \ldots (1.39)$$

Integrating from $t_D=0$ (where $p_D=0$) to $t_D$ and $p_D$, the result is

$$C_{sD}p_D=t_D. \quad \ldots \ldots (1.40)$$

Taking logarithms of both sides of the equation,

$$\log C_{sD}+\log p_D=\log t_D. \quad \ldots \ldots (1.41)$$

Thus, as long as $q_{sf}=0$, theory leads us to expect that a graph of log $p_D$ vs. log $t_D$ will have a slope of unity; it also leads us to expect that *any point* $(p_D, t_D)$ on this unit-slope line must satisfy the relation

$$\frac{C_{sD}p_D}{t_D}=1. \quad \ldots \ldots (1.42)$$

This observation is of major value in well test analysis.

### End of Wellbore Storage Distortion

When wellbore storage has ceased (i.e., when $q_{sf}\cong q$), we would expect the solution to the flow equations to be the same as if there had never been any wellbore storage – i.e., the same as for $C_{sD}=0$. In Fig. 1.6, note that the solutions for finite $C_{sD}$ and for $C_{sD}=0$ *do* become identical after sufficient elapsed time. One useful empirical observation is that this time (called the "end of wellbore storage distortion," $t_{wbs}$ occurs approximately one and a half log cycles after the disappearance of the unit-slope line. Another useful observation is that the dimensionless time at which wellbore storage distortion ceases is given by

$$t_D=(60+3.5\,s)C_{sD}. \quad \ldots \ldots (1.43)$$

These observations are also very useful in well test analysis.

### Linear Flow

Linear flow occurs in some petroleum reservoirs with long, highly conductive vertical fractures. For that reason, it is of interest to review one of the fundamental equations describing linear flow in a reservoir. Consider a situation with linear flow (in the $x$ direction, for convenience) of a slightly compressible fluid in an infinite, homogeneous reservoir, initially at uniform pressure, $p_i$. Fluid is produced at constant rate $qB$ over an area $A_f$ (square feet). [If the area $A_f$ represents a vertical fracture with two equal-length wings of length $L_f$ (feet) and height $h$ (feet), $A_f=4\,hL_f$, with flow entering both sides of each wing of the fracture.]

This situation is modeled by the diffusivity equation in the form

$$\frac{\partial^2 p}{\partial x^2}=\frac{\phi\mu c_t}{0.000264\,k}\frac{\partial p}{\partial t}. \quad \ldots \ldots (1.44)$$

For the conditions stated, the solution[8] to this equation at $x=0$ is

$$p_i-p_{wf}=16.26\frac{qB}{A_f}\left(\frac{\mu t}{k\phi c_t}\right)^{1/2}. \quad \ldots \ldots (1.45)$$

For linear flow into vertical fractures, $A_f=4\,hL_f$, and

$$p_i-p_{wf}=4.064\frac{qB}{hL_f}\left(\frac{\mu t}{k\phi c_t}\right)^{1/2}. \quad \ldots \ldots (1.46)$$

## 1.4 Radius of Investigation

The radius-of-investigation concept is of both quantitative and qualitative value in well test design and analysis. By radius of investigation, $r_i$, we mean the distance that a pressure transient has moved into a formation following a rate change in a well. We will show that this distance is related to formation rock and fluid properties and time elapsed since the rate change.

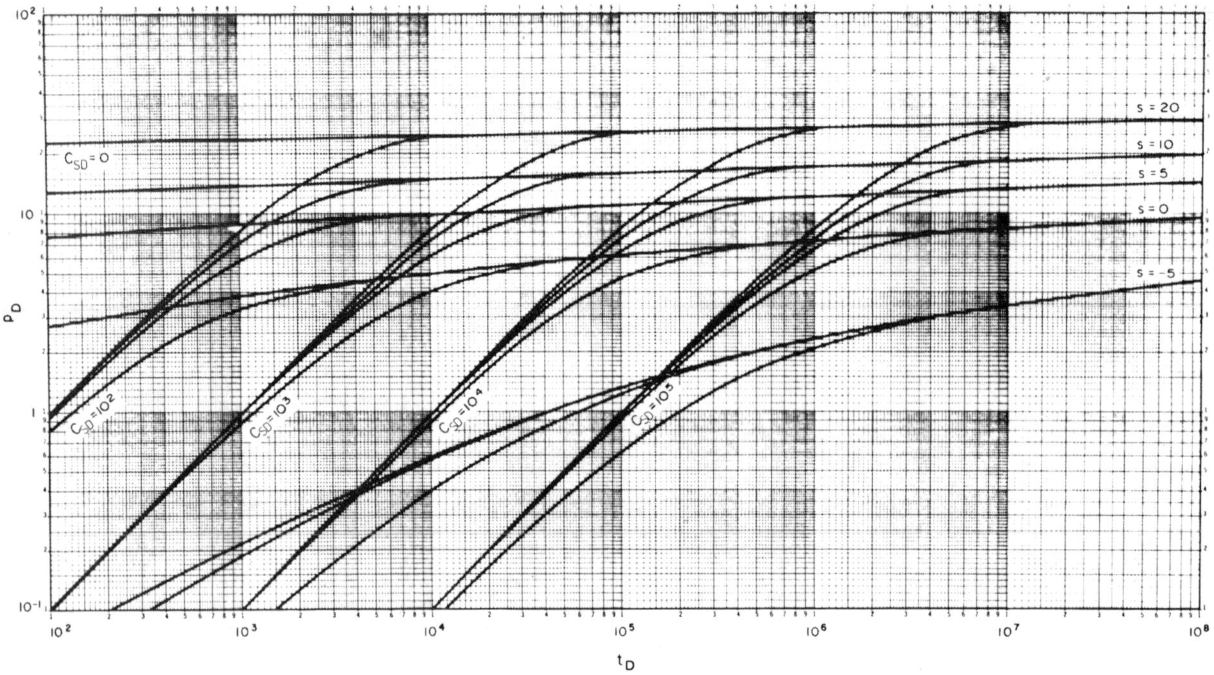


**Fig. 1.6**—Dimensionless pressure solutions with wellbore storage, skin factor. (After Earlougher.[10])

Before we develop a quantitative means of calculating $r_i$, however, we will examine pressure distributions at ever-increasing times to develop a feel for the movement of transients into a formation. Fig. 1.7 shows pressure as a function of radius for 0.1, 1.0, 10, and 100 hours after a well begins to produce from a formation originally at 2,000 psi. These pressure distributions were calculated using the *Ei*-function solution to the diffusivity equation for a well and a formation with these characteristics.

$$q = 177 \text{ STB/D},$$
$$\mu = 1 \text{ cp},$$
$$B = 1.2 \text{ RB/STB},$$
$$k = 10 \text{ md},$$
$$h = 150 \text{ ft},$$
$$\phi = 0.15,$$
$$c_t = 70.3 \times 10^{-6} \text{ psi}^{-1},$$
$$r_e = 3{,}000 \text{ ft},$$
$$r_w = 0.1 \text{ ft, and}$$
$$s = 0.$$

Two observations are particularly important:

1. The pressure in the wellbore (at $r=r_w$) decreases steadily with increasing flow time; likewise, pressures at other fixed values of $r$ also decrease with increasing flow time.

2. The pressure disturbance (or pressure transient) caused by producing the well moves further into the reservoir as flow time increases. For the range of flow times shown, there is always a point beyond which the drawdown in pressure from the original value is negligible.

Now consider a well into which we instantaneously inject a volume of liquid. This injection introduces a pressure disturbance into the formation; the

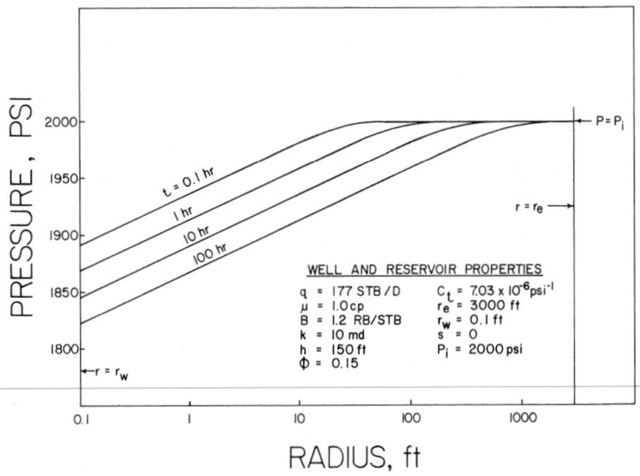


**Fig. 1.7** – Pressure distribution in formation near producing well.

disturbance at radius $r_i$ will reach its maximum at time $t_m$ after introduction of the fluid volume. We seek the relationship between $r_i$ and $t_m$. From the solution to the diffusivity equation for an instantaneous line source in an infinite medium,[9]

$$p - p_i = \frac{c_1}{t} e^{-r^2/4\eta t},$$

where $c_1$ is a constant, related to the strength of the instantaneous source. We find the time, $t_m$, at which the pressure disturbance is a maximum at $r_i$ by differentiating and setting equal to zero:

$$\frac{dp}{dt} = \frac{-c_1}{t^2} e^{-r^2/4\eta t} + \frac{c_1 r^2}{4\eta t^3} \cdot e^{-r^2/4\eta t} = 0.$$

Thus,

$$t_m = r_i^2/4\eta = 948\ \phi\mu c_t r_i^2/k.$$

Stated another way, in time $t$, a pressure disturbance reaches a distance $r_i$, which we shall call radius of investigation, as given by the equation

$$r_i = \left(\frac{kt}{948\ \phi\mu c_t}\right)^{1/2}. \quad \text{(1.47)}$$

The radius of investigation given by Eq. 1.47 also proves to be the distance a significant pressure disturbance is propagated by production or injection at a constant rate. For example, for the formation with pressure distributions shown in Fig. 1.7, application of Eq. 1.47 yields the following results.

| $t$ (hours) | $r_i$ (ft) |
|---|---|
| 0.1 | 32 |
| 1.0 | 100 |
| 10.0 | 316 |
| 100.0 | 1,000 |

Comparison of these results with the pressure distributions plotted shows that $r_i$ as calculated from Eq. 1.47 is near the point at which the drawdown in reservoir pressure caused by producing the well becomes negligible.

We also use Eq. 1.47 to calculate the radius of investigation achieved at any time after any rate change in a well. This is significant because the distance a transient has moved into a formation is approximately the distance from the well at which formation properties are being investigated at a particular time in a well test.

The radius of investigation has several uses in pressure transient test analysis and design. A qualitative use is to help explain the *shape* of a pressure buildup or pressure drawdown curve. For example, a buildup curve may have a difficult-to-interpret shape or slope at earliest times when the radius of investigation is in the zone of altered permeability, $k_s$, nearest the wellbore. Or, more commonly, a pressure buildup curve may change shape at long times when the radius of investigation reaches the general vicinity of a reservoir boundary (such as a sealing fault) or some massive reservoir heterogeneity. (In practice, we find that a heterogeneity or boundary influences pressure response in a well when the calculated radius of investigation is of the order of twice the distance to the heterogeneity.)

The radius-of-investigation concept provides a guide for well test design. For example, we may want to sample reservoir properties at least 500 ft from a tested well. How long a test shall be run? Six hours? Twenty-four hours? We are not forced to guess – or to run a test for an arbitrary length of time that could be either too short or too long. Instead, we can use the radius-of-investigation concept to estimate the time required to test to the desired depth in the formation.

The radius-of-investigation equation also provides a means of estimating the length of time required to achieve "stabilized" flow (i.e., the time required for a pressure transient to reach the boundaries of a tested reservoir). For example, if a well is centered in a cylindrical drainage area of radius $r_e$, then, setting $r_i = r_e$, the time required for stabilization, $t_s$, is found to be

$$t_s = 948\ \phi\mu c_t r_e^2/k. \quad \text{(1.48)}$$

It is no coincidence that this is the time at which pseudosteady-state flow begins (i.e., the time at which Eq. 1.12 becomes an accurate approximation of the exact solution to the diffusivity equation). A word of caution: For other drainage-area shapes, time to stabilize can be quite different, as illustrated in Example 1.3.

Useful as the radius-of-investigation concept is, we must caution the reader that it is no panacea. First, we note that it is exactly correct only for a homogeneous, isotropic, cylindrical reservoir – reservoir heterogeneities will decrease the accuracy of Eq. 1.47. Further, Eq. 1.47 is exact only for describing the time the maximum pressure disturbance reaches radius $r_i$ following an instantaneous burst of injection into or production from a well. Exact location of the radius of investigation becomes less well defined for continuous injection or production at constant rate following a change in rate. Limitations kept in mind, though, the radius-of-investigation concept can serve us well.

---

### *Example 1.4 – Calculation of Radius of Investigation*

**Problem.** We wish to run a flow test on an exploratory well for sufficiently long to ensure that the well will drain a cylinder of more than 1,000-ft radius. Preliminary well and fluid data analysis suggests that $k = 100$ md, $\phi = 0.2$, $c_t = 2 \times 10^{-5}$ psi$^{-1}$, and $\mu = 0.5$ cp. What length flow test appears advisable? What flow rate do you suggest?

**Solution.** The minimum length flow test would propagate a pressure transient approximately 2,000 ft from the well (twice the minimum radius of investigation for safety). Time required is

$$t = 948\ \phi\mu c_t r_i^2/k$$

$$= \frac{(948)(0.2)(0.5)(2 \times 10^{-5})(2{,}000)^2}{100}$$

$$= 75.8 \text{ hours.}$$

In principle, *any* flow rate would suffice – time required to achieve a particular radius of investigation is independent of flow rate. In practice, we require a flow rate sufficiently large that pressure change with time can be recorded with sufficient precision to be useful for analysis. What constitutes sufficient precision depends on the particular pressure gauge used in the test.

## 1.5 Principle of Superposition

At this point, the most useful solution to the flow equation, the *Ei*-function solution, appears to be applicable only for describing the pressure

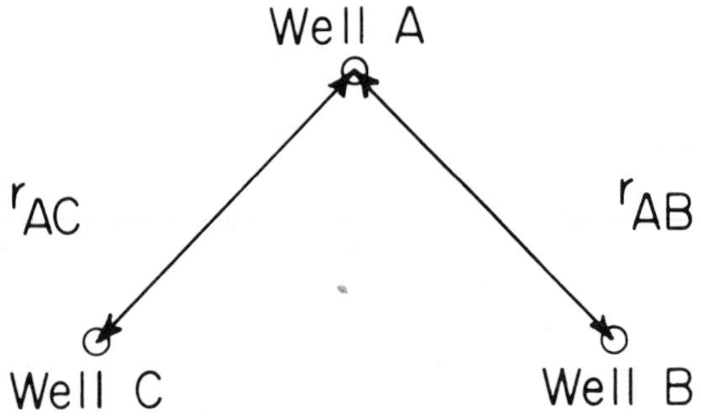


Fig. 1.8—Multiple-well system in infinite reservoir.

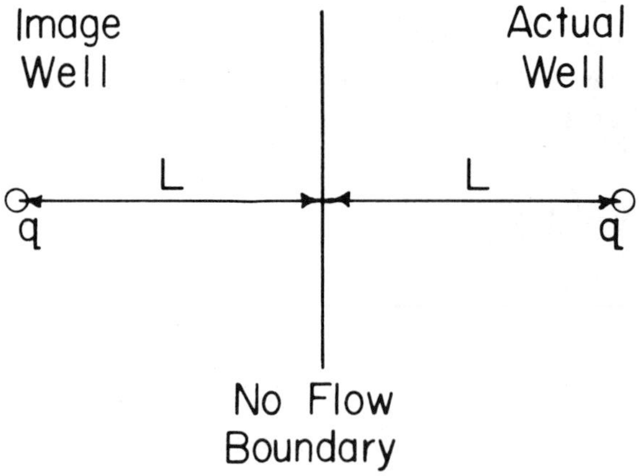


Fig. 1.9—Well near no-flow boundary illustrating use of imaging.

distribution in an infinite reservoir, caused by the production of a single well in the reservoir, and, most restrictive of all, production of the well at constant rate beginning at time zero. In this section, we demonstrate how application of the principle of superposition can remove some of these restrictions, and we conclude with examination of an approximation that greatly simplifies modeling a variable-rate well.

For our purposes, we state the principle of superposition in the following way: The total pressure drop at any point in a reservoir is the sum of the pressure drops at that point caused by flow in each of the wells in the reservoir. The simplest illustration of this principle is the case of more than one well in an infinite reservoir. As an example, consider three wells, Wells A, B, and C, that start to produce at the same time from an infinite reservoir (Fig. 1.8). Application of the principle of superposition shows that

$$(p_i - p_{wf})_{\text{total at Well A}} = (p_i - p)_{\text{due to Well A}} + (p_i - p)_{\text{due to Well B}} + (p_i - p)_{\text{due to Well C}}.$$

In terms of $Ei$ functions and logarithmic approximations,

$$(p_i - p_{wf})_{\text{total at Well A}} = -70.6 \frac{q_A B\mu}{kh}\left[\ln\left(\frac{1{,}688\ \phi\mu c_t r_{wA}^2}{kt}\right) - 2s_A\right]$$

$$-70.6 \frac{q_B B\mu}{kh} Ei\left(\frac{-948\ \phi\mu c_t r_{AB}^2}{kt}\right)$$

$$-70.6 \frac{q_C B\mu}{kh} Ei\left(\frac{-948\ \phi\mu c_t r_{AC}^2}{kt}\right), \quad \ldots (1.49)$$

where $q_A$ refers to the rate at which Well A produces; $q_B$, Well B; and $q_C$, Well C. Note that this equation includes a skin factor for Well A, but does not include skin factors for Wells B and C. Because most wells have a nonzero skin factor and because we are modeling pressure inside the zone of altered permeability near Well A, we must include its skin factor. However, the presence of nonzero skin factors for Wells B and C affects pressure only inside their zones of altered permeability and has no influence on pressure at Well A if Well A is not within the altered zone of either Well B or Well C.

Using this method, we can treat any number of wells flowing at constant rate in an infinite-acting reservoir. Thus, we can model so-called interference tests, which basically are designed to determine reservoir properties from the observed response in one well (such as Well A) to production from one or more other wells (such as Well B or Well C) in a reservoir. A relatively modern method of conducting interference tests, called pulse testing, is based on these ideas.[10]

Our next application of the principle of superposition is to simulate pressure behavior in bounded reservoirs. Consider the well in Fig. 1.9 a distance, $L$, from a single no-flow boundary (such as a sealing fault). Mathematically, this problem is identical to the problem of a well a distance $2L$ from an "image" well (i.e., a well that has the same production history as the actual well). The reason this two-well system simulates the behavior of a well near a boundary is that a line equidistant between the two wells can be shown to be a no-flow boundary – i.e., along this line the pressure gradient is zero, which means that there can be no flow. Thus, this is a simple two-well-in-an-infinite-reservoir problem:

$$p_i - p_{wf} = -70.6 \frac{qB\mu}{kh}\left(\ln \frac{1{,}688\ \phi\mu c_t r_w^2}{kt} - 2s\right)$$

$$-70.6 \frac{qB\mu}{kh} Ei\left(\frac{-948\ \phi\mu c_t (2L)^2}{kt}\right). \quad \ldots (1.50)$$

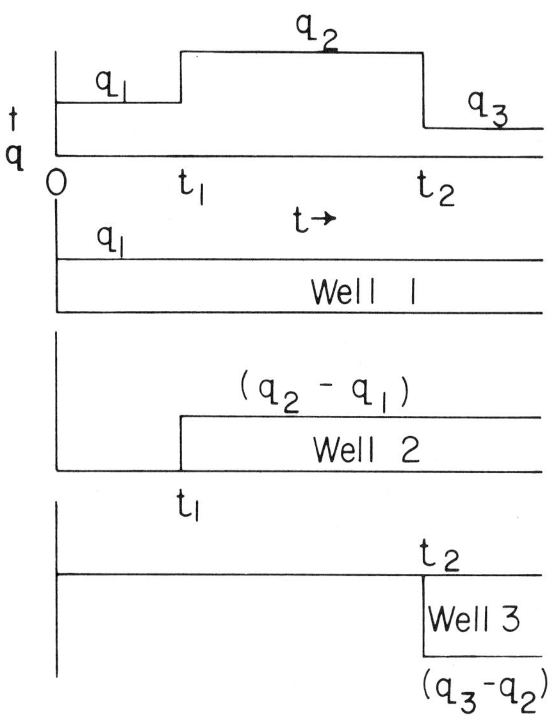


**Fig. 1.10**—Production schedule for variable-rate well.

Here again, note that whether the image well has a nonzero skin factor is immaterial. Its influence outside its zone of altered permeability is independent of whether this zone exists.

Extensions of the imaging technique also can be used, for example, to model (1) pressure distribution for a well between two boundaries intersecting at 90°; (2) the pressure behavior of a well between two parallel boundaries; and (3) pressure behavior for wells in various locations completely surrounded by no-flow boundaries in rectangular-shaped reservoirs. This last case has been studied quite completely; the study by Matthews *et al.*[11] is one of the methods most frequently used to estimate average drainage-area pressure from pressure buildup tests.

Our final and most important application of the superposition principle will be to model variable-rate producing wells. To illustrate this application, consider the case (Fig. 1.10) in which a well produces at rate $q_1$ from time 0 to time $t_1$; at $t_1$, the rate is changed to $q_2$; and at time $t_2$, the rate is changed to $q_3$. The problem that we wish to solve is this: At some time $t>t_2$, what is the pressure at the sandface of the well? To solve this problem, we will use superposition as before, but, in this case, each well that contributes to the total pressure drawdown will be at the same position in the reservoir – the wells simply will be "turned on" at different times.

The first contribution to a drawdown in reservoir pressure is by a well producing at rate $q_1$ starting at $t=0$. This well, in general, will be inside a zone of altered permeability; thus, its contribution to drawdown of reservoir pressure is

$$(\Delta p)_1=(p_i-p_{wf})_1=-70.6\frac{\mu q_1 B}{kh}\cdot\left[\ln\left(\frac{1{,}688\;\phi\mu c_t r_w^2}{kt}\right)-2s\right].$$

Starting at time $t_1$, the new total rate is $q_2$. We introduce a Well 2, producing at rate $(q_2-q_1)$ starting at time $t_1$, so that the total rate after $t_1$ is the required $q_2$. Note that total elapsed time since this well started producing is $(t-t_1)$; note further that this well is still inside a zone of altered permeability. Thus, the contribution of Well 2 to drawdown of reservoir pressure is

$$(\Delta p)_2=(p_i-p_{wf})_2=-70.6\frac{\mu(q_2-q_1)B}{kh}\cdot\left\{\ln\left[\frac{1{,}688\;\phi\mu c_t r_w^2}{k(t-t_1)}\right]-2s\right\}.$$

Similarly, the contribution of a third well is

$$(\Delta p)_3=(p_i-p_{wf})_3=-70.6\frac{\mu(q_3-q_2)B}{kh}\cdot\left\{\ln\left[\frac{1{,}688\;\phi\mu c_t r_w^2}{k(t-t_2)}\right]-2s\right\}.$$

Thus, the total drawdown for the well with two changes in rate is

$$\begin{aligned}p_i-p_{wf}&=(\Delta p)_1+(\Delta p)_2+(\Delta p)_3\\&=-70.6\frac{\mu q_1 B}{kh}\left[\ln\left(\frac{1{,}688\;\phi\mu c_t r_w^2}{kt}\right)-2s\right]\\&\quad-70.6\frac{\mu(q_2-q_1)B}{kh}\cdot\left\{\ln\left[\frac{1{,}688\;\phi\mu c_t r_w^2}{k(t-t_1)}\right]-2s\right\}\\&\quad-70.6\frac{\mu(q_3-q_2)B}{kh}\cdot\left\{\ln\left[\frac{1{,}688\;\phi\mu c_t r_w^2}{k(t-t_2)}\right]-2s\right\}.\qquad\text{.....(1.51)}\end{aligned}$$

Proceeding in a similar way, we can model an actual well with dozens of rate changes in its history; we also can model the rate history for a well with a continuously changing rate (with a sequence of constant-rate periods at the average rate during the period) – but, in many such cases, this use of superposition yields a lengthy equation, tedious to use in hand calculations. Note, however, that such an equation is valid only if Eq. 1.11 is valid for the total time elapsed since the well began to flow at its initial rate – i.e., for time $t$, $r_i$ must be $\leq r_e$.

*Example 1.5 – Use of Superposition*

**Problem.** A flowing well is completed in a reservoir that has the following properties.

$$\begin{aligned} p_i &= 2{,}500 \text{ psia}, \\ B &= 1.32 \text{ RB/STB}, \\ \mu &= 0.44 \text{ cp}, \\ k &= 25 \text{ md}, \\ h &= 43 \text{ ft}, \\ c_t &= 18\times10^{-6} \text{psi}^{-1}, \text{ and} \\ \phi &= 0.16. \end{aligned}$$

What will the pressure drop be in a shut-in well 500 ft from the flowing well when the flowing well has been shut in for 1 day following a flow period of 5 days at 300 STB/D?

**Solution.** We must superimpose the contributions of two wells because of the rate change:

$$p_i-p=-\frac{70.6\,\mu B}{kh}\left\{q_1\,Ei\left(\frac{-948\,\phi\mu c_t r^2}{kt}\right)\right.$$

$$\left.+\,(q_2-q_1)\,Ei\left[\frac{-948\,\phi\mu c_t r^2}{k(t-t_1)}\right]\right\}.$$

Now,

$$\frac{948\,\phi\mu c_t r^2}{k}=\left[(948)(0.16)(0.44)(1.8\times10^{-5})\right.$$

$$\left.\cdot(500)^2\right]/25=12.01.$$

Then,

$$p_i-p=-\frac{(70.6)(0.44)(1.32)}{(25)(43)}$$

$$\cdot\left\{(300)\,Ei\left[\frac{-12.01}{(6)(24)}\right]\right.$$

$$\left.+\,(0-300)\,Ei\left[\frac{-12.01}{(1)(24)}\right]\right\}$$

$$=11.44\left[-Ei(-0.0834)+Ei(-0.5)\right]$$

$$=11.44\,(1.989-0.560)$$

$$=16.35 \text{ psi}.$$

## 1.6 Horner's Approximation

In 1951, Horner[12] reported an approximation that can be used in many cases to avoid the use of superposition in modeling the production history of a variable-rate well. With this approximation, we can replace the sequence of *Ei* functions, reflecting rate changes, with a single *Ei* function that contains a single producing time and a single producing rate. The single rate is the most recent nonzero rate at which the well was produced; we call this rate $q_{\text{last}}$ for now. The single producing time is found by dividing cumulative production from the well by the most recent rate; we call this producing time $t_p$, or pseudoproducing time:

$t_p$(hours) =

$$24\,\frac{\text{cumulative production from well, } N_p\text{(STB)}}{\text{most recent rate, } q_{\text{last}}\text{(STB/D)}}\quad \ldots (1.52)$$

Then, to model pressure behavior at any point in a reservoir, we can use the simple equation

$$p_i-p=-\frac{70.6\,\mu q_{\text{last}} B}{kh}\,Ei\left(\frac{-948\,\phi\mu c_t r^2}{kt_p}\right).\quad (1.53)$$

Two questions arise logically at this point: (1) What is the basis for this approximation? (2) Under what conditions is it applicable?

The basis for the approximation is not rigorous, but intuitive, and is founded on two criteria: (1) If we use a single rate in the approximation, the clear choice is the most recent rate; such a rate, maintained for any significant period, determines the pressure distribution nearest the wellbore and approximately out to the radius of investigation achieved with that rate. (2) Given the single rate to use, intuition suggests that we choose an effective production time such that the product of the rate and the production time results in the correct cumulative production. In this way, material balances will be maintained accurately.

But when is the approximation adequate? If we maintain a most-recent rate for too brief a time interval, previous rates will play a more important role in determining the pressure distribution in a tested reservoir. We can offer two helpful guidelines. First, if the most recent rate is maintained sufficiently long for the radius of investigation achieved at this rate to reach the drainage radius of the tested well, then Horner's approximation is always sufficiently accurate. This rule is quite conservative, however. Second, we find that, for a new well that undergoes a series of rather rapid rate changes, it is usually sufficient to establish the last constant rate for at least twice as long as the previous rate. When there is any doubt about whether these guidelines are satisfied, the safe approach is to use superposition to model the production history of the well.

---

*Example 1.6 – Application of Horner's Approximation*

**Problem.** Following completion, a well is produced for a short time and then shut in for a buildup test. The production history was as follows.

| Production Time (hours) | Total Production (STB) |
|---|---|
| 25 | 52 |
| 12 | 0 |
| 26 | 46 |
| 72 | 68 |

1. Calculate the pseudoproducing time, $t_p$.
2. Is Horner's approximation adequate for this case? If not, how should the production history for this well be simulated?

**Solution.**

1.

$$q_{\text{last}} = \frac{68 \text{ STB}}{72 \text{ hours}} \times \frac{24 \text{ hours}}{\text{day}} = 22.7 \text{ STB/D}.$$

Then,

$$t_p = \frac{24 \text{ (cumulative production, STB)}}{q_{\text{last}}, \text{ STB/D}}$$

$$= \frac{(24)(166)}{(22.7)} = 176 \text{ hours}.$$

2. In this case,

$$\frac{\Delta t_{\text{last}}}{\Delta t_{\text{next-to-last}}} = \frac{72}{26} = 2.77 > 2.$$

Thus, Horner's approximation is probably adequate for this case. It should not be necessary to use superposition, which is required when Horner's approximation is not adequate.

## Exercises

1.1 Compare values of $Ei(-x)$ and $\ln(1.781x)$ for the following values of $x$: 0.01, 0.02, 0.1, and 1. What do you conclude about the accuracy of the logarithmic approximation? About its range of applicability?

1.2 A well has flowed for 10 days at a rate of 350 STB/D. Rock and fluid properties include $B = 1.13$ RB/STB; $p_i = 3{,}000$ psia; $\mu = 0.5$ cp; $k = 25$ md (uniform to wellbore – i.e., $s = 0$); $h = 50$ ft; $c_t = 2 \times 10^{-5}$ psi$^{-1}$; $\phi = 0.16$; and $r_w = 0.333$ ft. Calculate pressures at radii of 0.333, 1, 10, 100, 1,000, and 3,160 ft, and plot the results as pressure vs. the logarithm of radius. What minimum drainage radius have you assumed in this calculation?

1.3 For the well described in Exercise 1.2, plot pressure in the wellbore vs. logarithm of time at times of 0.1, 1, and 10 days. What minimum drainage radius have you assumed in this calculation?

1.4 Calculate (a) elapsed time required for the *Ei*-function solution to be valid for the conditions described in Exercise 1.2; (b) time required for the logarithmic approximation of the *Ei* function to apply for calculations at the wellbore; and (c) time required for the logarithmic approximation to apply for calculations at a radius of 1,000 ft. Is the logarithmic approximation valid by the time the *Ei* function itself is a valid solution to the flow equation at the wellbore? At a radius of 1,000 ft?

1.5 Estimate the radius of investigation achieved after 10 days flow time for the reservoir described in Exercise 1.2. Compare this estimate with the extrapolation to 3,000 psi of the straight line passing through radii of 0.333 and 100 ft on the plot of pressure vs. logarithm of radius.

On this plot, how far into the formation has a significant pressure disturbance been propagated? What is the size of the pressure disturbance at the radius of investigation calculated from Eq. 1.47?

1.6 If the drainage radius of the well described in Exercise 1.2 were 3,160 ft, and if the flow rate at the well suddenly was changed from 350 to 500 STB/D, how long would it take for the well to stabilize at the new rate?

1.7 Suppose the well described in Exercise 1.2 flowed at a rate of 700 STB/D for 10 days. Prepare a plot of pressure vs. logarithm of radius for this situation on the same graph as the plot developed for a rate of 350 STB/D. Is the radius of investigation calculated from Eq. 1.47 affected by change in flow rate? Does the extrapolation of the straight line referred to in Exercise 1.5 change? What is the effect of increased rate?

1.8 Write an equation similar to Eq. 1.49 for the case in which Wells, A, B, and C begin to produce at different times from one another. What do you assume about the location of reservoir boundaries when you write this equation?

1.9 (a) Suppose a well is 250 ft due west of a north-south trending fault. From pressure transient tests, the skin factor, $s$, of this well has been found to be 5.0. Suppose further that the well has been flowing for 8 days at 350 B/D; reservoir and well properties are those given in Exercise 1.2. Calculate pressure at the flowing well.

(b) Suppose there is a shut-in well 500 ft due north of the producing well. Calculate the pressure at the shut-in well at the end of 8 days.

1.10 A reservoir has the following properties.

$p_i$ = 2,500 psia,
$B$ = 1.32 RB/STB,
$\mu$ = 0.44 cp,
$k$ = 25 md,
$h$ = 43 ft,
$c_t$ = $18 \times 10^{-6}$ psi$^{-1}$, and
$\phi$ = 0.16.

In this reservoir, a well is opened to flow at 250 STB/D for 1 day. The second day its flow is increased to 450 B/D and the third to 500 B/D. What is the pressure in a shut-in well 660 ft away after the third day?

1.11 In Example 1.6, Application of Horner's Approximation, what influence did the 12-hour shut-in time have on the calculation? How would the influence of this shut-in period have changed had the shut-in period been 120 hours? How do you suggest that the calculation procedure be modified to take into account long shut-in periods prior to producing at the final rate?

1.12 Consider a well and formation with the following properties.

$B$ = 1.0 RB/STB,
$\mu$ = 1.0 cp,
$h$ = 10 ft,
$k$ = 25 md,
$\phi$ = 0.2,
$p_i$ = 3,000 psig,
$c_t$ = $10 \times 10^{-6}$ psi$^{-1}$,
$s$ = 0, and
$r_w$ = 1.0 ft.

The well produced 100 STB/D for 3.0 days, was shut-in for the next 1.0 day, produced 150 STB/D for the next 2.0 days, produced 50 STB/D for the next 1.0 day, and produced 200 STB/D for next 2.0 days.

(a) Calculate the pseudoproducing time, $t_p$. Compare this with the actual total producing time.

(b) Calculate and plot the pressure distribution in the reservoir at the end of 9 days using Horner's approximation.

(c) On the same graph, plot the pressure distribution at the end of 9 days using superposition.

(d) What do you conclude about the adequacy of Horner's approximation in this particular case?

1.13 A well and reservoir have the following properties.

$1.742 \times 10^6$ sq ft

$$A = 17.42 \times 10^6 \text{ sq ft (40 acres)},$$
$$\phi = 0.2,$$
$$\mu = 1 \text{ cp},$$
$$c_t = 10 \times 10^{-6} \text{ psi}^{-1},$$
$$k = 100 \text{ md},$$
$$h = 10 \text{ ft},$$
$$s = 3.0,$$
$$r_w = 0.3 \text{ ft, and}$$
$$B = 1.2 \text{ RB/STB}.$$

5 areas on handout

For each of the drainage areas in Table 1.2, determine (a) the time (hours) up to which the reservoir is infinite-acting; (b) the time (hours) beyond which the pseudosteady-state solution is an adequate approximation; (c) PI of the well; and (d) stabilized production rate with 500-psi drawdown.

# References

1. Matthews, C.S. and Russell, D.G.: *Pressure Buildup and Flow Tests in Wells*, Monograph Series, SPE, Dallas (1967) **1**.
2. van Everdingen, A.F. and Hurst, W.: "The Application of the Laplace Transformation to Flow Problems in Reservoirs," *Trans.*, AIME (1949) **186**, 305-324.
3. Slider, H.C.: *Practical Petroleum Reservoir Engineering Methods*, Petroleum Publishing Co., Tulsa (1976) 70.
4. Hawkins, M.F. Jr.: "A Note on the Skin Effect," *Trans.*, AIME (1956) **207**, 356-357.
5. Odeh, A.S.: "Pseudosteady-State Flow Equation and Productivity Index for a Well With Noncircular Drainage Area," *J. Pet. Tech.* (Nov. 1978) 1630-1632.
6. Agarwal, R.G., Al-Hussainy, R., and Ramey, H.J. Jr.: "An Investigation of Wellbore Storage and Skin Effect in Unsteady Liquid Flow – I. Analytical Treatment," *Soc. Pet. Eng. J.* (Sept. 1970) 279-290; *Trans.*, AIME, **249**.
7. Wattenbarger, R.A. and Ramey, H.J. Jr.: "An Investigation of Wellbore Storage and Skin Effect in Unsteady Liquid Flow – II. Finite-Difference Treatment," *Soc. Pet. Eng. J.* (Sept. 1970) 291-297; *Trans.*, AIME, **249**.
8. Katz, D.L. *et al.*: *Handbook of Natural Gas Engineering*, McGraw-Hill Book Co. Inc., New York (1959) 411.
9. Carslaw, H.S. and Jaeger, J.C.: *Conduction of Heat in Solids*, second ed., Oxford at the Clarendon Press (1959) 258.
10. Earlougher, R.C. Jr.: *Advances in Well Test Analysis*, Monograph Series, SPE, Dallas (1977) **5**.
11. Matthews, C.S., Brons, F., and Hazebroek, P.: "A Method for Determination of Average Pressure in a Bounded Reservoir," *Trans.*, AIME (1954) **201**, 182-191.
12. Horner, D.R.: "Pressure Build-Up in Wells," *Proc.*, Third World Pet. Cong., The Hague (1951) Sec. II, 503-523.

# Chapter 2
# Pressure Buildup Tests

## 2.1 Introduction

This chapter discusses the most frequently used pressure transient test, the pressure buildup test. Basically, the test is conducted by producing a well at constant rate for some time, shutting the well in (usually at the surface), allowing the pressure to build up in the wellbore, and recording the pressure (usually downhole) in the wellbore as a function of time. From these data, it is frequently possible to estimate formation permeability and current drainage-area pressure, and to characterize damage or stimulation and reservoir heterogeneities or boundaries.

The analysis method discussed in this chapter is based largely on a plotting procedure suggested by Horner.[1] While this procedure is strictly correct only for infinite-acting reservoirs, these plots also can be interpreted correctly for finite reservoirs,[2] so only this plotting method is emphasized. Another important analysis technique for buildup tests, using type curves, is discussed in Chap. 4.

The chapter begins with a derivation of the Horner plotting technique and the equation for calculating skin factor. Differences in actual and idealized test behavior then are discussed, followed by comments on dealing with deviations from assumptions made in developing the Horner plotting technique. We then examine qualitatively the behavior of actual tests in common reservoir situations. The chapter next develops in detail a systematic analysis procedure for buildup tests: (1) effects and duration of afterflow (continued production into the wellbore following surface shut-in), (2) determination of permeability, (3) well damage and stimulation, (4) determination of pressure level in the surrounding formation, and (5) reservoir limits tests.

Up to this point, the analysis procedure discussed is applicable only to single-phase flow of a slightly compressible liquid. The chapter concludes with a discussion of how the procedure can be modified to analyze tests in gas wells and in wells with two or three phases flowing simultaneously.

## 2.2 The Ideal Buildup Test

In this section we derive an equation describing an ideal pressure buildup test. By ideal test we mean a test in an infinite, homogeneous, isotropic reservoir containing a slightly compressible, single-phase fluid with constant fluid properties. Any wellbore damage or stimulation is considered to be concentrated in a skin of zero thickness at the wellbore; at the instant of shut-in, flow into the wellbore ceases totally. No actual buildup test is modeled exactly by this idealized description, but the analysis methods developed for this case prove useful for more realistic situations if we recognize the effect of deviation from some of these assumptions on actual test behavior.

Assume that (1) a well is producing from an infinite-acting reservoir (one in which no boundary effects are felt during the entire flow and later shut-in period), (2) the formation and fluids have uniform properties, so that the *Ei* function (and, thus, its logarithmic approximation) applies, and (3) that Horner's pseudoproducing time approximation is applicable. If the well has produced for a time $t_p$ at rate $q$ before shut-in, and if we call time elapsed since shut-in $\Delta t$, then, using superposition (Fig. 2.1), we find that following shut-in

$$p_i - p_{ws} = -70.6\frac{qB\mu}{kh}\left\{\ln\left[\frac{1{,}688\,\phi\mu c_t r_w^2}{k(t_p+\Delta t)}\right] - 2s\right\}$$

$$-70.6\frac{(-q)\,B\mu}{kh}\left[\ln\left(\frac{1{,}688\,\phi\mu c_t r_w^2}{k\Delta t}\right) - 2s\right],$$

which becomes

$$p_{ws} = p_i - 70.6\frac{qB\mu}{kh}\ln\left[(t_p+\Delta t)/\Delta t\right],$$

or

$$p_{ws} = p_i - 162.6\frac{qB\mu}{kh}\log\left[(t_p+\Delta t)/\Delta t\right]. \quad \ldots (2.1)$$

The form of Eq. 2.1 suggests that shut-in BHP, $p_{ws}$, recorded during a pressure buildup test should plot as a straight-line function of log $[(t_p+\Delta t)/\Delta t]$. Further, the slope $m$ of this straight line should be

$$m = -162.6\frac{qB\mu}{kh}.$$

It is convenient to use the absolute value of $m$ in test

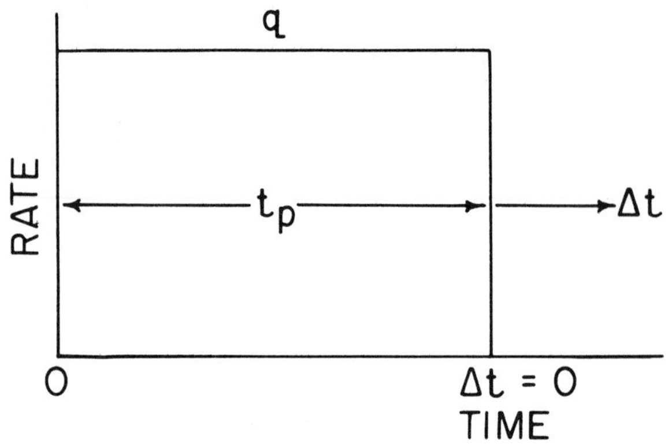


Fig. 2.1 – Rate history for ideal pressure buildup test.

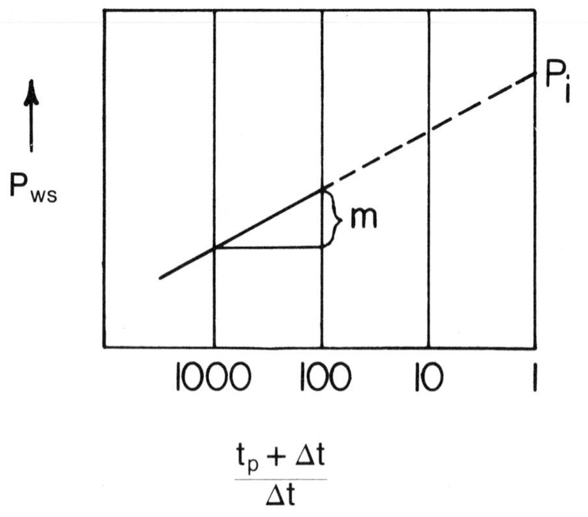


Fig. 2.2 – Plotting technique for pressure buildup test.

analysis; accordingly, in this text we will use the convention that $m$ is considered a positive number and that

$$m = 162.6\frac{qB\mu}{kh}. \quad \text{(2.2)}$$

Thus, formation permeability, $k$, can be determined from a buildup test by measuring the slope $m$. In addition, if we extrapolate this straight line to infinite shut-in time [i.e., $(t_p + \Delta t)/\Delta t = 1$] the pressure at this time will be the original formation pressure $p_i$.

Conventional practice in the industry is to plot $p_{ws}$ vs. $(t_p + \Delta t)/\Delta t$ (Fig. 2.2) on semilogarithmic paper with values of $(t_p + \Delta t)/\Delta t$ decreasing from left to right. The slope $m$ on such a plot is found by simply subtracting the pressures at any two points on the straight line that are one cycle (i.e., a factor of 10) apart on the semilog paper.

We also can determine skin factor $s$ from the data available in the idealized pressure buildup test. At the instant a well is shut in, the flowing BHP, $p_{wf}$, is

$$p_{wf} = p_i + 70.6\frac{qB\mu}{kh}\left[\ln\left(\frac{1{,}688\ \phi\mu c_t r_w^2}{kt_p}\right) - 2s\right]$$

$$= p_i + 162.6\frac{qB\mu}{kh}$$

$$\cdot\left[\log\left(\frac{1{,}688\ \phi\mu c_t r_w^2}{kt_p}\right) - 0.869\,s\right]$$

$$= p_i + m\left[\log\left(\frac{1{,}688\ \phi\mu c_t r_w^2}{kt_p}\right) - 0.869\,s\right].$$

At shut-in time $\Delta t$ in the buildup test,

$$p_{ws} = p_i - m\log[(t_p + \Delta t)/\Delta t].$$

Combining these equations and solving for the skin factor $s$, we have

$$s = 1.151\left(\frac{p_{ws} - p_{wf}}{m}\right) + 1.151\log\left(\frac{1{,}688\ \phi\mu c_t r_w^2}{k\Delta t}\right)$$

$$+ 1.151\log\left(\frac{t_p + \Delta t}{t_p}\right). \quad \text{(2.3)}$$

It is conventional practice in the petroleum industry to choose a fixed shut-in time, $\Delta t$, of 1 hour and the corresponding shut-in pressure, $p_{1\,hr}$, to use in this equation (although *any* shut-in time and the corresponding pressure would work just as well). The pressure, $p_{1\,hr}$, must be *on* the straight line or its extrapolation. We usually can assume further that log $(t_p + \Delta t)/t_p$ is negligible. With these simplifications,

$$s = 1.151\left[\frac{(p_{1\,hr} - p_{wf})}{m} - \log\left(\frac{k}{\phi\mu c_t r_w^2}\right) + 3.23\right]. \quad \text{(2.4)}$$

Note again that the slope $m$ is considered to be a positive number in this equation.

In summary, from the ideal buildup test, we can determine formation permeability (from the slope $m$ of the plotted test results), original reservoir pressure, $p_i$, and skin factor, $s$, which is a measure of damage or stimulation.

---

### *Example 2.1 – Analysis of Ideal Pressure Buildup Test*

**Problem.** A new oil well produced 500 STB/D for 3 days; it then was shut in for a pressure buildup test, during which the data in Table 2.1 were recorded.

For this well, net sand thickness, $h$, is 22 ft; formation volume factor, $B_o$, is 1.3 RB/STB; porosity, $\phi$, is 0.2; total compressibility, $c_t$, is $20 \times 10^{-6}$; oil viscosity, $\mu_o$, is 1.0 cp; and wellbore radius, $r_w$, is 0.3 ft. From these data, estimate formation permeability, $k$, initial reservoir pressure, $p_i$, and skin factor, $s$.

**TABLE 2.1 – IDEAL PRESSURE BUILDUP DATA**

| Time After Shut-In $\Delta t$ (hours) | $p_{ws}$ (psig) |
|---|---|
| 0 | 1,150 |
| 2 | 1,794 |
| 4 | 1,823 |
| 8 | 1,850 |
| 16 | 1,876 |
| 24 | 1,890 |
| 48 | 1,910 |

**TABLE 2.2 – BUILDUP TEST DATA FOR HORNER PLOT**

| $\Delta t$ (hours) | $\frac{t_p+\Delta t}{\Delta t}$ | $p_{ws}$ (psig) |
|---|---|---|
| 2 | 37.0 | 1,794 |
| 4 | 19.0 | 1,823 |
| 8 | 10.0 | 1,850 |
| 16 | 5.5 | 1,876 |
| 24 | 4.0 | 1,890 |
| 48 | 2.5 | 1,910 |

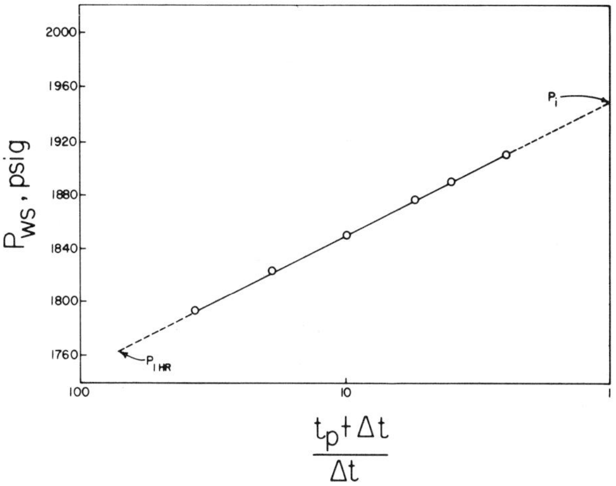


**Fig. 2.3** – Ideal pressure buildup test graph.

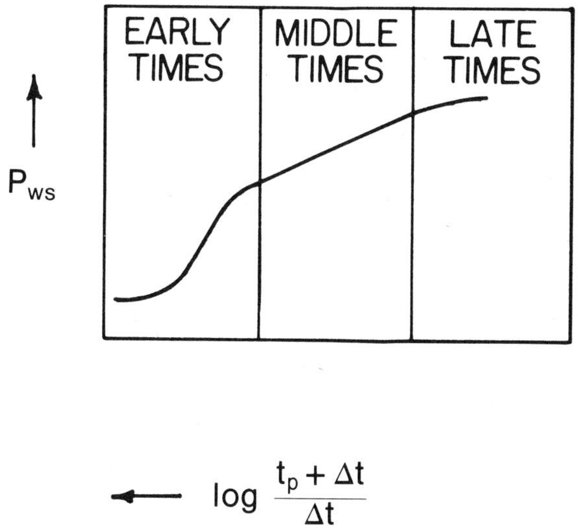


**Fig. 2.4** – Actual buildup test graph.

**Solution.** To estimate permeability, original reservoir pressure, and skin factor, we must plot shut-in BHP, $p_{ws}$, vs. log $(t_p+\Delta t)/\Delta t$; measure the slope $m$ and use Eq. 2.2 to calculate formation permeability, $k$; extrapolate the curve to $[(t_p+\Delta t)/\Delta t]=1$ and read original reservoir pressure, $p_i$; and use Eq. 2.4 to calculate the skin factor $s$. (See Fig. 2.3.)

Producing time, $t_p$, is given to be 3 days, or 72 hours (in this case, Horner's approximation is *exact* because the well was produced at constant rate since time zero). Thus, we develop Table 2.2.

We plot these data, and they fall along a straight line suggested by ideal theory. The slope $m$ of the straight line is $1{,}950-1{,}850=100$ psi (units are actually psi/cycle). The formation permeability $k$ is

$$k=162.6\frac{qB\mu}{mh}=\frac{(162.6)(500)(1.3)(1.0)}{(100)(22)}=48 \text{ md.}$$

From extrapolation of the buildup curve to $[(t_p+\Delta t)/\Delta t]=1$, $p_i=1{,}950$ psig. The skin factor $s$ is found from Eq. 2.4:

$$s=1.151\left[\frac{(p_{1\,\mathrm{hr}}-p_{wf})}{m}-\log\left(\frac{k}{\phi\mu c_t r_w^2}\right)+3.23\right].$$

The value for $p_{ws}$ is $p_{1\,\mathrm{hr}}$ on the ideal straight line at $(t_p+\Delta t)/\Delta t=(72+1)/1=73$; this value is $p_{1\,\mathrm{hr}}=$ 1,764 psig. Thus,

$$s=1.151\left[\frac{(1{,}764-1{,}150)}{(100)}\right.$$

$$\left.-\log\left(\frac{48}{(0.2)(1.0)(2\times10^{-5})(0.3)^2}\right)+3.23\right]$$

$$=1.43.$$

This means the well has a flow restriction.

## 2.3 Actual Buildup Tests

Encouraged by the simplicity and ease of application of the ideal buildup test theory, we may test an actual well and obtain a most discouraging result: Instead of a single straight line for all times, we obtain a curve with a complicated shape. To explain what went wrong, the radius-of-investigation concept is useful. Based on this concept, we logically can divide a buildup curve into three regions (Fig. 2.4): (1) an early-time region during which a pressure transient is moving through the formation nearest the wellbore; (2) a middle-time region during which the pressure transient has moved away from the wellbore and into the bulk formation; and (3) a late-time region, in which the radius of investigation has reached the well's drainage boundaries. Let us examine each region in more detail.

### Early-Time Region

As we have noted, most wells have altered permeability near the wellbore. Until the pressure transient caused by shutting in the well for the buildup test moves through this region of altered permeability, there is no reason to expect a straight-line slope that is related to formation permeability. (We should note that the ideal buildup curve – i.e., one with a single straight line over virtually all time – is possible for a damaged well only when the damage is concentrated in a very thin skin at the sandface.)

There is another complication at earliest times in a pressure buildup test. Continued movement of fluid into a wellbore (afterflow, a form of wellbore storage) following the usual surface shut-in compresses the fluids (gas, oil, and water) in the wellbore. Why should this affect the character of a buildup curve at earliest times? Perhaps the clearest answer lies in the observation that the idealized theory leading to the equation $p_{ws} = p_i - m \log [(t_p + \Delta t)/\Delta t]$ explicitly assumed that, at $\Delta t = 0$, flow rate abruptly changed from $q$ to zero. In practice, $q$ declines toward zero but, at the instant of surface shut-in, the downhole rate is, in fact, still $q$. (See Fig. 2.5.) Thus, one of the assumptions we made in deriving the buildup equation is violated in the actual test, and another question arises. Does afterflow *ever* diminish to such an extent that data obtained in a pressure buildup test can be analyzed as in the ideal test? The answer is yes, fortunately, but the important problem of finding the point at which afterflow ceases distorting buildup data remains. This is the point at which the early-time region usually ends, because afterflow frequently lasts longer than the time required for a transient to move through the altered zone near a well. We will deal with this problem more completely when we discuss a systematic analysis procedure for pressure buildup tests.

### Middle-Time Region

When the radius of investigation has moved beyond the influence of the altered zone near the tested well, and when afterflow has ceased distorting the pressure buildup test data, we usually observe the ideal straight line whose slope is related to formation permeability. This straight line ordinarily will continue until the radius of investigation reaches one or more reservoir boundaries, massive heterogeneities, or a fluid/fluid contact.

Systematic analysis of a pressure buildup test using the Horner method of plotting $[p_{ws}$ vs. $\log (t_p + \Delta t)/\Delta t]$ requires that we recognize this middle-time line and that, in particular, we do not confuse it with false straight lines in the early- and late-time regions. As we have seen, determination of reservoir permeability and skin factor depends on recognition of the middle-time line; estimation of average drainage-area pressure for a well in a developed field also requires that this line be defined.

### Late-Time Region

Given enough time, the radius of investigation eventually will reach the drainage boundaries of a well. In this late-time region pressure behavior is influenced by boundary configuration, interference from nearby wells, significant reservoir heterogeneities, and fluid/fluid contacts.

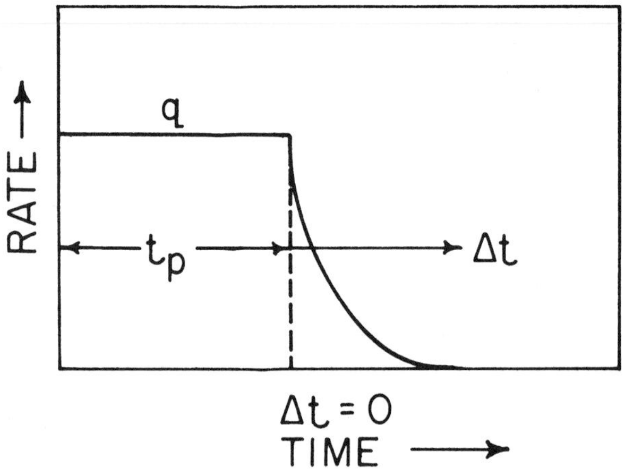


Fig. 2.5 – Rate history for actual pressure buildup test.

## 2.4 Deviations From Assumptions in Ideal Test Theory

In suggesting that tests logically can be divided into early-, middle-, and late-time regions, we have recognized that several assumptions made in developing the theory of ideal buildup test behavior are not valid for actual tests. In this section, we examine further the implications of three over-idealized assumptions: (1) the infinite-reservoir assumption; (2) the single-phase liquid assumption; and (3) the homogeneous reservoir assumption.

### Infinite Reservoir Assumption

In developing the equation suggesting the Horner plot, we assumed that the reservoir was infinite acting during both the production period preceding the buildup test and the buildup test itself. Frequently, the reservoir is at pseudosteady-state before shut-in; if so, neither the *Ei*-function solution nor its logarithmic approximation should be used to describe the pressure drawdown caused by the producing well:

$$(p_i - p_{wf})_{\text{prod well}} \neq -70.6 \frac{qB\mu}{kh} \cdot \left\{ \ln\left[ \frac{1{,}688\, \phi\mu c_t r_w^2}{k(t_p + \Delta t)} \right] - 2s \right\}.$$

Instead, if the well is centered in a cylindrical reservoir,

$$(p_i - p_{wf})_{\text{prod well}} = 141.2 \frac{qB\mu}{kh} \cdot \left[ \frac{0.000527\, k(t_p + \Delta t)}{\phi\mu c_t r_e^2} + \ln\left( \frac{r_e}{r_w} \right) - \frac{3}{4} \right].$$

Thus, we must conclude that in principle, the Horner plot is *incorrect* when the reservoir is not infinite acting during the flow period preceding the buildup test. Boundaries become important as $r_i \rightarrow r_e$.

The problem is compounded when $r_i \rightarrow r_e$ after shut-in. Then, too, the Horner plot is incorrect in principle.

This difficulty is resolved in different ways by

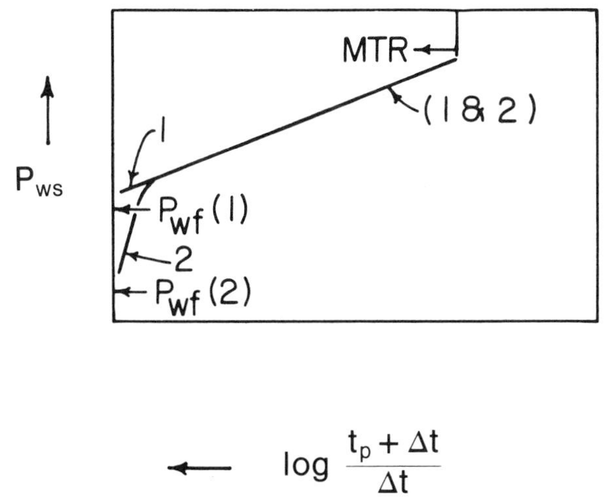


**Fig. 2.6** – Buildup test with no afterflow: (1) without wellbore damage and (2) with wellbore damage.

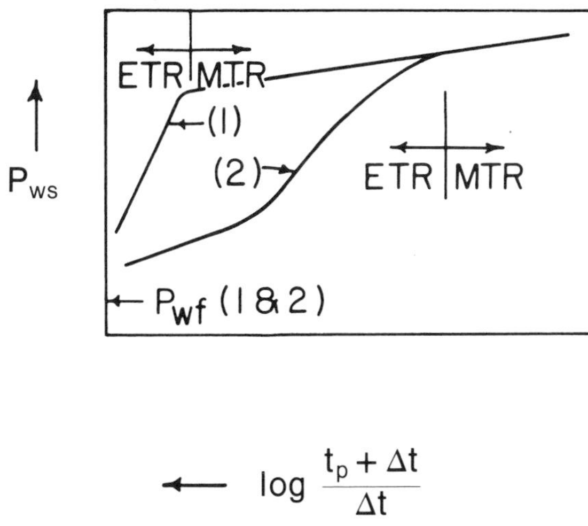


**Fig. 2.7** – Buildup test with formation damage: (1) without afterflow and (2) with afterflow.

different analysts. In this text, we will use a method supported by the research of Cobb and Smith.[2] We will use the Horner plot for all tests (even when the reservoir has reached pseudosteady-state during the production period preceding the test) for the following reasons.

1. This method of plotting is correct *theoretically* for an infinite-acting reservoir (i.e., one for which, at time $t_p + \Delta t$, $r_i < r_e$).

2. The Horner plot offers a convenient means of extrapolating to $\Delta t \rightarrow \infty$ not found in some other plots; the pressure at this shut-in time is a useful checkpoint for the test analyst.

3. For finite-acting reservoirs, formation permeability can be determined accurately from the slope of the Horner plot at even greater shut-in times than from a plotting method developed specifically for reservoirs at pseudosteady state at shut-in.[2] The curve *will* begin to deviate from the ideal slope before $r_i$ during shut-in reaches reservoir boundaries. However, the middle-time region still can be identified, except for long early-time regions.

Other analysis methods for finite-acting reservoirs are discussed by Miller, Dyes, and Hutchinson[3] (MDH) and Slider.[4] Many analysts use the data-plotting method suggested by MDH because it is simpler than the Horner method. Consider a buildup test with a middle-time region described by Eq. 2.1:

$$p_{ws} = p_i - m \log\,(t_p + \Delta t)/\Delta t$$

$$= p_i - m \log\,(t_p + \Delta t) + m \log \Delta t.$$

If $t_p \gg \Delta t$ during the range of shut-in time values examined, then $\log\,(t_p + \Delta t) \simeq \log t_p = \text{constant}$, and

$$p_{ws} = \text{constant} + m \log \Delta t.$$

This leads to the plotting technique suggested by MDH: $p_{ws}$ vs. $\log \Delta t$. It has the same slope $m$ as the Horner plot (in the time range of applicability). Further insight into this plotting technique is provided by Exercise 2.2.

### Single-Phase Liquid Assumption

The assumption that a petroleum reservoir contains only a single-phase liquid must be modified. Even reservoirs in which only oil *flows* contain an immobile water saturation; many also contain an immobile gas saturation. Also, in many cases, compressibility of the formation cannot be ignored. These factors are taken into account if we use total compressibility, $c_t$, in solutions to flow equations:

$$c_t = c_o S_o + c_w S_w + c_g S_g + c_f. \qquad (1.4)$$

Even in single-phase flow, when $S_g \neq 0$, evaluation of oil compressibility, $c_o$, and water compressibility, $c_w$, are somewhat complicated:

$$c_o = \frac{-1}{B_o}\frac{dB_o}{dp} + \frac{B_g}{B_o}\frac{dR_s}{dp}. \qquad (2.5)$$

$$c_w = \frac{-1}{B_w}\frac{dB_w}{dp} + \frac{B_g}{B_w}\frac{dR_{sw}}{dp}. \qquad (2.6)$$

These compressibility relationships allow us to model single-phase flow of oil. Later sections in this chapter discuss modifications to the flow equations required for (1) single-phase gas flow and (2) simultaneous flow of two or three phases. Appendix D summarizes correlations useful for calculating compressibilities and other fluid properties needed in analysis of well tests.

### Homogeneous Reservoir Assumption

No reservoir is homogeneous, yet solutions to the flow equations are valid only for homogeneous reservoirs. The solutions prove to be adequate for most real reservoirs, particularly early in time while conditions nearest the tested well dominate test behavior. Rate of pressure change is dominated by *average* rock and fluid properties. When massive heterogeneities are encountered (particularly in a localized portion of the reservoir), the simple solutions to flow equations lose accuracy. Examples include changes of depositional environment, with

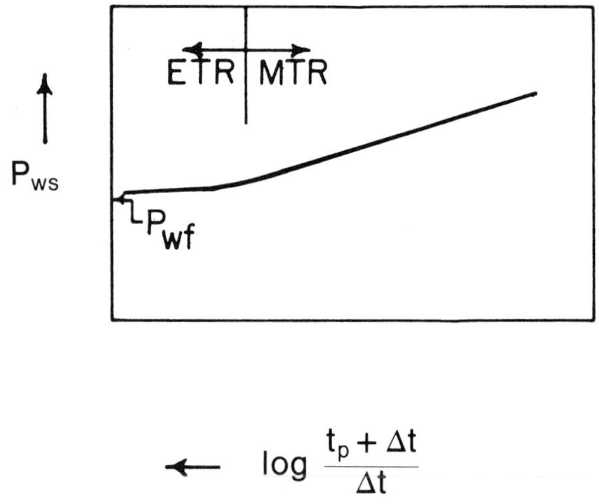


Fig. 2.8 – Buildup test in hydraulically fractured well.

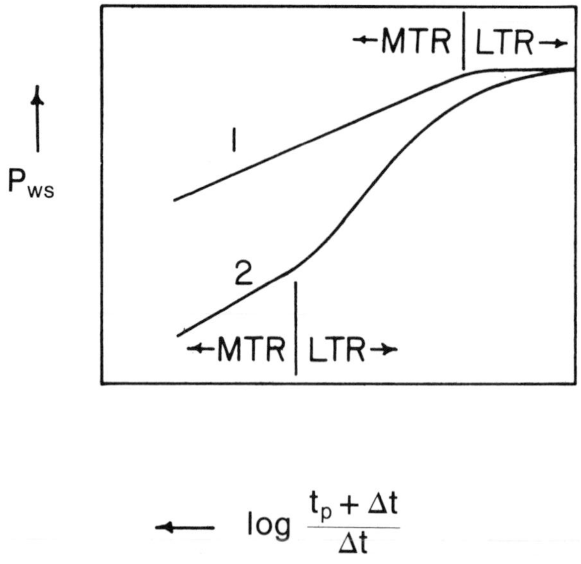


Fig. 2.9 – Boundary effects in pressure buildup test: (1) well centered in drainage area and (2) well off-center in drainage area.

resultant changes in permeability or thickness, and some fluid/fluid contacts. The longer a test is run, the higher the probability that a significant heterogeneity will be encompassed within the radius of investigation and thus influence the test.

Modifications to the simple reservoir models have been developed for some important reservoir heterogeneities. Still, in actual heterogeneous reservoirs, the test analyst must be aware constantly of the possibility of an unknown or improperly modeled heterogeneity. These heterogeneities make analysis of late-time data in transient tests more difficult – reservoirs are rarely uniform cylinders or parallelepipeds, and the analysis technique that is based on these assumptions for treatment of late-time data can be difficult to apply without ambiguity.

What is the test analyst to do with late-time data? Opinions vary. One frequent approach is to use analysis techniques suggested by published simple models – but to try to find other models that also fit the observed data. One then chooses the most probable reservoir description, and recognizes that the analysis may be *absolutely incorrect*.

## 2.5 Qualitative Behavior of Field Tests

We now have developed the background required to understand the qualitative behavior of commonly occurring pressure buildup curves. There is an important reason for this examination of behavior. It provides a convenient means of introducing some factors that influence these curves and that can obscure interpretation unless they are recognized. In the figures that follow, the early-, middle-, and late-time regions are designated by ETR, MTR, and LTR, respectively. In these curves, the most important region is the MTR. Interpretation of the test using the Horner plot [$p_{ws}$ vs. log $(t_p + \Delta t)/\Delta t$] is usually impossible unless the MTR can be recognized.

Fig. 2.6 illustrates the ideal buildup test, in which the MTR spans almost the entire range of the plotted data. Such a curve is possible for an undamaged well (Curve 1, with the level of $p_{wf}$, the flowing pressure at shut-in, is shown for reference) and for a damaged well with an altered zone concentrated at the wellbore. This latter situation, shown in Curve 2, is indicated by a rapid rise in pressure from flowing pressure at shut-in to the pressures along the MTR. Neither case is observed often in practice with a surface shut-in because afterflow usually distorts the early data that would fall on the straight line.

Fig. 2.7 illustrates the pressure buildup test obtained for a damaged well. Curve 1 would be obtained with a shut-in near the perforations (minimizing the duration of afterflow); Curve 2 would be obtained with the more conventional surface shut-in. Note in this figure that the flowing BHP at shut-in, $p_{wf}$, is the same for either case, but that the afterflow that appears with the surface shut-in (1) completely obscures information reflecting near-well conditions in the ETR and (2) delays the beginning of the MTR. A further complication introduced by afterflow is that several apparent straight lines appear on the buildup curve. The question arises, how do we find *the* straight line (the MTR line) whose slope is related to formation permeability? We will deal with this question shortly.

Fig. 2.8 shows characteristic behavior of a buildup test for a fractured well without afterflow. For such a well, the pressure builds up slowly at first; the MTR develops only when the pressure transient has moved beyond the region influenced by the fracture. In a buildup test for a fractured well, there is a possibility that boundary effects will appear before the ETR has ended (i.e., that there will be no MTR at all).

Fig. 2.9 illustrates two different types of behavior in the LTR of a buildup test plot. Curve 1 illustrates middle- and late-time behavior for a well reasonably

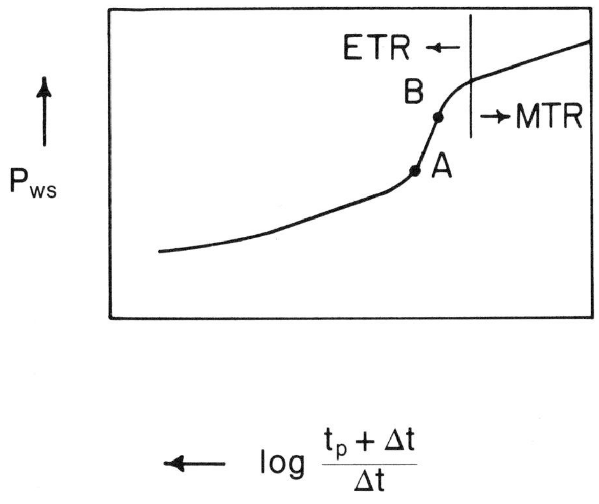


Fig. 2.10 – Characteristic influence of afterflow on Horner graph.

centered in its drainage area; Curve 2 illustrates behavior for a well highly off center in its drainage area. For simplicity, the ETR is not shown in either case.

Many curve shapes other than those discussed above appear in practice, of course. Still, these few examples illustrate the need for a systematic analysis procedure that allows us to determine the end of the ETR (usually, the time at which afterflow ceases distorting the test data) and the beginning of the LTR.

Without this procedure, there is a high probability of choosing the incorrect straight-line segment and using it to estimate permeability and skin factor.

## 2.6 Effects and Duration of Afterflow

In our discussion thus far, we have noted several problems that afterflow causes the buildup test analyst. Summarizing, these problems include (1) delay in the beginning of the MTR, making its recognition more difficult; (2) total lack of development of the MTR in some cases, with relatively long periods of afterflow and relatively early onset of boundary effects; and (3) development of several false straight lines, any one of which could be mistaken for the MTR line. We note further that recognition of the middle-time line is essential for successful buildup curve analysis based on the Horner plotting method $\{p_{ws}$ vs. log $[(t_p+\Delta t)/\Delta t]\}$, because the line must be identified to estimate reservoir permeability, to calculate skin factor, and to estimate static drainage-area pressure. The need for methods to determine when (if ever) afterflow ceased distorting a buildup test is clear; this section fills that need.

The characteristic influence of afterflow on a pressure buildup test plot is a lazy S-shape at early times, as shown in Fig. 2.10. In some tests, parts of the S-shape may be missing in the time range during which data have been recorded – e.g., data before Time A may be missing, or data for times greater than Time B may be absent. Thus, the shape of the buildup test alone is not sufficient to indicate the presence or absence of afterflow – it is merely a clue that sometimes indicates presence of afterflow.

A log-log graph of pressure change, $p_{ws}-p_{wf}$, in a buildup test vs. shut-in time, $\Delta t$, is an even more diagnostic indicator of the end of afterflow distortion. Fig. 1.6, based on solutions to the flow equations for constant-rate production with wellbore storage distortion, describes pressure buildup tests, as we discuss in some detail in Chap. 4. For use of this figure for buildup tests, dimensionless pressure, $p_D$, is defined as

$$p_D = \frac{0.00708\, kh(p_{ws}-p_{wf})}{qB\mu} \quad \text{. . . . . . . . . . . . (2.7)}$$

Dimensionless time, $t_D$, and dimensionless wellbore storage constant, $C_{sD}$, are defined essentially as for constant-rate production:

$$t_D = \frac{0.000264\, k\Delta t_e}{\phi\mu c_t r_w^2}, \quad \text{. . . . . . . . . . . . . . . . . . (2.8)}$$

$$C_{sD} = \frac{0.894\, C_s}{\phi c_t r_w^2 h}, \quad \text{. . . . . . . . . . . . . . . . . . . (2.9)}$$

where

$$C_s = 25.65 \frac{A_{wb}}{\rho}$$

for a well with a rising liquid/gas interface in the wellbore, and

$$C_s = c_{wb} V_{wb}$$

for a wellbore containing only single-phase fluid (liquid or gas).

We define[5]

$$\Delta t_e = \Delta t/(1+\Delta t/t_p). \quad \text{. . . . . . . . . . . . . . . . . (2.10)}$$

As noted in Chap. 1, wellbore storage distortion (afterflow in the case of a buildup test) has ceased when the graphed solutions for finite $C_{sD}$ become identical to those for $C_{sD}=0$. Also, a line with unit slope (45° line) appears at early times for most values of $C_{sD}$ and $s$. The meaning of this line in a buildup test is that the rate of afterflow is identical to the flow rate just before shut-in.

If the unit-slope line is present, the end of afterflow distortion occurs at approximately one and a half log cycles after the disappearance of the unit-slope line. Regardless of whether the unit-slope line is present, the end of afterflow distortion can be determined by overlaying the log-log plot of the test data onto the Ramey solution (Fig. 1.6) – plotted on graph or tracing paper with a scale identical in size to the Ramey graph – finding *any* preplotted curve that matches the test data, and noting when the preplotted curve for finite value of $C_{sD}$ becomes identical to the curve for $C_{sD}=0$. This point, on the actual data plot, is the end of afterflow or wellbore storage distortion.

If the unit-slope line is present, we can use a relationship developed in Chap. 1 to establish the value of $C_{sD}$ that characterizes the actual test. There,

**TABLE 2.3 – OILWELL PRESSURE BUILDUP TEST DATA**

| $\Delta t$ (hours) | $\frac{t_p+\Delta t}{\Delta t}$ | $\Delta t_e=\Delta t/\left(1+\frac{\Delta t}{t_p}\right)$ (hours) | $p_{ws}$ (psia) | $p_{ws}-p_{wf}$ (psia) |
|---|---|---|---|---|
| 0 | – | – | 3,534 | 0 |
| 0.15 | 90,900 | 0.15 | 3,680 | 146 |
| 0.2 | 68,200 | 0.2 | 3,723 | 189 |
| 0.3 | 45,400 | 0.3 | 3,800 | 266 |
| 0.4 | 34,100 | 0.4 | 3,866 | 332 |
| 0.5 | 27,300 | 0.5 | 3,920 | 386 |
| 1 | 13,600 | 1 | 4,103 | 569 |
| 2 | 6,860 | 2 | 4,250 | 716 |
| 4 | 3,410 | 4 | 4,320 | 786 |
| 6 | 2,270 | 6 | 4,340 | 806 |
| 7 | 1,950 | 7 | 4,344 | 810 |
| 8 | 1,710 | 8 | 4,350 | 816 |
| 12 | 1,140 | 12 | 4,364 | 830 |
| 16 | 853 | 16 | 4,373 | 839 |
| 20 | 683 | 20 | 4,379 | 845 |
| 24 | 569 | 24 | 4,384 | 850 |
| 30 | 455 | 29.9 | 4,393 | 859 |
| 40 | 342 | 39.9 | 4,398 | 864 |
| 50 | 274 | 49.8 | 4,402 | 868 |
| 60 | 228 | 59.7 | 4,405 | 871 |
| 72 | 190 | 71.6 | 4,407 | 873 |

we noted that any point on the unit-slope line must satisfy the relationship

$$\frac{C_{sD}p_D}{t_D}=1, \qquad (1.42)$$

which, in terms of variables with dimensions, leads to

$$C_s=\frac{qB}{24}\frac{\Delta t}{\Delta p}, \qquad (2.11)$$

where $\Delta t$ and $\Delta p$ are values read from a point on the unit-slope line. If we can establish $C_s$ in this way (a less acceptable alternative is to use the actual mechanical properties of the well – e.g., $C_s=25.65\ A_{wb}/\rho_{wb}$ for a well with a rising liquid/gas interface), we then can establish $C_{sD}$ from Eq. 2.9 and thus determine the proper curve on Fig. 1.6 on which to attempt a curve match. (It is difficult to interpolate between values of $C_{sD}$ on this curve; accordingly, many test analysts prefer to find a match with the preplotted value of $C_{sD}$ closest in value to the calculated value.) With $C_{sD}$ established, and permeability, $k$, and skin factor, $s$, determined from complete analysis of the test, we can use the empirical relationships below to verify the time, $t_{wbs}$, marking the end of wellbore storage distortion.

$$t_D \cong 50\ C_{sD}e^{0.14s}, \qquad (2.12)$$

or

$$t_{wbs} \cong \frac{170{,}000\ C_s e^{0.14s}}{(kh/\mu)}. \qquad (2.13)$$

We will illustrate (1) the application of the basic curve-matching procedure in Example 2.2; (2) the check provided by Eq. 2.13 in Example 2.4; and (3) complete, quantitative curve-matching procedures in Chap. 4.

In this discussion of the qualitative application of curve matching, we also should note that appearance of boundary effects or the effects of heterogeneities frequently can be verified from the curves. Fig. 1.6 is a solution to the flow equations for an infinite-acting, homogeneous reservoir; when, in an actual reservoir, a pressure transient reaches a boundary or important heterogeneity, the actual test data plot will deviate from Fig. 1.6. This characteristic of curve matching is illustrated in Example 2.2.

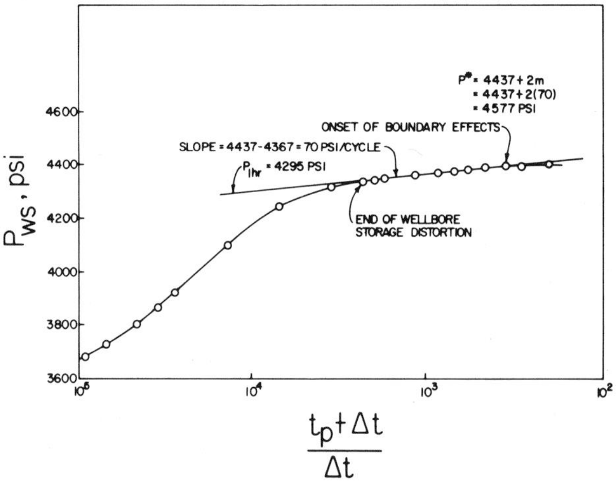


Fig. 2.11 – Semilog graph of example buildup test data.

## *Example 2.2 Finding the End of Wellbore Storage Distortion*

**Problem.** The data in Table 2.3 were obtained in a pressure buildup test on an oil well producing above the bubble point.

The well was produced for an effective time of 13,630 hours at the final rate (i.e., $t_p=13{,}630$ hours). Other data include the following.

$q_o$ = 250 STB/D,
$\mu_o$ = 0.8 cp,
$\phi$ = 0.039,
$B$ = 1.136 RB/STB,

$c_t$ = $17 \times 10^{-6}$ psi$^{-1}$,
$r_w$ = 0.198 ft,
$r_e$ = 1,489 ft (well is centered in square drainage area, 2,640 × 2,640 ft; $r_e$ is radius of circle with same area),
$\rho_o$ = 53 lbm/cu ft,
$A_{wb}$ = 0.0218 sq ft,
$h$ = 69 ft, and
rising liquid level in well during shut-in.

A semilog graph [$p_{ws}$ vs. log $(t_p+\Delta t)/\Delta t$] of these data is shown in Fig. 2.11, and a log-log graph ($p_{ws}-p_{wf}$ vs. $\Delta t_e$) in Fig. 2.12. From these graphs, answer the following questions.

1. At what shut-in time ($\Delta t$) does afterflow cease distorting the pressure buildup test data?

2. At what shut-in time ($\Delta t$) do boundary effects appear?

**Solution.** From the semilog graph (Fig. 2.11), it seems plausible that afterflow distortion disappears at $(t_p+\Delta t)/\Delta t \cong 2{,}270$ or $\Delta t = 6$ hours because of the end of the characteristic lazy-S-shaped curve. However, other reservoir features can lead to this same shape, so we confirm the result with the log-log graph. After plotting $\Delta p = p_{ws} - p_{wf}$ vs. $\Delta t_e = \Delta t/(1+\Delta t/t_p)$ on log-log paper, we find that the actual data fit well* curves for $s=5$ for several values of $C_{sD}$ (e.g., $C_{sD} = 10^3$, $10^4$, and$10^5$). In each case, the curve fitting the earliest data coincides with the $C_{sD}=0$ curve for $s=5$ at $\Delta t_e = \Delta t = 4$ to 6 hours. This, then, is the end of wellbore effects: $t_{wbs} \cong 6$ hours. The data begin to deviate from the semilog straight line at $(t_p+\Delta t)/\Delta t \cong 274$ or $\Delta t = 50$ hours. On the log-log graph, data begin falling below the fitting curve at $\Delta t = \Delta t_e \cong 40$ hours, consistent with the semilog graph.

In summary, basing our quantitative judgment on the more sensitive semilog graph, we say that the MTR spans the time range of $\Delta t = 6$ hours to $\Delta t = 50$ hours. This judgment is verified qualitatively by the log-log graph curve matching. Even though the semilog graph is more sensitive (i.e., can be read with greater accuracy), it alone is not sufficient to determine the beginning and end of the MTR: matching Ramey's solution is a critically important part of the analysis.

The log-log curve-matching analysis was performed without knowledge of $C_{sD}$. Note that $C_{sD}$ *can* be established in this case, at least approximately: from the curve match, we note that the data are near the unit-slope line on the graph of Ramey's solution; the point $\Delta p = 100$, $\Delta t = 0.1$ is essentially on this line. Thus, from Eq. 2.11,

$$C_s \cong \frac{qB}{24}\frac{\Delta t}{\Delta p} = \frac{(250)(1.136)}{24}\frac{(0.1)}{(100)}$$

$$= 0.0118 \text{ bbl/psi.}$$

Alternatively (and, in general, less accurately),

$$C_s = \frac{25.65\,A_{wb}}{\rho} = \frac{(25.65)(0.0218)}{53}$$

$$= 0.0106 \text{ bbl/psi.}$$

*Data are plotted on 3 x 5 cycle log-log graph paper (11 x 16½ in.) and matched with the Ramey solution (such as provided in the SPE type-curve package) plotted on the same size scale.

Then, from Eq. 2.9,

$$C_{sD} = \frac{0.894\,C_s}{\phi c_t h r_w^2}$$

$$= \frac{(0.894)(0.0118)}{(0.039)(1.7\times 10^{-5})(69)(0.198)^2} = 5{,}880.$$

Thus, matching should be attempted in the range $10^3 < C_{sD} < 10^4$.

## 2.7 Determination of Permeability

In this section, we examine techniques for the next step in the systematic analysis of a pressure buildup or falloff test: determining bulk-formation permeability. Because bulk-formation permeability is obtained from the slope of the MTR line, correct selection of this region is critical. Average permeability, $k_J$, also can be estimated from information available in buildup tests.

The first problem is identification of the MTR. This region *cannot* begin until afterflow ceases distorting the data; indeed, cessation of afterflow effects *usually* determines the beginning of the MTR. If the altered zone is unusually deep (as with a hydraulic fracture), passage of the transient through the region of the drainage area influenced by the fracture will determine the beginning of the MTR.

Predicting the time at which the MTR ends is more difficult than predicting when it begins. Basically, the middle-time line ends when the radius of investigation begins to detect drainage boundaries of the tested well; at this time, the pressure buildup curve begins to bend. The problem is that the time at which the middle region ends depends on (1) the distance from the tested well to the reservoir boundaries, (2) the geometry of the area drained by the well, and (3) the duration of the flow period as well as the shut-in period. Cobb and Smith[2] present charts that allow the analyst to predict the shut-in time $\Delta t$ at which the MTR should end if drainage-area geometry and producing time are known. If this information is available and if the reservoir is sufficiently homogeneous that, until the LTR begins, it behaves in the ideal way required by Cobb and Smith's theory, their charts can be used to check results.

One useful generalization can be made from their results. If a well was at pseudosteady-state before shut-in, the time $\Delta t$ at which the LTR begins* is approximately $\Delta t_{\ell t} \simeq 38\,\phi\mu c_t A/k$ for a well centered in a square or circular drainage area. In the equation, $A$ (sq ft) is the drainage area of the tested well. If the well was not at pseudosteady-state, $\Delta t_{\ell t}$ is larger than calculated by the rule above. In many cases, we simply assume that the straight line spanning the times between the end of afterflow distortion and a later bend of the Horner plot constitutes the MTR. Use of the log-log graph and curve matching, as in Example 2.2, can help confirm this assumption.

The calculated radius of investigation ($r_i$) at the assumed end of the MTR provides a *qualitative*

*Choice of time at which LTR begins is somewhat arbitrary. The rule stated is based on a 10% deviation in slope of the Horner plot from the true MTR.

estimate only of the radius of the infinite-acting drainage area in the reservoir.

In summary, the procedure for determining bulk-formation permeability is as follows.

1. Determine the probable beginning of the MTR by estimating when afterflow effects disappear.

2. Assume that the probable end of the MTR occurs when the Horner plot becomes nonlinear, verified by a deviation from a curve fitting early- and middle-time data on a log-log graph using the curve-matching technique.

3. If there *is* an apparent linear MTR, calculate the slope of the middle-time line, and estimate permeability from

$$k=162.6\frac{qB\mu}{mh}.$$

4. Radius-of-investigation estimates at the beginning and end of the assumed MTR may help establish its plausibility but should be viewed as qualitative only.

5. If there is no clear-cut MTR or if it is so short that its slope cannot be determined with confidence, bulk-formation permeability estimates from quantitative type-curve analysis (i.e., curve-matching) can be used and probably are essential.

It can be helpful in MTR analysis to calculate average permeability, $k_J$, from data obtained in a buildup test. From Eq. 1.19, which is valid only if pseudosteady-state is reached during the production period,

$$k_J=\frac{141.2\,qB\mu\left[\ln\,(r_e/r_w)-\frac{3}{4}\right]}{h(\bar{p}-p_{wf})}.$$

For a well that is neither damaged nor stimulated, $k_J$ should equal bulk-formation permeability, $k$, determined from the slope of the MTR; for a damaged well, $k_J<k$; and for a stimulated well, $k_J>k$. Average permeability is valuable in checking consistency in buildup test analysis. If $k$ and $k_J$ do not bear the proper relationship to each other, something is wrong – possibly an incorrect analysis due to incorrect choice of MTR.

---

### *Example 2.3 – Estimating Formation Permeability*

**Problem.** For the test discussed in Example 2.2, determine formation permeability.

**Solution.** In Example 2.2, we established that the MTR spans the time range of $\Delta t\simeq 6$ hours to $\Delta t\cong 50$ hours $[2{,}270\geq(t_p+\Delta t)/\Delta t\geq 274]$. From Fig. 2.11, note that the slope $m$ of this straight line is

$$m=4{,}437-4{,}367=70\text{ psi/cycle}.$$

Then,

$$k=162.6\frac{qB\mu}{mh}=\frac{(162.6)(250)(1.136)(0.8)}{(70)(69)}$$

$$=7.65\text{ md}.$$

It is of interest to determine the portion of the reservoir sampled during the MTR; that region is given roughly by the radius of investigation achieved by the shut-in transient at the start and end of the MTR. From Eq. 1.47,

$$r_i=\left(\frac{kt}{948\,\phi\mu c_t}\right)^{1/2}.$$

Thus, at $\Delta t=6$ hours,

$$r_i=\left[\frac{(7.65)(6)}{(948)(0.039)(0.8)(1.7\times 10^{-5})}\right]^{1/2}$$

$$=302\text{ ft},$$

and, at $\Delta t=50$ hours,

$$r_i=302\left(\frac{50}{6}\right)^{1/2}=872\text{ ft}.$$

Thus, a significant fraction of the well's drainage area has been sampled; its permeability is 7.65 md.

## 2.8 Well Damage and Stimulation

This section shows how to use data available in a buildup or falloff test to estimate damage or stimulation quantitatively. The basic technique when analyzing data from a plot of $p_{ws}$ vs. $(t_p+\Delta t)/\Delta t$ is called the skin factor method. It involves calculating the skin factor and translating it into a more easily visualized characterization of a well. We will deal with another method of analyzing damage or stimulation in a later chapter – a method using type-curve analysis.

Before we examine the skin factor method, it is useful to consider briefly the physical reasons for damage or stimulation of a well. Wellbore damage is a descriptive term applied when permeability is reduced near a wellbore. This permeability reduction usually occurs during drilling or completion operations. Causes include plugging of pores with fine material in the drilling fluid and reaction of the formation with filtrate from the drilling fluid (e.g., swelling of clays in the formation resulting from contact with low-salinity filtrate). Completion fluids can cause similar permeability reduction as they enter the formation.

Stimulation usually results from deliberate attempts to improve a well's productivity. Common techniques include acidization and hydraulic fracturing. Acidization is dissolving plugging materials and the formation near the wellbore with acid injection through the perforations. Hydraulic fracturing is creating seams in the formation with high-pressure injection of special fluids, usually accompanied by sand or some other agent that will prop the fracture open when the pressure creating the fracture is removed.

Eq. 2.4 shows how we can calculate skin factor, $s$, once the MTR is identified and bulk-formation permeability is estimated:

$$s=1.151\left[\frac{(p_{1\,hr}-p_{wf})}{m}-\log\left(\frac{k}{\phi\mu c_t r_w^2}\right)+3.23\right]. \quad (2.4)$$

We recall that $p_{1\,hr}$ is the value of $p_{ws}$ at shut-in

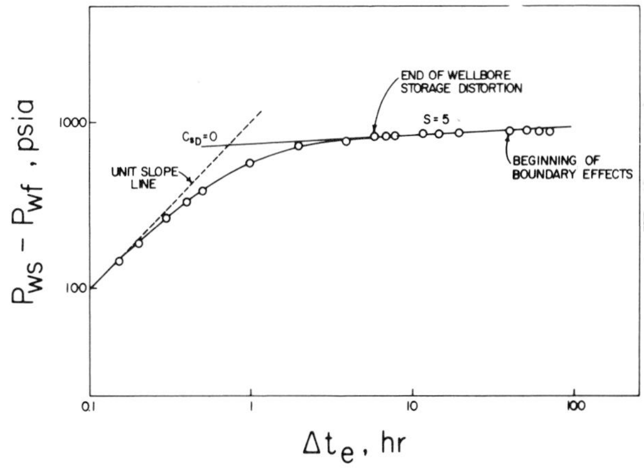


**Fig. 2.12** – Log-log graph of example buildup test data

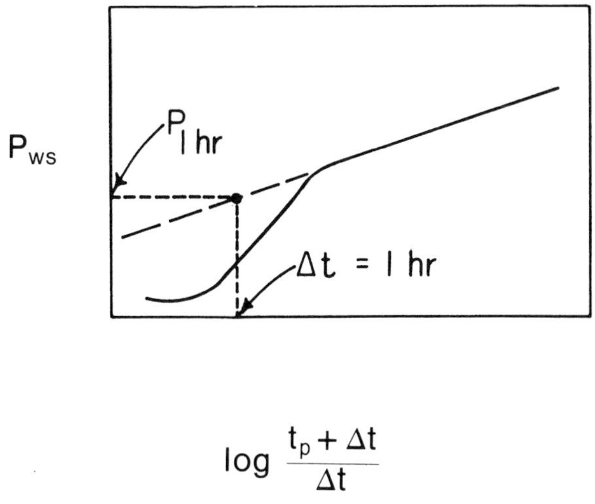


**Fig. 2.13** – Determination of $p_{1\,hr}$.

time $\Delta t$ of 1 hour on the middle-time line, or its extrapolation as shown in Fig. 2.13. It is not possible to calculate the skin factor until the middle-time line has been established because values of $k$, $m$, and $p_{1\,hr}$ are found from this line. If an accurate skin factor is to be calculated from a buildup test, the flowing pressure $p_{wf}$ must be measured before shut-in.

Interpretation of a given numerical value of the skin factor can be summarized as follows.

1. A positive skin factor indicates a flow restriction (e.g., wellbore damage); the larger the skin factor, the more severe the restriction.

2. A negative skin factor indicates stimulation; the larger the absolute value of the skin factor, the more effective the stimulation.

3. Conditions other than wellbore damage can cause an apparent skin factor. The reason is that any deviation from purely radial flow near a well, which results in total well production squeezing through a smaller vertical thickness near the well than away from the well, increases the pressure drop near the well. This is precisely the same effect that wellbore damage has; damage also results in an increased pressure drop near the well. The basic equation used in constructing our theory of pressure buildup and falloff test behavior, Eq. 1.7, is based on the assumption that flow is radial throughout the drainage area of the well up to the sandface; a deviation from this assumption invalidates the equation, but Eqs. 1.11 and 1.16 are usually excellent approximations when the nonradial flow occurs near the wellbore only.

Conditions leading to nonradial flow near the wellbore include (1) when the well does not completely penetrate the productive interval, and (2) when the well is perforated only in a portion of the interval (e.g., the top 10 ft of a 50-ft sand). In these cases, the analyst will calculate a positive skin factor even for an undamaged well. (In addition, the perforations themselves – their size, spacing, and depth – also can affect the skin factor.) We will examine results of this nonradial flow after completing our consideration of skin factor solely caused by formation damage or stimulation.

We now turn our attention to methods for translating values of $s$ into less abstract characterizations of the wellbore. We consider three methods: estimation of effective wellbore radius, $r_{wa}$; calculation of additional pressure drop near the wellbore; and calculation of flow efficiency.

### Estimation of Effective Wellbore Radius

The effective wellbore radius $r_{wa}$ is defined as

$$r_{wa} = r_w e^{-s}. \quad \text{(2.14)}$$

To understand the significance of this quantity, note that from Eq. 1.11,

$$p_i - p_{wf} = -70.6\frac{qB\mu}{kh}\left[\ln\left(\frac{1{,}688\ \phi\mu c_t r_w^2}{kt}\right) - 2s\right]$$

$$= -70.6\frac{qB\mu}{kh}\left[\ln\left(\frac{1{,}688\ \phi\mu c_t r_w^2}{kt}\right) + \ln(e^{-2s})\right]$$

$$= -70.6\frac{qB\mu}{kh}\left[\ln\left(\frac{1{,}688\ \phi\mu c_t r_w^2 e^{-2s}}{kt}\right)\right]$$

$$= -70.6\frac{qB\mu}{kh}\ln\left(\frac{1{,}688\ \phi\mu c_t r_{wa}^2}{kt}\right).$$

This shows that the effect of $s$ on total pressure drawdown is the same as that of a well with no altered zone but with a wellbore radius of $r_{wa}$.

Calculation of effective wellbore radius is of special value for analyzing wells with vertical fractures. Model studies have shown that for highly conductive vertical fractures with two equal-length wings of length $L_f$,

$$L_f \simeq 2\,r_{wa}. \quad \text{(2.15)}$$

Thus, calculation of skin factor from a pressure

buildup or falloff test can lead to an estimate of fracture length – useful in a postfracture analysis. However, this analysis technique for a fractured well is frequently oversimplified; more complete methods are discussed later.

### Calculation of Additional Pressure Drop Near Wellbore

We defined additional pressure drop $(\Delta p)_s$ across the altered zone in Eq. 1.9 in terms of the skin factor $s$:

$$(\Delta p)_s = 141.2\frac{qB\mu}{kh}s.$$

In terms of the slope $m$ of the middle-time line,

$$(\Delta p)_s = 0.869\, m(s). \quad \text{(2.16)}$$

Calculation of this additional pressure drop across the altered zone can be a meaningful way of translating the abstract skin factor into a concrete characterization of the tested well. For example, a well may be producing 100 STB/D oil with a drawdown of 1,000 psi. Analysis of a buildup test might show that $(\Delta p)_s$ is 900 psi and, thus, that 900 psi of the total drawdown occurs across the altered zone. This implies that if the damage were removed, the well could produce much more fluid with the same drawdown or, alternatively, could produce the same 100 STB/D with a much smaller drawdown.

### Calculation of Flow Efficiency

The final method that we will examine for translating $s$ into a physically meaningful characterization of a well is by calculation of the flow efficiency, $E$. We define flow efficiency as the ratio of actual or observed PI of a tested well to its ideal PI (i.e., the PI it would have if the permeability were unaltered all the way to the sandface of the well). In mathematical terms,

$$E = \frac{J_{\text{actual}}}{J_{\text{ideal}}}$$

$$= \frac{\bar{p} - p_{wf} - (\Delta p)_s}{\bar{p} - p_{wf}}. \quad \text{(2.17)}$$

For rapid analysis of a pressure buildup or falloff test, Eq. 2.17 can be written in approximate form as

$$E \simeq \frac{p^* - p_{wf} - (\Delta p)_s}{p^* - p_{wf}}, \quad \text{(2.18)}$$

where $p^*$, the extrapolation of the middle-time line to $(t_p + \Delta t)/\Delta t = 1$, is found more readily than $\bar{p}$, which can require lengthy analysis. Flow efficiency is actually time dependent unless a well reaches pseudosteady state during the producing period (only then is $\bar{p} - p_{wf}$ in Eq. 2.17 constant).

Flow efficiency is unity for a well that is neither damaged nor stimulated. For a damaged well, flow efficiency is less than one; for a stimulated well, flow efficiency is greater than one. A damaged well with a calculated flow efficiency of 0.1 is producing about 10% as much fluid with a given pressure drawdown as it would if the damage were removed; a stimulated well with a calculated flow efficiency of two is producing about twice as much fluid with a given pressure drawdown as it would had the well not been stimulated.

Use of the skin factor method is illustrated in Example 2.4.

---

### *Example 2.4 – Damage Analysis*

**Problem.** Consider the buildup test described in Examples 2.2 and 2.3. Make the following calculations with those data.

1. Calculate the skin factor for the tested well.
2. Calculate the effective wellbore radius $r_{wa}$.
3. Calculate the additional pressure drop near the wellbore caused by the damage that is present.
4. Calculate the flow efficiency.
5. Verify the end of wellbore storage distortion using Eq. 2.13.

**Solution.**

1. *Skin Factor.* In the skin factor equation, we need $p_{1\,\text{hr}}$ from extrapolation of the middle-time line to a shut-in time of 1 hour. At $\Delta t = 1$ hour, $(t_p + \Delta t)/\Delta t = 13{,}631$. From an extrapolation (Fig. 2.11) of the middle-time line to this time $p_{1\,\text{hr}} = 4{,}295$ psi. (Note how different this is from the *actual* pressure at $\Delta t = 1$ hour: 4,103 psi.) Then, because $k/\phi\mu c_t = 1.442 \times 10^7$,

$$s = 1.151\left[\frac{(p_{1\,\text{hr}} - p_{wf})}{m} - \log\left(\frac{k}{\phi\mu c_t r_w^2}\right) + 3.23\right]$$

$$= 1.151\left\{\frac{(4{,}295 - 3{,}534)}{70}\right.$$

$$\left. - \log\left[\frac{1.442 \times 10^7}{(0.198)^2}\right] + 3.23\right\} = 6.37.$$

In Example 2.2, we found from curve matching that $s \cong 5$, which is good agreement.

2. *Effective Wellbore Radius.* From Eq. 2.14,

$$r_{wa} = r_w e^{-s}$$

$$= (0.198)e^{-6.37}$$

$$= 0.00034 \text{ ft}.$$

The physical interpretation of this result is that the tested well is producing 250 STB/D oil with the same pressure drawdown as would a well with a wellbore radius of 0.00034 ft and permeability unaltered up to the sandface.

3. *Additional Pressure Drop Near the Wellbore.* From Eq. 2.16,

$$(\Delta p)_s = 0.869\, m(s)$$

$$= (0.869)(70)(6.37)$$

$$= 387 \text{ psi}.$$

Thus, of the total drawdown of approximately 4,420 – 3,534 = 886 psi, about 387 psi is caused by damage. Much of this additional drawdown could be avoided if the skin resulted from formation damage (rather than from partial penetration, for example) and if the well were stimulated.

4. *Flow Efficiency.* To calculate flow efficiency, we need $p^*$, the value of $p_{ws}$ on the middle-time line at $(t_p+\Delta t)/\Delta t=1$. We cannot extrapolate directly on our plot because there are no values of $(t_p+\Delta t)/\Delta t$ less than 100, but we note that the pressure increases by 70 psi over each cycle; thus, we can add 2 (70) psi to the value of $p$ at $(t_p+\Delta t)/\Delta t=100$:

$$p^*=4{,}437+2\,(70)=4{,}577 \text{ psi}.$$

Then, from Eq. 2.18,

$$E=\frac{p^*-p_{wf}-(\Delta p)_s}{p^*-p_{wf}}$$

$$=\frac{4{,}577-3{,}534-387}{4{,}577-3{,}534}$$

$$=0.629.$$

This means that the well is producing about 62% as much fluid with the given drawdown as an undamaged well in a completely perforated interval would produce.

5. *End of Wellbore Storage Distortion.* From Eq. 2.13 and Examples 2.2 and 2.3,

$$t_{wbs}=\frac{170{,}000\,C_s e^{0.14s}}{kh/\mu}$$

$$=\frac{(170{,}000)(0.0118)e^{(0.14)(6.37)}}{(7.65)(69)/0.8}$$

$$=7.42 \text{ hours}.$$

This agrees closely with the results of Example 2.2.

### Effect of Incompletely Perforated Interval

When the completed interval is less than total formation thickness, the pressure drop near the well is increased and the apparent skin factor becomes increasingly positive. In a review of technology in this area, Saidikowski[6] found that total skin factor, $s$, determined from a pressure transient test is related to true skin factor, $s_d$, caused by formation damage and apparent skin factor, $s_p$, caused by an incompletely perforated interval. The relationship between these skin factors is

$$s=\frac{h_t}{h_p}s_d+s_p, \quad\text{.......................}\quad (2.19)$$

where $h_t$ is total interval height (ft) and $h_p$ is the perforated interval (ft).

Saidikowski also verified that $s_p$ can be estimated from the equation

$$s_p=\left(\frac{h_t}{h_p}-1\right)\left[\ln\left(\frac{h_t}{r_w}\sqrt{\frac{k_H}{k_V}}\right)-2\right], \quad\text{.....}\quad (2.20)$$

where $k_H$ is horizontal permeability (md) and $k_V$ is vertical permeability (md). Use of these equations is best illustrated with an example.

---

## *Example 2.5 – Incompletely Perforated Interval*

**Problem.** A well with disappointing productivity is perforated in 10 ft of a total formation thickness of 50 ft. Vertical and horizontal permeabilities are believed to be equal. A pressure buildup test was run on the well; results and basic properties are summarized as follows.

$$\begin{aligned}
p_{wf} &= 1{,}190 \text{ psi},\\
p_{1\,\text{hr}} &= 1{,}940 \text{ psi},\\
\phi &= 0.20,\\
m &= 50 \text{ psi/cycle},\\
\mu &= 0.5 \text{ cp},\\
r_w &= 0.25 \text{ ft},\\
c_t &= 15\times10^{-6} \text{ psi}^{-1}, \text{ and}\\
k &= 3.35 \text{ md}.
\end{aligned}$$

Calculate $s$, $s_d$, and $s_p$; on the basis of these results, determine whether the productivity problem results from formation damage or from other causes.

**Solution.** From Eq. 2.4,

$$s=1.151\left[\frac{p_{1\,\text{hr}}-p_{wf}}{m}-\log\left(\frac{k}{\phi\mu c_t r_w^2}\right)+3.23\right]$$

$$=1.151\left[\frac{(1{,}940-1{,}190)}{50}\right.$$

$$\left.-\log\left(\frac{3.35}{(0.2)(0.5)(1.5\times10^{-5})(0.25)^2}\right)+3.23\right]$$

$$=12.3.$$

The contribution of an incompletely perforated interval to the total skin factor is, from Eq. 2.20,

$$s_p=\left(\frac{h_t}{h_p}-1\right)\left[\ln\left(\frac{h_t}{r_w}\sqrt{\frac{k_H}{k_V}}\right)-2\right]$$

$$=\left(\frac{50}{10}-1\right)\left[\ln\left(\frac{50}{0.25}\sqrt{\frac{3.35}{3.35}}\right)-2\right]$$

$$=13.2.$$

Rearranging Eq. 2.19, skin factor, $s_d$, resulting from formation damage is

$$s_d=\frac{h_p}{h_t}(s-s_p)$$

$$=\frac{10}{50}(12.3-13.2)$$

$$=-0.18.$$

As a practical matter, the well is neither damaged nor stimulated. The observed productivity problem is caused entirely by the effects of an incompletely perforated interval.

### Analysis of Hydraulically Fractured Wells

Type curves provide a general method of analyzing hydraulically fractured wells – particularly because

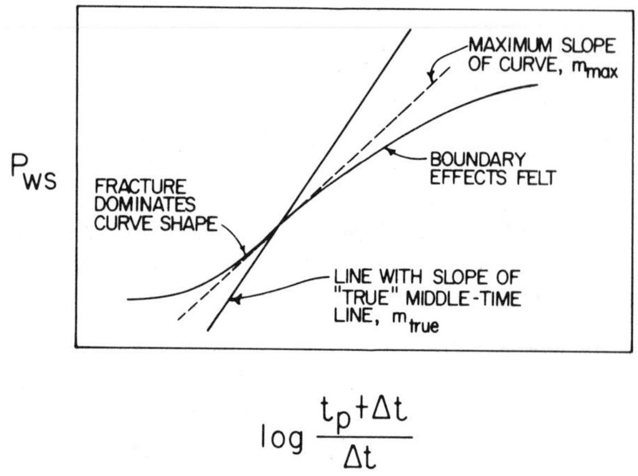


**Fig. 2.14** – Buildup curve for hydraulically fractured well, bounded reservoir.

**TABLE 2.4 – BUILDUP TEST SLOPES FOR HYDRAULICALLY FRACTURED WELLS**

| $L_f/r_e$ | $m_{max}/m_{true}$ |
|---|---|
| 0.1 | 0.87 |
| 0.2 | 0.70 |
| 0.4 | 0.46 |
| 0.6 | 0.32 |
| 1.0 | 0.28 |

finite conductivity can be considered. Some conventional methods are also of value for infinite-conductivity fractures. This section summarizes some of the useful conventional methods.

When fractures are highly conductive (i.e., when there is little pressure drop in the fracture itself) and when there is uniform flux of fluid into the fracture, linear flow theory describes well behavior at earliest times in a buildup test. (Uniform flux means identical flow rates of formation fluid into the fracture per unit cross-sectional area at all points along the fracture.) From Eq. 1.46, for constant-rate production,

$$p_i - p_{wf} = 4.064 \frac{qB}{hL_f}\left(\frac{\mu t}{k\phi c_t}\right)^{1/2}.$$

For a buildup test, for $t_p \gg \Delta t$,

$$p_{ws} - p_{wf} = 4.064 \frac{qB}{hL_f}\left(\frac{\mu \Delta t}{k\phi c_t}\right)^{1/2}.$$

Thus, the slope $m_L$ of a $p_{ws}$ vs. $\sqrt{\Delta t}$ plot is

$$m_L = 4.064 \frac{qB}{hL_f}\left(\frac{\mu}{k\phi c_t}\right)^{1/2}. \quad \ldots\ldots\ldots\ldots (2.21)$$

From measurements of this slope, fracture length, $L_f$, can be estimated. This procedure requires that an independent estimate of permeability be available – from a prefracture pressure buildup test on the well, for example.

When linear flow cannot be recognized (i.e., when there is no early straight-line relationship between $p_{ws}$ and $\sqrt{\Delta t}$), we can use the observation that $L_f = 2r_{wa}$ for infinitely conductive fractures to estimate fracture length. Rather than calculate $s$ directly, we can note that

$$s = 1.151\left[\frac{(p_{1\,hr} - p_{wf})}{m} - \log\left(\frac{k}{\phi\mu c_t r_w^2}\right) + 3.23\right];$$

and, because

$$r_{wa} = \frac{L_f}{2} = r_w e^{-s},$$

then

$$s = -\ln\left(\frac{L_f}{2r_w}\right) = -2.303 \log\left(\frac{L_f}{2r_w}\right).$$

Thus,

$$-2\log\left(\frac{L_f}{2r_w}\right) = -2\log\frac{L_f}{2} - 2\log\left(\frac{1}{r_w}\right)$$

$$= \frac{(p_{1\,hr} - p_{wf})}{m} - \log\left(\frac{k}{\phi\mu c_t}\right)$$

$$-2\log\left(\frac{1}{r_w}\right) + 3.23.$$

This simplifies to

$$\log(L_f) = \frac{1}{2}\left[\left(\frac{p_{wf} - p_{1\,hr}}{m}\right) + \log\frac{k}{\phi\mu c_t} - 2.63\right]. \quad \ldots\ldots\ldots\ldots (2.22)$$

Using Eq. 2.22, fracture length, $L_f$, can be estimated *if* the MTR can be recognized, which allows $m$, $p_{1\,hr}$, and $k$ to be determined.

In buildup tests from some hydraulically fractured wells, the true middle-time line does not appear, as illustrated in Fig. 2.14. (Afterflow can cause the same curve shape.) This situation arises because, at earliest times, the depth of investigation is in a region dominated by the fracture; at later times, the depth of investigation reaches a point dominated by boundary effects. (See Fig. 2.14.) When the length $L_f$ of a vertical fracture is greater than one-tenth of the drainage radius $r_e$ of a well centered in its drainage area, it has been found[7] that boundary effects begin before the influence of the fracture disappears. For a given drainage radius, the greater the fracture length, the greater the discrepancy between the maximum slope achieved on a buildup test and the slope of the true middle-time line. Table 2.4 summarizes the ratio of the maximum slope attained in a buildup test to the slope of the true middle-time line (from the work of Russell and Truitt[7]) for infinitely conductive fractures.

The implication is that if the test analyst simply does the best he can, and finds the maximum slope on a buildup test from a hydraulically fractured well and assumes that this maximum slope is an adequate approximation to the slope of the true middle-time

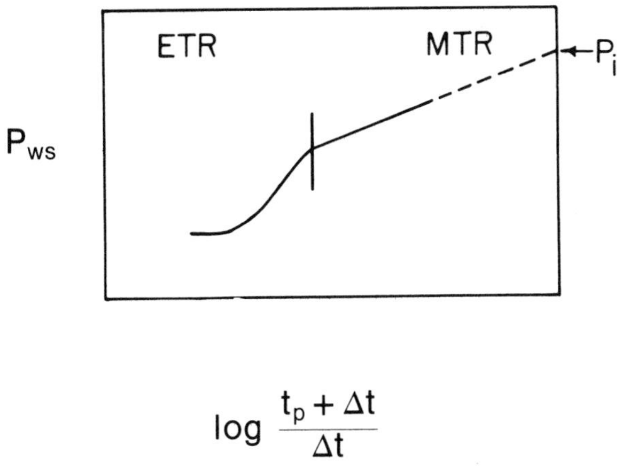


**Fig. 2.15** – Buildup test graph for infinite-acting reservoir.

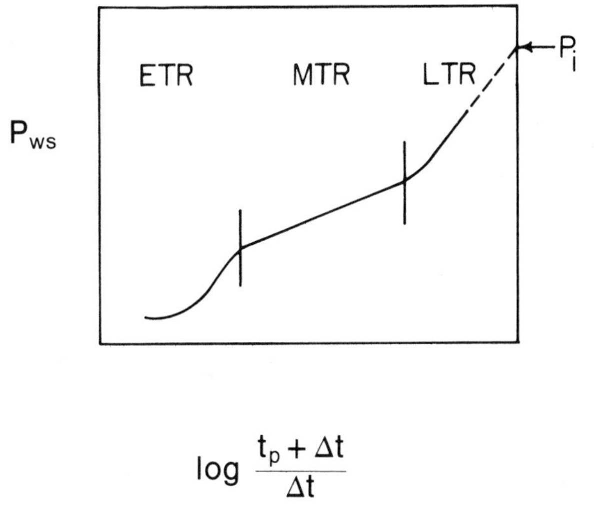


**Fig. 2.16** – Buildup test graph for well near reservoir limit(s).

line, then the permeability, skin factor, and fracture estimates will be in error, with the error growing as fracture length increases.

Correlation of reservoir model results by Russell and Truitt[7] showed that an equation similar to Eq. 2.22 can be used to estimate true fracture length even when $L_f > 0.1\, r_e$.

We again emphasize that all methods in this section assume highly conductive, vertical fractures with two equal-length wings. When fracture conductivity is not high, fracture length estimated by these methods will be too small.

## 2.9 Pressure Level in Surrounding Formation

A pressure buildup test can be used to determine average drainage-area pressure in the formation surrounding a tested well. We have seen that ideal pressure buildup theory suggests a method for estimating *original* reservoir pressure in an infinite-acting reservoir – that is, extrapolating the buildup test to infinite shut-in time $[(t_p + \Delta t)/\Delta t = 1]$ and reading the pressure there. For wells with partial pressure depletion, extrapolation of a buildup test to infinite shut-in time provides an estimate of $p^*$, which is related to, but *is not equal to*, current average drainage-area pressure. In this section, we will examine methods for estimating original and current average drainage-area pressures.

### Original Reservoir Pressure

For a well with an uncomplicated drainage area, original reservoir pressure, $p_i$, is found as suggested by ideal theory. We simply identify the middle-time line, extrapolate it to infinite shut-in time, and read the pressure, which is original reservoir pressure (Fig. 2.15). This technique is possible only for a well in a new reservoir (i.e., one in which there has been negligible pressure depletion). Strictly speaking, this is true only for tests in which the radius of investigation does not encounter any reservoir boundary during production.

For a reservoir with one or more boundaries relatively near a tested well (and encountered by the radius of investigation during the production period), the late-time line must be extrapolated (Fig. 2.16). (This can be quite complex for multiple boundaries near a well.) Note that our discussion is still restricted to reservoirs in which there has been negligible pressure depletion. Thus, even in the case under consideration, the well must be relatively far from boundaries in at least one direction.

### Static Drainage-Area Pressure

For a well in a reservoir in which there has been some pressure depletion, we do not obtain an estimate of original reservoir pressure from extrapolation of a buildup curve. Our usual objective is to estimate the average pressure in the drainage area of the well; we will call this pressure static drainage-area pressure. We will examine two useful methods for making these estimates: (1) the Matthews-Brons-Hazebroek (MBH)[8] $p^*$ method and (2) the modified Muskat method.[9]

The $p^*$ method was developed by Matthews *et al.* by computing buildup curves for wells at various positions in drainage areas of various shapes and then, from the plotted buildup curves, comparing the pressure ($p^*$) on an extrapolated middle-time line with the static drainage-area pressure ($\bar{p}$), which is the value at which the pressure will stabilize given sufficient shut-in time. The buildup curves were computed using imaging techniques and the principle of superposition. The results of the investigation are summarized in a series of plots of $kh(p^* - \bar{p})/70.6\, q\mu B$ vs. $0.000264\, kt_p/\phi\mu c_t A$. [Note that $kh(p^* - \bar{p})/70.6\, q\mu B$ can be written more compactly as $2.303\,(p^* - \bar{p})/m$. Also, the group $0.000264\, kt_p/\phi\mu c_t A$ is a dimensionless time and is symbolized by $t_{DA}$. The group $kh(p^* - \bar{p})/70.6\, q\mu B$ is a dimensionless pressure and is given the symbol $p_{D\,\mathrm{MBH}}$]. The only new symbol in these expressions is the drainage area, $A$, of the tested well expressed in square feet. Figs. 2.17A through 2.17G (reproduced from the Mat-

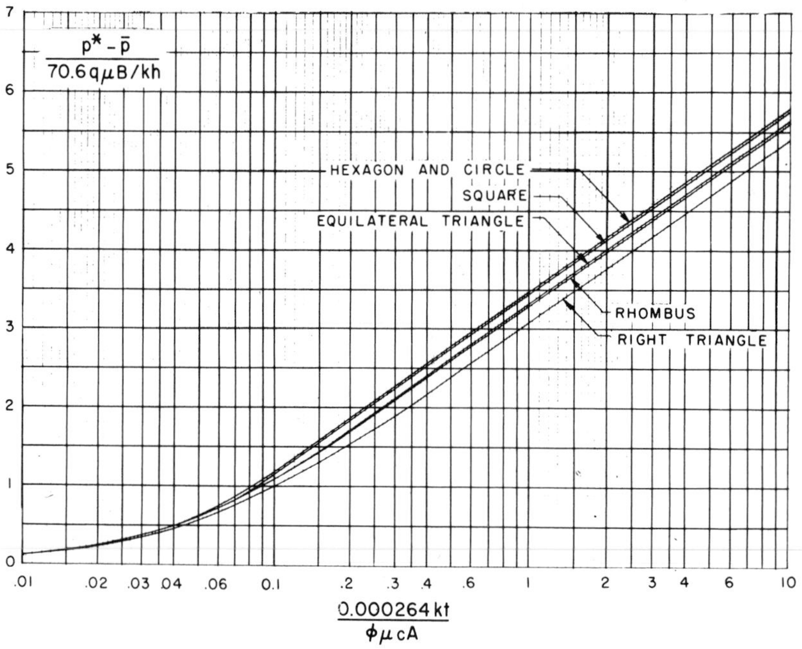


**Fig. 2.17A** – MBH pressure function for well in center of equilateral figures.

thews and Russell monograph[10]) summarize the results.

The steps in application of this method are as follows.

(1) Extrapolate the middle-time line to $(t_p + \Delta t)/\Delta t = 1$, and read the extrapolated pressure, $p^*$.

(2) Estimate the drainage area shape, which is frequently reasonably symmetrical, particularly for a well in a field developed on regular spacing.*

(3) Choose the proper curve from Figs. 2.17A through 2.17 G for the drainage-area shape of the tested well.

(4) Estimate $0.000264\ kt_p/\phi\mu c_t A$, and find $2.303(p^* - \bar{p})/m = p_{D\mathrm{MBH}}$. (It is essential that $t_p$ used in this calculation be the same used in the Horner plot.)

(5) Calculate $\bar{p} = p^* - mp_{D\mathrm{MBH}}/2.303$.

The advantages of the $p^*$ method for estimating static drainage-area pressure are that it does not require data beyond the MTR and that it is applicable to a wide variety of drainage-area shapes. Its disadvantages are that it requires knowledge of drainage-area size and shape and estimates of reservoir and fluid properties such as $c_t$ and $\phi$, which are not always known with great accuracy. We will compare these advantages and disadvantages with those of the modified Muskat method. First, we present an example illustrating application of the $p^*$ method.

---

## *Example 2.6 – p* Method To Calculate Average Pressure in Drainage Area*

**Problem.** Consider the buildup test described in Examples 2.2 and 2.3. Estimate the average pressure in the well's drainage area using the $p^*$ method.

**Solution.** In our use of the $p^*$ method, we assume that the well is centered in a square drainage area; thus, Fig. 2.17A applies. To use this figure, we first calculate the group

$$\frac{0.000264\ kt_p}{\phi\mu c_t A} = \left[(0.000264)(1.442\times 10^7)\right.$$

$$\left.\cdot(1.363\times 10^4)\right]/(2\times 1{,}320)^2$$

$$= 7.44.$$

From Fig. 2.17A,

$$\frac{2.303(p^* - \bar{p})}{m} = 5.45.$$

Therefore,

$$\bar{p} = p^* - \frac{5.45\ m}{2.303} = 4{,}577 - \frac{(5.45)(70)}{2.303}$$

$$= 4{,}411 \text{ psi}.$$

Accuracy of the $p^*$ method may be improved by using $t_{pss}$, producing time required to achieve pseudosteady state, in the Horner plot *and* in the abscissa in the MBH figures.[11] In principle, results should be identical for any $t_p > t_{pss}$; in practice, use of the smallest possible $t_p$ may reduce the error arising from a lengthy extrapolation. Time to reach pseudosteady state can be calculated after formation permeability, $k$, has been established, given drainage-area size and shape. Values are given in Table 1.2 for common drainage areas. As an example, for conditions in Example 2.6, in the column "Exact for $t_{DA}$," we read 0.1 for a well centered in a square drainage area. This means that

*Ref. 10 explains a method for estimating drainage-area size and shape for use when the symmetrical-drainage-area assumption is inadequate. The method can be tedious because it can involve verifying by trial-and-error planimetering that drainage volumes are proportional to flow rates in individual wells.

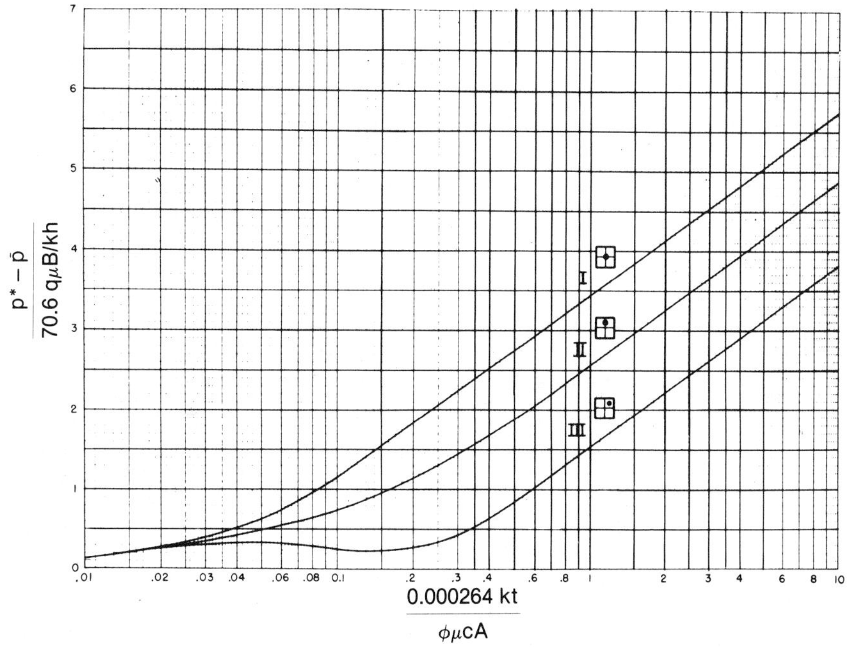


Fig. 2.17B – MBH pressure function for different well locations in a square boundary.

$$t_{DA} = \frac{0.000264\ kt_{pss}}{\phi\mu c_t A} = 0.1,$$

or, for this case (with $A = 160$ acres $= 6.97 \times 10^6$ sq ft),

$$t_{pss} = 183 \text{ hours.}$$

The reader can verify that use of $t_{pss}$ in the Horner plot and in the $p^*$ method leads to the same results as in Examples 2.3 and 2.6.

### Shape Factors for Reservoirs

Brons and Miller[12] observed that reservoirs with general drainage-area shapes follow equations of the form

$$\bar{p} - p_{wf} = 141.2 \frac{qB\mu}{kh}\left[\frac{1}{2}\ln\left(\frac{10.06\ A}{C_A r_w^2}\right) - \frac{3}{4} + s\right], \quad \text{(1.20)}$$

after pseudosteady state flow is achieved. Values for the shape factor $C_A$ can be derived from the $p_{D\text{MBH}}$ vs. $t_{DA}$ charts for the various reservoir geometries in Figs. 2.17A through 2.17G. Note that the definition of $p^*$ implies that

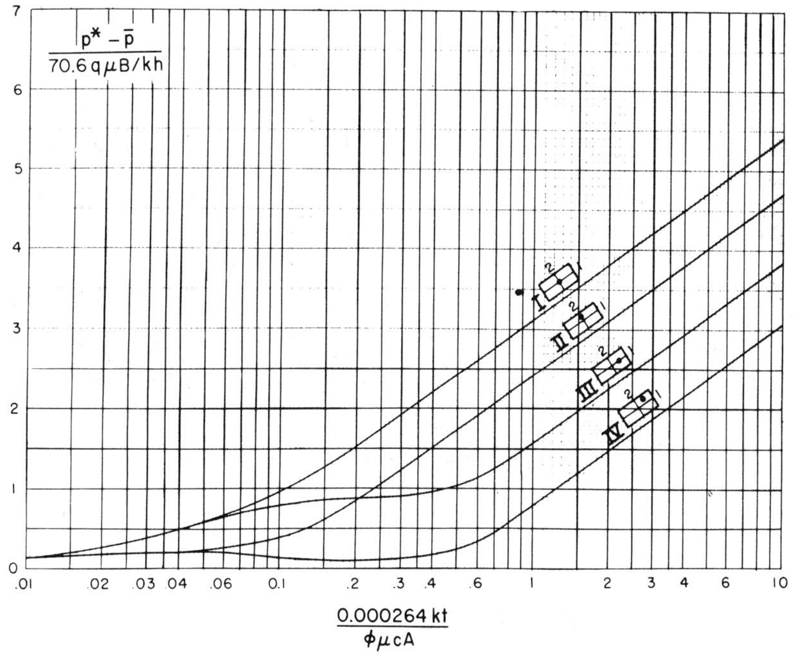


Fig. 2.17C – MBH pressure function for different well locations in a 2:1 rectangular boundary

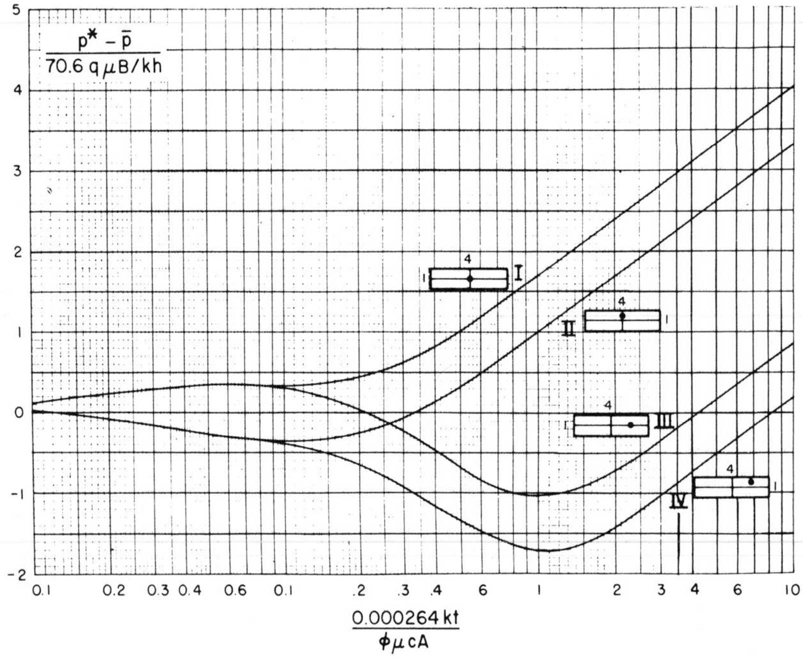


**Fig. 2.17D** – MBH pressure function for different well locations in a 4:1 rectangular boundary.

$$p^* - p_{ws} = 70.6 \frac{qB\mu}{kh} \ln(t_p + \Delta t)/\Delta t,$$

and that, at the instant of shut-in ($\Delta t = 0$),

$$p^* - p_{wf} = 70.6 \frac{qB\mu}{kh} \left[ \ln\left( \frac{kt_p}{1{,}688\, \phi\mu c_t r_w^2} \right) + 2s \right].$$

These relationships result from replacing $p_i$ with $p^*$ in Eqs. 1.11 and 2.1, which are valid for infinite-acting reservoirs only.

Eliminating $p_{wf}$ between equations,

$$p^* - \bar{p} = 70.6 \frac{qB\mu}{kh} \left[ \ln\left( \frac{kt_p}{1{,}688\, \phi\mu c_t r_w^2} \right) \right.$$

$$\left. - \ln\left( \frac{10.06\, A}{C_A r_w^2} \right) + 1.5 \right]$$

$$= 70.6 \frac{qB\mu}{kh} \ln\left( \frac{0.000264\, kt_p C_A}{\phi\mu c_t A} \right)$$

$$= 70.6 \frac{qB\mu}{kh} \ln(C_A t_{DA}),$$

or

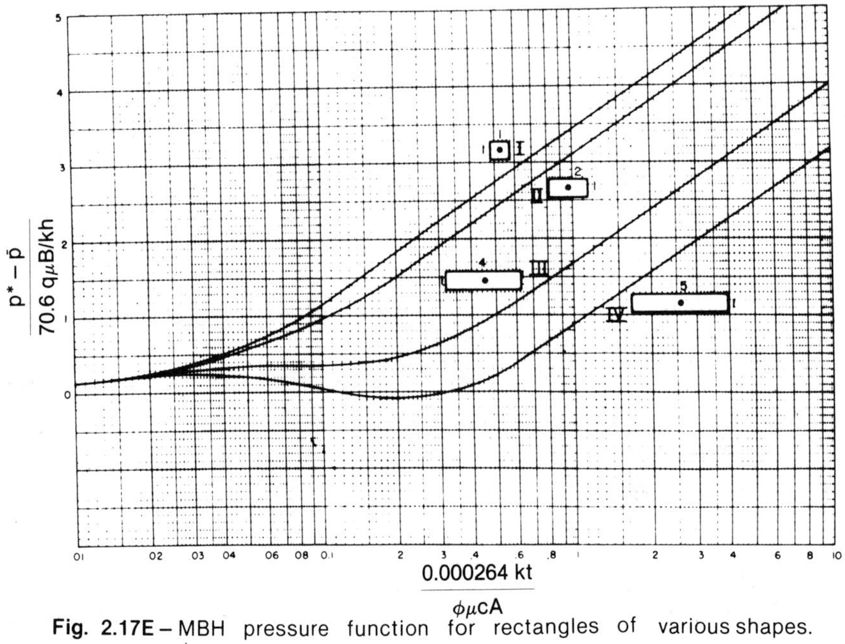


**Fig. 2.17E** – MBH pressure function for rectangles of various shapes.

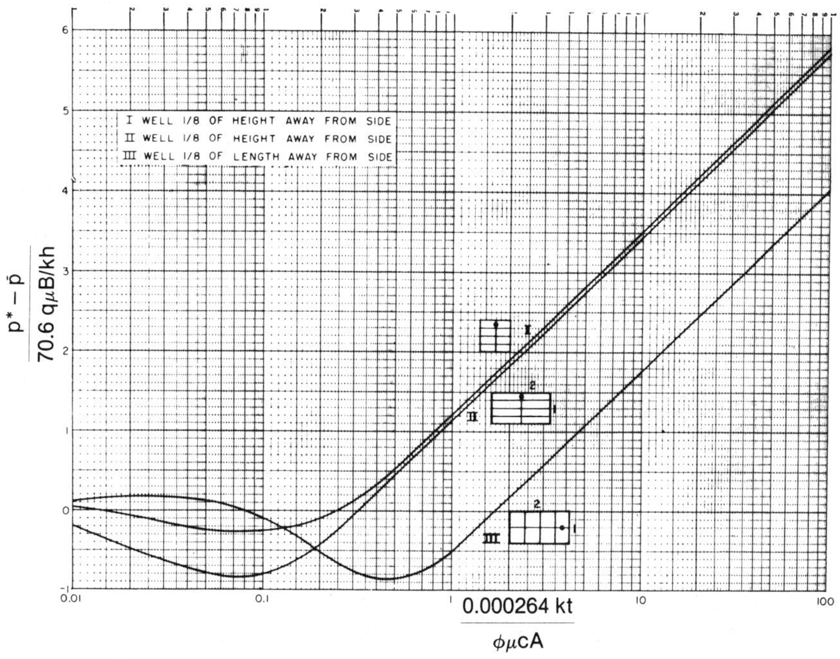


Fig. 2.17F – MBH pressure function in a square and in 2:1 rectangles.

$$p_{D\,\mathrm{MBH}} = \frac{p^* - \bar{p}}{70.6\, qB\mu/kh} = \ln(C_A t_{DA}).$$

This equation implies a linear relationship between $p_{D\,\mathrm{MBH}}$ and $t_{DA}$ after pseudosteady state flow has been achieved. Indeed, inspection of the curves in Figs. 2.17A through 2.17G shows that, for sufficiently large $t_{DA}$, a linear relation *does* develop. Further, any point on this straight line can be used to determine $C_A$. For example, consider a circular drainage area with a centered well.

From Fig. 2.17A, for

$$t_{DA} = \frac{0.000264\, kt_p}{\phi\mu c_t A} = 1,$$

$$p_{D\,\mathrm{MBH}} = \frac{p^* - \bar{p}}{70.6\, qB\mu/kh} = 3.454.$$

Thus, ln $[C_A(1)] = 3.454$ or $C_A = 31.6$ – essentially

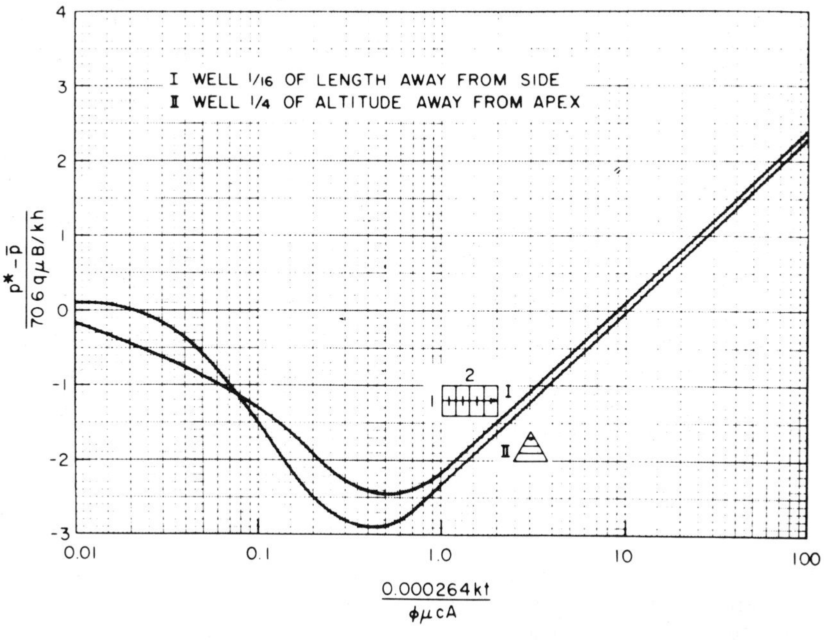


Fig. 2.17G – MBH pressure function on a 2:1 rectangle and equilateral triangle.

TABLE 2.5 – APPLICATION OF MODIFIED MUSKAT METHOD

| $\Delta t$ (hours) | $p_{ws}$ (psi) | $\bar{p}-p_{ws}$ (psi) | | |
|---|---|---|---|---|
| | | $\bar{p}=4{,}408$ | $\bar{p}=4{,}412$ | $\bar{p}=4{,}422$ |
| 30 | 4,393 | 15 | 19 | 29 |
| 40 | 4,398 | 10 | 14 | 24 |
| 50 | 4,402 | 6 | 10 | 20 |
| 60 | 4,405 | 3 | 7 | 17 |
| 72 | 4,407 | 1 | 5 | 15 |

the value given in Table 1.2. Note also that the linear relationship between $p_{D\text{MBH}}$ and $t_{DA}$ begins at $t_{DA} \simeq 0.1$; in Table 1.2, this is the value at which the pseudosteady-state flow equation becomes exact (in the column "Exact for $t_{DA}>$").

**Modified Muskat Method**

The modified Muskat method is based on a limiting form of Eq. 1.6, which is a solution to the flow equations for a well producing from a closed, cylindrical reservoir at constant rate. Using superposition to simulate a buildup following stabilized flow (depth of investigation has reached reservoir boundaries) and noting that, once boundary effects are felt in the buildup, the equation can be approximated as

$$\bar{p}-p_{ws}=118.6\frac{qB\mu}{kh}\exp(-0.00388\,k\Delta t/\phi\mu c_t r_e^2). \quad (2.23)$$

For analysis of buildup tests, we usually express this equation as

$$\log(\bar{p}-p_{ws})=\log\left(118.6\frac{qB\mu}{kh}\right)-\frac{0.00168\,k\Delta t}{\phi\mu c_t r_e^2}. \quad (2.24)$$

Approximations used in developing this equation are valid in the shut-in time range

$$\frac{250\,\phi\mu c_t r_e^2}{k}\le\Delta t\le\frac{750\,\phi\mu c_t r_e^2}{k}.$$

Note that Eq. 2.24 has the form

$$\log(\bar{p}-p_{ws})=A+B\Delta t,$$

where $A$ and $B$ are constants. This form of the equation suggests how it is applied. We assume a value for $\bar{p}$ and plot $\log(\bar{p}-p_{ws})$ vs. $\Delta t$ until a straight line results; when it does, the correct value of $\bar{p}$ (the static drainage-area pressure of the tested well) has been found. Experience with the method indicates that it is quite sensitive. An assumed value of $\bar{p}$ that is too low produces a plot of $\log(\bar{p}-p_{ws})$ vs. $\Delta t$ that has a noticeable downward curvature for data in the time range

$$\frac{250\,\phi\mu c_t r_e^2}{k}\le\Delta t\le\frac{750\,\phi\mu c_t r_e^2}{k}.$$

An assumed value of $\bar{p}$ that is too high produces a noticeable upward curvature in this same time range. The correct assumed value of $\bar{p}$ produces a straight line in (and only in) the proper time range. (See Fig. 2.18.)

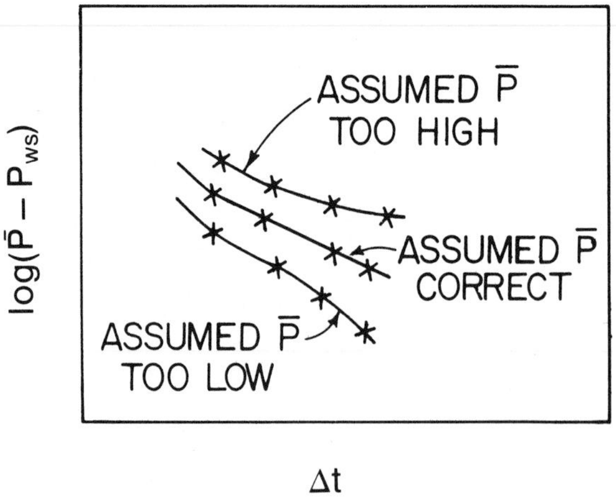


Fig. 2.18 – Schematic graph for modified Muskat method.

The modified Muskat method has two important advantages over the $p^*$ method: (1) it requires no estimates of reservoir properties when it is used to establish $\bar{p}$ (except to choose the correct portion of data for analysis); and (2) it has been found to provide satisfactory $\bar{p}$ estimates for hydraulically fractured wells and for wells with layers of different permeability that communicate only at the wellbore. In these cases, the $p^*$ method fails. The Muskat method has serious disadvantages, too: (1) it fails when the tested well is not reasonably centered in its drainage area (although the drainage area need not be cylindrical, as implied in derivation of this method); and (2) the required shut-in times of $(250\,\phi\mu c_t r_e^2)/k$ to $(750\,\phi\mu c_t r_e^2)/k$ are frequently impractically long, particularly in low-permeability reservoirs.

We now illustrate this method with an example.

---

## *Example 2.7 – Modified Muskat Method To Calculate Average Pressure in Drainage Area*

**Problem.** Consider the buildup test described in Examples 2.2 and 2.3. Estimate the average pressure in the well's drainage area by using the modified Muskat method.

**Solution.** The data that can be examined by this method are those in the range $\Delta t=(250\,\phi\mu c_t r_e^2)/k$ to $\Delta t=(750\,\phi\mu c_t r_e^2)/k$. In this case we are fortunate to have estimates of $k$ and $r_e$, so we can eliminate data outside the time range of interest. Often, of course, we do not have these estimates – a situation that does not limit the applicability of the method, but one that slightly complicates the trial-and-error nature of the calculations. Here,

$$\frac{250\,\phi\mu c_t r_e^2}{k}=\frac{(250)(1{,}320)^2}{(1.442\times10^7)}=30.2\text{ hours},$$

and

$$\frac{750\,\phi\mu c_t r_e^2}{k}=90.6\text{ hours}.$$

Thus, we can examine data from $\Delta t=30$ hours until

**TABLE 2.6 – BUILDUP DATA FOR WELL NEAR BOUNDARY**

| $\Delta t$ (hours) | $p_{ws}$ (psia) | $\Delta t$ (hours) | $p_{ws}$ (psia) | $\Delta t$ (hours) | $p_{ws}$ (psia) |
|---|---|---|---|---|---|
| 0 | 3,103 | 8 | 4,085 | 30 | 4,614 |
| 1 | 3,488 | 10 | 4,172 | 36 | 4,700 |
| 2 | 3,673 | 12 | 4,240 | 42 | 4,770 |
| 3 | 3,780 | 14 | 4,298 | 48 | 4,827 |
| 4 | 3,861 | 16 | 4,353 | 54 | 4,882 |
| 5 | 3,936 | 20 | 4,435 | 60 | 4,931 |
| 6 | 3,996 | 24 | 4,520 | 66 | 4,975 |

we stop recording pressures at $\Delta t = 72$ hours.

We make Table 2.5 for three trial values of $\bar{p}$. From plots of $(\bar{p} - p_{ws})$, we see that $\bar{p} = 4{,}412$ psi is clearly the best choice; we also note the sensitivity of this method. There is a noticeable curvature for $\bar{p} = 4{,}408$ psi and $\bar{p} = 4{,}422$ psi (Fig. 2.19).

This example application of the modified Muskat method is intended to illustrate only the mechanics of using the method. The pressure had built up to within 5 psi of its static value at a shut-in time of 72 hours; in such a case, there is little value in applying either this method or the MBH $p^*$ method. Both methods are of value only when the final pressure is a significant distance from stabilization.

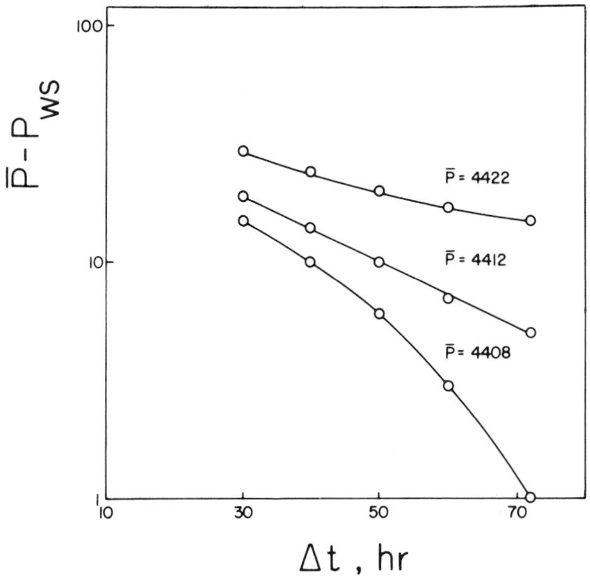


Fig. 2.19 – Modified Muskat method applied to example buildup test.

## 2.10 Reservoir Limits Test

In this section, we deal briefly with techniques for estimating reservoir size and distance to boundaries. These comments are introductory only, and deal only with the simplest cases. The intent is to illustrate an approach to these problems rather than the state of the art. The techniques presented are based on *buildup*-test data analysis. Much technology based on *drawdown* test analysis has been developed and is summarized by Earlougher.[13]

We demonstrate that a single boundary near a well causes the slope of a buildup curve to double, and then develop a method for estimating distance from a tested well to a single boundary.

In Chap. 1, when we illustrated application of the superposition principle, we showed that the flowing pressure in a well a distance $L$ from a no-flow boundary (such as a sealing fault) is given by Eq. 1.50. Rearranged, it becomes

$$p_i - p_{wf} = -70.6\frac{qB\mu}{kh}\left[\ln\left(\frac{1{,}688\,\phi\mu c_t r_w^2}{kt_p}\right) - 2s\right]$$

$$-70.6\frac{qB\mu}{kh}Ei\left[\frac{-948\,\phi\mu c_t (2L)^2}{kt_p}\right].$$

We can develop an equation describing a buildup test for such a well. Note that

$$p_i - p_{ws} = -70.6\frac{qB\mu}{kh}\left\{\ln\left[\frac{1{,}688\,\phi\mu c_t r_w^2}{k(t_p+\Delta t)}\right] - 2s\right\}$$

$$-70.6\,(-q)\,\frac{B\mu}{kh}\left[\ln\left(\frac{1{,}688\,\phi\mu c_t r_w^2}{k\Delta t}\right) - 2s\right]$$

$$-70.6\frac{qB\mu}{kh}Ei\left[\frac{-3{,}792\,\phi\mu c_t L^2}{k(t_p+\Delta t)}\right]$$

$$-70.6\,(-q)\,\frac{B\mu}{kh}Ei\left(\frac{-3{,}792\,\phi\mu c_t L^2}{k\Delta t}\right). \quad \text{(2.25)}$$

For a shut-in time sufficiently large that the logarithmic approximation is accurate for the *Ei* functions, the equation becomes

$$p_i - p_{ws} = 70.6\frac{qB\mu}{kh}\left[\ln\left(\frac{t_p+\Delta t}{\Delta t}\right) + \ln\left(\frac{t_p+\Delta t}{\Delta t}\right)\right]$$

$$= 141.2\frac{qB\mu}{kh}\ln\left(\frac{t_p+\Delta t}{\Delta t}\right).$$

This can be written as

$$p_{ws} = p_i - 325.2\frac{qB\mu}{kh}\log[(t_p+\Delta t)/\Delta t]. \quad \text{(2.26)}$$

Two observations can be made: (1) for a well near a single boundary, such as a sealing fault, Eq. 2.26 shows that the slope of a buildup curve will eventually double (compare Eq. 2.26 with Eq. 2.1) and (2) the time required for the slope to double can be long – specifically, $3{,}792\ \phi\mu c_t L^2/k\Delta t < 0.02$, or $\Delta t > 1.9\times10^5\ \phi\mu c_t L^2/k$. For large values of $L$ or small values of permeability, shut-in time required for the logarithmic approximation to be valid can be longer than the time ordinarily available for a buildup test. For this reason, awaiting a doubling in slope on a buildup test is not necessarily a satisfactory method of identifying a no-flow boundary near a well and estimating distance from the well to the boundary. Consequently, some analysts prefer to use Eq. 2.25 more directly, noting that for $t_p \gg \Delta t$ it may be rearranged as

$$p_{ws} \simeq p_i - 162.6\frac{qB\mu}{kh}\left[\log\left(\frac{t_p+\Delta t}{\Delta t}\right) - 0.434\,Ei\left(\frac{-3{,}792\,\phi\mu c_t L^2}{kt_p}\right)\right] - 70.6\frac{qB\mu}{kh}Ei\left(\frac{-3{,}792\,\phi\mu c_t L^2}{k\Delta t}\right). \quad \ldots (2.27)$$

Reasons for arranging the equation in this form are as follows.

1. The term $162.6\,(qB\mu/kh)\{\log[(t_p+\Delta t)/\Delta t] - 0.434\,Ei(-3{,}792\,\phi\mu c_t L^2/kt_p)]\}$ determines the position of the middle-time line. Note that the $Ei$ function is a constant; thus, it affects only the position of the MTR and has no effect on slope.
2. At earliest shut-in times in a buildup test, $Ei(-3{,}792\,\phi\mu c_t L^2/k\Delta t)$ is negligible. Physically, this means that the radius of investigation has not yet encountered the no-flow boundary and, mathematically, that the late-time region in the buildup test has not yet begun.

These observations suggest a method for analyzing the buildup test (Fig. 2.20):

1. Plot $p_{ws}$ vs. $\log\,(t_p+\Delta t)/\Delta t$.
2. Establish the middle-time region.
3. Extrapolate the MTR into the LTR.
4. Tabulate the differences, $\Delta p^*_{ws}$, between the buildup curve and extrapolated MTR for several points ($\Delta p^*_{ws} = p_{ws} - p_{MT}$).
5. Estimate $L$ from the relationship implied by Eq. 2.28:

$$\Delta p^*_{ws} = 70.6\frac{qB\mu}{kh}\left[-Ei\left(\frac{-3{,}792\,\phi\mu c_t L^2}{k\Delta t}\right)\right]. \quad \ldots\ldots\ldots\ldots (2.28)$$

$L$ is the only unknown in this equation, so it can be solved directly. Remember, though, that accuracy of this equation requires that $\Delta t \ll t_p$; when this condition is not satisfied, a computer history match using Eq. 2.25 in its complete form is required to determine $L$.

This calculation implied in Eq. 2.28 should be made for several values of $\Delta t$. If the apparent value of $L$ tends to increase or to decrease systematically with time, there is a strong indication that the model does not describe the reservoir adequately (i.e., the well is not behaving as if it were in a reservoir of uniform thickness and porosity, and much nearer one boundary than any others).

The following example illustrates this computational technique.

---

## *Example 2.8 – Estimating Distance to a No-Flow Boundary*

**Problem.** Geologists suspect a fault near a newly drilled well. To confirm this fault and to estimate distance from it, we run a pressure buildup test. Data from the test are given in Table 2.6. Well and reservoir data include the following.

$\phi$ = 0.15,
$\mu_o$ = 0.6 cp,
$c_t$ = $17\times10^{-6}$ psi$^{-1}$,
$r_w$ = 0.5 ft,
$A_{wb}$ = 0.00545 sq ft,
$\rho_o$ = 54.8 lbm/cu ft,
$q_o$ = 1,221 STB/D,
$B_o$ = 1.310 RB/STB, and
$h$ = 8 ft.

The well produced only oil and dissolved gas. Before shut-in, a total of 14,206 STB oil had been produced.

Analysis of these data show that afterflow distorts none of the data recorded at shut-in times of 1 hour or more. Based on the slope (650 psi/cycle) of the earliest straight line, permeability, $k$, appears to be 30 md. Depth of investigation at a shut-in time of 1 hour is 144 ft, lending confidence to the choice of the middle-time line. Pseudoproducing time, $t_p$, is 279.2 hours.

From these data, determine whether the buildup test data indicate that the well is behaving as if it were near a single fault and estimate distance to an apparent fault from buildup data at several times in the LTR.

**Solution.** Our attack is to plot $p_{ws}$ vs. $(t_p+\Delta t)/\Delta t$; extrapolate the middle-time line on into the LTR; read pressures, $p_{MT}$, from this extrapolated line; subtract those pressures from observed values of $p_{ws}$ in the LTR ($\Delta p^*_{ws} = p_{ws} - P_{MT}$); estimate values of $L$ from Eq. 2.28; and assume that, if calculated values of $L$ are fairly constant, the well is indeed near a single sealing fault.

From Fig. 2.21, we obtain the data in Table 2.7. We now estimate $L$ from Eq. 2.28. Note that

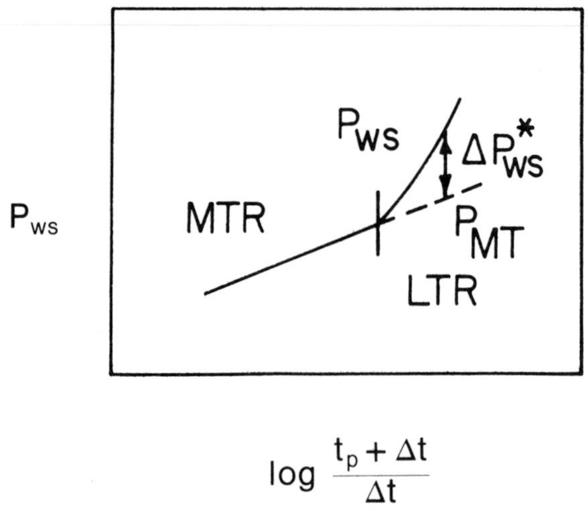


Fig. 2.20 – Buildup test graph for well near reservoir boundary.

$$\frac{3{,}792\,\phi\mu c_t}{k}=\frac{(3{,}792)(0.15)(0.6)(1.7\times10^{-5})}{30}$$

$$=1.934\times10^{-4}.$$

We first estimate $L$ at $\Delta t=10$ hours, which assumes that the approximation $t_p \simeq t_p+\Delta t$ is adequate in this case.

$$\Delta p^*_{ws}=52=-70.6\frac{qB\mu}{kh}Ei\left(\frac{-3{,}792\,\phi\mu c_t L^2}{k\Delta t}\right)$$

$$=\frac{-(70.6)(1{,}221)(1.310)(0.6)}{(30)(8)}$$

$$\cdot Ei\left(\frac{-3{,}792\,\phi\mu c_t L^2}{k\Delta t}\right).$$

$$-Ei\left(\frac{-1.934\times10^{-4}\,L^2}{10}\right)=0.184.$$

$$L^2=\frac{(1.107)(10)}{1.934\times10^{-4}}=5.72\times10^4,$$

or

$$L=239\text{ ft}.$$

For larger values of shut-in time, the approximation $t_p \simeq t_p+\Delta t$ becomes decreasingly accurate, and no terms in Eq. 2.25 can be neglected, but $L\simeq 240$ ft satisfies the equation for all values of shut-in time.

For the case in which the slope of the buildup test has time to double, estimation of distance from well to boundary is easier. From the buildup tests plot, we find the time, $\Delta t_x$, at which the two straight-line sections intersect (Fig. 2.22). Gray[14] suggests that the distance $L$ from the well to the fault can be calculated from

$$L=\sqrt{\frac{0.000148\,k\Delta t_x}{\phi\mu c_t}}\quad\ldots\ldots\ldots\ldots\ldots\ldots(2.29)$$

In Fig. 2.21, the slope did double, and the figure shows that $(t_p+\Delta t_x)/\Delta t_x=17$, from which $\Delta t_x=17.45$ hours. Eq. 2.29 then shows that $L\simeq 225$ ft, in reasonable agreement with our previous calculation.

The results of pressure buildup tests sometimes can be used to estimate reservoir size. The basic idea is to compare average static reservoir pressure before and after production of a known quantity of fluid from a closed, volumetric reservoir, with constant compressibility, $c_t$. If $V_R$ is the reservoir volume (barrels), $\Delta N_p$ is the stock-tank barrels of oil produced between Times 1 and 2, and $\bar{p}_1$ and $\bar{p}_2$ are the average reservoir pressures before and after oil production, then a material balance on the reservoir shows that

$$\bar{p}_2=\bar{p}_1-\frac{(\Delta N_p)(B_o)}{V_R c_t \phi},$$

**TABLE 2.7 – ANALYSIS OF DATA FROM WELL NEAR BOUNDARY**

| $\Delta t$ (hours) | $(t_p+\Delta t)/\Delta t$ | $p_{ws}$ (psi) | $p_{MT}$ (psi) | $(p_{ws}-p_{MT})=\Delta p^*_{ws}$ (psi) |
|---|---|---|---|---|
| 6 | 47.5 | 3,996 | 3,980 | 16 |
| 8 | 35.9 | 4,085 | 4,051 | 34 |
| 10 | 28.9 | 4,172 | 4,120 | 52 |
| 12 | 24.3 | 4,240 | 4,170 | 70 |
| 14 | 20.9 | 4,298 | 4,210 | 88 |
| 16 | 18.5 | 4,353 | 4,250 | 103 |
| 20 | 15.0 | 4,435 | 4,300 | 135 |
| 24 | 12.6 | 4,520 | 4,355 | 165 |
| 30 | 10.3 | 4,614 | 4,410 | 204 |
| 36 | 8.76 | 4,700 | 4,455 | 245 |
| 42 | 7.65 | 4,770 | 4,495 | 275 |
| 48 | 6.82 | 4,827 | 4,525 | 302 |
| 54 | 6.17 | 4,882 | 4,552 | 330 |
| 60 | 5.65 | 4,931 | 4,578 | 353 |
| 66 | 5.23 | 4,975 | 4,600 | 375 |

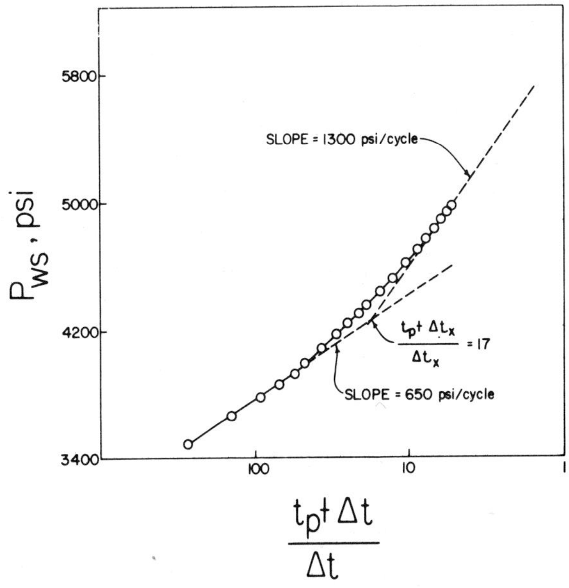


Fig. 2.21 – Estimating distance to a no-flow boundary.

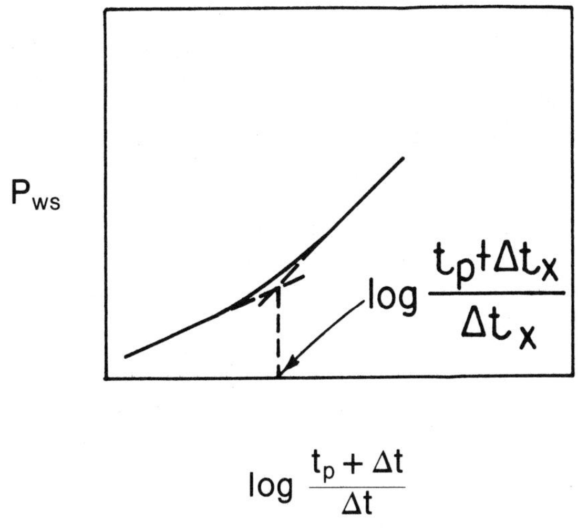


Fig. 2.22 – Distance to boundary from slope doubling.

or

$$V_R = \frac{(\Delta N_p)(B_o)}{(\bar{p}_1 - \bar{p}_2)c_t\phi}. \quad \text{(2.30)}$$

---

*Example 2.9 – Estimating Reservoir Size*

**Problem.** Two pressure buildup tests are run on the only well in a closed reservoir. The first test indicates an average pressure of 3,000 psi, the second indicates 2,100 psi. The well produced an average of 150 STB/D of oil in the year between tests. Average oil formation volume factor, $B_o$, is 1.3 RB/STB; total compressibility, $c_t$, is $10\times10^{-6}$ psi$^{-1}$; porosity, $\phi$, is 22%; and average sand thickness, $h$, is 10 ft. Estimate area, $A_R$, of the reservoir in acres.

**Solution.** From Eq. 2.30,

$$V_R = \frac{\Delta N_p B_o}{(\bar{p}_1 - \bar{p}_2)c_t\phi}$$

$$= \frac{(qt)B_o}{(\bar{p}_1 - \bar{p}_2)(c_t\phi)}$$

$$= \frac{(150 \text{ STB/D})(365 \text{ days})(1.3 \text{ RB/STB})}{(3{,}000-2{,}100)\text{psi}\,(10\times10^{-6}\text{ psi}^{-1})(0.22)}.$$

Thus,

$$V_R = 35.9\times10^6 \text{ bbl}$$

$$= \frac{43{,}560 A_R h}{5.615},$$

and

$$A_R = \frac{(35.9\times10^6 \text{ bbl})(5.615 \text{ cu ft/bbl})}{(10 \text{ ft})(43.56\times10^3 \text{ sq ft/acre})}$$

$$= 463 \text{ acres}.$$

## 2.11 Modifications for Gases

This section presents modifications of the basic drawdown and buildup equations so that they can be applied to analysis of gas reservoirs. These modifications are based on results obtained with the gas pseudopressure,[15] although a more complete discussion of that subject is left to Chap. 5.

Wattenbarger and Ramey[16] have shown that for some gases at pressures above 3,000 psi, flow in an infinite-acting reservoir can be modeled accurately by the equation

$$p_{wf} = p_i + \frac{162.6\, q_g B_{gi}\mu_i}{kh}\left[\log\left(\frac{1{,}688\,\phi\mu_i c_{ti}}{kt_p}\right) - \frac{(s+Dq_g)}{1.151}\right]. \quad \text{(2.31)}$$

This equation has the same form as the equation for a slightly compressible liquid, but there are some important differences:

1. Gas production rate, $q_g$, is conveniently expressed in thousands of standard cubic feet per day (Mscf/D), and gas formation volume factor, $B_g$, is then expressed in reservoir barrels per thousand standard cubic feet (RB/Mscf), so that the product $q_g B_g$ is in reservoir barrels per day (RB/D) as in the analogous equation for slightly compressible liquids.

2. All gas properties ($B_g$, $\mu$, and $c_g$) are evaluated at original reservoir pressure, $p_i$. (More generally, these properties should be evaluated at the uniform pressure in the reservoir before initiation of flow.) In Eq. 2.31,

$$B_{gi} = \frac{178.1\, z_i T p_{sc}}{p_i T_{sc}} \text{ (RB/Mscf)},$$

$$c_{ti} = c_{gi}S_g + c_w S_w + c_f \simeq c_{gi}S_g.$$

3. The factor $D$ is a measure of non-Darcy or turbulent pressure loss (i.e., a pressure drop in addition to that predicted by Darcy's law). It cannot be calculated separately from the skin factor from a single buildup or drawdown test; thus, the concept of apparent skin factor, $s' = s + Dq_g$, is sometimes convenient since it can be determined from a single test.

For many cases at pressures below 2,000 psi, flow in an infinite acting reservoir can be modeled by

$$p_{wf}^2 = p_i^2 + \frac{1{,}637\, q_g\mu_i z_i T}{kh} \cdot \left[\log\left(\frac{1{,}688\,\phi\mu c_{ti}}{kt_p}\right) - \frac{(s+Dq_g)}{1.151}\right]. \quad \text{(2.32)}$$

Using these basic drawdown equations, we can use superposition to develop equations describing a buildup test for gas wells.

For $p > 3{,}000$ psi,

$$p_{ws} = p_i - 162.6\frac{q_g B_{gi}\mu_i}{kh}\left[\log\left(\frac{t_p+\Delta t}{\Delta t}\right)\right], \quad \text{(2.33)}$$

and

$$s' = s + D(q_g) = 1.151\left[\frac{(p_{1\,\text{hr}} - p_{wf})}{m} - \log\left(\frac{k}{\phi\mu_i c_{ti} r_w^2}\right) + 3.23\right]. \quad \text{(2.34)}$$

For $p < 2{,}000$ psi,

$$p_{ws}^2 = p_i^2 - 1{,}637\frac{q_g\mu_i z_i T}{kh}\log\left(\frac{t_p+\Delta t}{\Delta t}\right), \quad \text{(2.35)}$$

and

$$s' = s + D(q_g) = 1.151\left[\frac{(p_{1\,\text{hr}}^2 - p_{wf}^2)}{m''} - \log\left(\frac{k}{\phi\mu_i c_{ti} r_w^2}\right) + 3.23\right], \quad \text{(2.36)}$$

where $m''$ is the slope of the plot $p_{ws}^2$ vs. $\log[(t_p+\Delta t)/\Delta t]$, which is $1{,}637\, q_g\mu_i z_i T/kh$.

An obvious question is, what technique should be used to analyze gas reservoirs with pressures in the range 2,000 $<p<$ 3,000 psi? One approach is to use equations written in terms of the gas pseudopressure instead of either pressure or pressure squared. This is at least somewhat inconvenient, so an alternative approach is to use equations written in terms of either $p_{ws}$ or $p_{ws}^2$ and accept the resultant inaccuracies, which, in real, heterogeneous reservoirs, may be far from the most significant oversimplification on which the test analysis procedure is based. The smaller the pressure drawdown during the test, the less the inaccuracy in this approach.

---

### *Example 2.10 – Gas Well Buildup Test Analysis*

**Problem.** A gas well is shut in for a pressure buildup test. Test data include the following.

$q_g$ = 5,256 Mscf/D,
$T$ = 181°F = 641°R,
$h$ = 28 ft,
$\mu_i$ = 0.028 cp,
$S_w$ = 0.3,
$\phi$ = 0.18,
$z_i$ = 0.85,
$r_w$ = 0.3 ft,
$c_{gi}$ = $0.344\times10^{-3}$ psi$^{-1}$,
$p_i$ = 2,906 psia, and
$p_{wf}$ = 1,801 psia.

Most of the test data fall in the intermediate pressure range, 2,000 $<p<$ 3,000 psia. On a plot of $p_{ws}$ vs. log $(t_p+\Delta t)/\Delta t$, the MTR had a slope, $m$, of 81 psi/cycle; on this plot, $p_{1\,hr}$ was found to be 2,525 psia. Alternatively, on a plot of $p_{ws}^2$ vs. log $(t_p+\Delta t)/\Delta t$, the MTR had a slope of $0.48\times10^6$ psi$^2$/cycle and $p_{1\,hr}^2$ of $7.29\times10^6$ psi$^2$.

From these data, estimate apparent values of $k$ and $s'$ (1) based on characteristics of the $p_{ws}$ plot and (2) based on characteristics of the $p_{ws}^2$ plot.

**Solution.** From results of the $p_{ws}$ plot, for standard conditions of 14.7 psia and 60°F,

$$B_{gi}=178.1\frac{z_iTp_{sc}}{p_iT_{sc}}$$

$$=\frac{(178.1)(0.85)(641)(14.7)}{(2{,}906)(520)}$$

$$=0.944 \text{ RB/Mscf},$$

$$k=162.6\frac{q_gB_{gi}\mu_i}{mh}$$

$$=\frac{(162.6)(5{,}256)(0.944)(0.028)}{(81)(28)}$$

$$=9.96 \text{ md},$$

$$c_{ti}\simeq c_{gi}S_g=(3.44\times10^{-4})(0.7)$$

$$=2.41\times10^{-4}\text{ psi}^{-1},$$

and

$$s'=s+D(q_g)=1.151\left[\frac{(p_{1\,hr}-p_{wf})}{m}-\log\left(\frac{k}{\phi\mu_ic_{ti}r_w^2}\right)+3.23\right]$$

$$=1.151\left\{\frac{(2{,}525-1{,}801)}{81}-\log\left[\frac{9.96}{(0.18)(0.028)(2.41\times10^{-4})(0.3)^2}\right]+3.23\right\}=4.84.$$

From results of the $p_{ws}^2$ plot,

$$k=1{,}637\frac{q_g\mu_iz_iT}{m''h}$$

$$=\frac{(1{,}637)(5{,}256)(0.028)(0.85)(641)}{(4.8\times10^5)(28)}$$

$$=9.77 \text{ md},$$

and

$$s'=1.151\left[\frac{(p_{1\,hr}^2-p_{wf}^2)}{m''}-\log\frac{k}{\phi\mu_ic_{ti}r_w^2}+3.23\right]$$

$$=1.151\left\{\frac{[7.29\times10^6-(1{,}801)^2]}{4.8\times10^5}-\log\left[\frac{9.77}{(0.18)(0.028)(2.41\times10^{-4})(0.3)^2}\right]+3.23\right\}$$

$$=4.27.$$

Neither set of results ($k$ and $s'$) is necessarily more accurate than the other in the *general* case; as in this *particular* case, use of an analysis procedure based on gas pseudopressure can be used to improve accuracy if disagreement in results from $p_{ws}$ and $p_{ws}^2$ plots is unacceptably large.

## 2.12 Modifications for Multiphase Flow

Basic buildup and drawdown equations can be modified to model multiphase flow.[17,18] For an infinite-acting reservoir, the drawdown equation becomes

$$p_{wf}=p_i+162.6\frac{q_{Rt}}{\lambda_th}\left[\log\left(\frac{1{,}688\,\phi c_tr_w^2}{\lambda_tt}\right)-\frac{s}{1.151}\right], \quad \text{(2.37)}$$

and the buildup equation becomes

$$p_{ws}=p_i-162.6\frac{q_{Rt}}{\lambda_t h}\log\left(\frac{t_p+\Delta t}{\Delta t}\right). \quad \ldots\ldots (2.38)$$

In these equations, the total flow rate $q_{Rt}$ is in reservoir barrels per day (neglecting solution gas liberated from produced water),

$$q_{Rt}=q_oB_o+\left(q_g-\frac{q_oR_s}{1{,}000}\right)B_g+q_wB_w, \quad \ldots\ldots (2.39)$$

and total mobility, $\lambda_t$, is

$$\lambda_t=\frac{k_o}{\mu_o}+\frac{k_w}{\mu_w}+\frac{k_g}{\mu_g}. \quad \ldots\ldots (2.40)$$

Total compressibility, $c_t$, was defined in Eq. 1.4.

These equations imply that it is possible to determine $\lambda_t$ from the slope $m$ of a buildup test run on a well that produces two or three phases simultaneously:

$$\lambda_t=162.6\frac{q_{Rt}}{mh}. \quad \ldots\ldots (2.41)$$

Perrine[17] has shown that it is also possible to estimate the permeability to *each phase* flowing from the same slope, $m$:

$$k_o=162.6\frac{q_oB_o\mu_o}{mh}, \quad \ldots\ldots (2.42)$$

$$k_g=162.6\frac{\left(q_g-\frac{q_oR_s}{1{,}000}\right)B_g\mu_g}{mh}, \quad \ldots\ldots (2.43)$$

and

$$k_w=162.6\frac{q_wB_w\mu_w}{mh}. \quad \ldots\ldots (2.44)$$

The term $(q_g-q_oR_s/1{,}000)B_g$, which appears in Eqs. 2.39 and 2.43, is the *free* gas flow rate in the reservoir. It is found by subtracting the dissolved gas rate $(q_oR_s/1{,}000)$ from the total surface gas rate $(q_g)$ and converting to a reservoir-condition basis.

Simultaneous solution of Eqs. 2.37 and 2.38 results in the following expression for the skin factor $s$.

$$s=1.151\left[\frac{p_{1\,\mathrm{hr}}-p_{wf}}{m}-\log\left(\frac{\lambda_t}{\phi c_t r_w^2}\right)+3.23\right]. \quad \ldots\ldots (2.45)$$

Static drainage-area pressure, $\bar{p}$, is calculated just as for a single-phase reservoir. In use of the MBH charts to determine $\bar{p}$ (and in the Horner plot itself), the effective production time $t_p$ is best estimated by dividing cumulative *oil* production by the *oil* production rate just before shut-in.

An important assumption required for accurate use of these equations for multiphase flow analysis is that saturations of each phase remain essentially uniform throughout the drainage area of the tested well.

---

## *Example 2.12 – Multiphase Buildup Test Analysis*

**Problem.** A buildup test is run in a well that produces oil, water, and gas simultaneously. Well, rock, and fluid properties evaluated at average reservoir pressure during the test include the following.

$S_o$ = 0.58,
$S_g$ = 0.08,
$S_w$ = 0.34,
$c_w$ = $3.6\times10^{-6}$ psi$^{-1}$,
$c_f$ = $3.5\times10^{-6}$ psi$^{-1}$,
$c_g$ = $0.39\times10^{-3}$ psi$^{-1}$,
$\mu_o$ = 1.5 cp,
$\mu_w$ = 0.7 cp,
$\mu_g$ = 0.03 cp,
$B_o$ = 1.3 RB/STB,
$B_w$ = 1.02 RB/STB,
$B_g$ = 1.480 RB/Mscf,
$R_s$ = 685 scf/STB,
$\phi$ = 0.17,
$r_w$ = 0.3 ft, and
$h$ = 38 ft.

From plots of $B_o$ vs. $p$ and $R_s$ vs. $p$ at average pressure in the buildup test,

$$\frac{dR_s}{dp}=0.0776 \text{ scf/STB/psi},$$

and

$$\frac{dB_o}{dp}=2.48\times10^{-6} \text{ RB/STB/psi}.$$

The production rates prior to the buildup test were $q_o$ = 245 STB/D, $q_w$ = 38 STB/D, and $q_g$ = 489 Mscf/D.

A plot of $p_{ws}$ vs. $\log(t_p+\Delta t)/\Delta t$ shows that the slope of the MTR, $m$, is 78 psi/cycle and that $p_{1\,\mathrm{hr}}$ = 2,466 psia. Flowing pressure, $p_{wf}$, at the instant of shut-in was 2,028 psia.

From these data, estimate $\lambda_t$, $k_o$, $k_w$, $k_g$, and $s$.

**Solution.** Permeabilities to each phase can be determined from the slope $m$ of the MTR:

$$k_o=162.6\frac{q_oB_o\mu_o}{mh}$$

$$=\frac{(162.6)(245)(1.5)(1.3)}{(78)(38)}=26.2 \text{ md},$$

$$k_w=162.6\frac{q_wB_w\mu_w}{mh}$$

$$=\frac{(162.6)(38)(0.7)(1.02)}{(78)(38)}=1.49 \text{ md},$$

$$k_g=162.6\frac{\left(q_g-\frac{q_oR_s}{1{,}000}\right)B_g\mu_g}{mh}$$

$$= 162.6 \frac{\left[489 - \frac{(245)(685)}{(1{,}000)}\right](1.480)(0.03)}{(78)(38)}$$

$$= 0.782 \text{ md.}$$

To calculate total mobility, $\lambda_t$, we first need total flow rate, $q_{Rt}$:

$$q_{Rt} = q_o B_o + \left(q_g - \frac{q_o R_s}{1{,}000}\right) B_g + q_w B_w$$

$$= (245)(1.3) + \left[489 - \frac{(245)(685)}{1{,}000}\right]$$

$$\cdot (1.480) + (38)(1.02)$$

$$= 833 \text{ RB/D.}$$

Then,

$$\lambda_t = 162.6 \frac{q_{Rt}}{mh} = \frac{(162.6)(833)}{(78)(38)}$$

$$= 45.7 \text{ md/cp.}$$

To calculate skin factor, $s$, we first need $c_o$ and $c_t$:

$$c_o = \frac{B_g}{B_o} \frac{dR_s}{dp} - \frac{1}{B_o} \frac{dB_o}{dp}$$

$$= \frac{(1.480 \text{ RB/Mscf})(0.0776 \text{ scf})}{(1.3 \text{ RB/STB})(\text{STB-psi})}$$

$$\cdot \frac{1 \text{ Mscf}}{1{,}000 \text{ scf}} - \frac{(2.48 \times 10^{-6})}{1.3 \text{ (psi)}}$$

$$= 86.4 \times 10^{-6} \text{ psi}^{-1}.$$

Then,

$$c_t = S_o c_o + S_g c_g + S_w c_w + c_f$$

$$= (0.58)(86.4 \times 10^{-6}) + (0.08)(0.39 \times 10^{-3})$$

$$+ (0.34)(3.6 \times 10^{-6}) + 3.5 \times 10^{-6}$$

$$= 86.0 \times 10^{-6} \text{ psi}^{-1},$$

and

$$s = 1.151 \left[\frac{p_{1\text{ hr}} - p_{wf}}{m} - \log\left(\frac{\lambda_t}{\phi c_t r_w^2}\right) + 3.23\right]$$

$$= 1.151 \left\{ \frac{2{,}466 - 2{,}028}{78} \right.$$

$$\left. - \log\left[\frac{45.7}{(0.17)(86.0 \times 10^{-6})(0.3)^2}\right] + 3.23 \right\}$$

$$= 1.50.$$

## Exercises

2.1. In Example 2.1, what error arises because we used Eq. 2.4 to calculate skin factor instead of the more exact Eq. 2.3? What difference would it have made in the value of $s$ had we used a shut-in time of 10 hours in Eq. 2.3 and the corresponding value of $p_{ws}$? What assumption have we made about distance from tested well to reservoir boundaries in Example 2.1?

2.2. Prove that the slope of a plot of shut-in BHP vs. log $(t_p + \Delta t)/\Delta t$ is, as asserted in the text, the difference in pressure at two points one cycle apart. Also prove that, for $\Delta t \ll t_p$, we obtain the same slope on a plot of $p_{ws}$ vs. log $\Delta t$. Finally, prove that on a plot of $p_{ws}$ vs. log $\Delta t$, we obtain the same slope regardless of the units used for shut-in time, $\Delta t$, on the plot (i.e., that $\Delta t$ can be expressed in minutes, hours, or days without affecting the slope of the plot).

2.3. A well producing only oil and dissolved gas has produced 12,173 STB. The well has not been stimulated, nor is there any reason to believe that there is a significant amount of formation damage. A pressure buildup test is run with the primary objective of estimating static drainage-area pressure. During buildup, there is a rising liquid level in the wellbore. Well and reservoir data are:

$\phi$ = 0.14,
$\mu$ = 0.55 cp,
$c_t$ = $16 \times 10^{-6}$ psi$^{-1}$,
$r_w$ = 0.5 ft,
$A_{wb}$ = 0.0218 sq ft,
$r_e$ = 1,320 ft (well centered in cylindrical drainage area),
$\rho$ = 54.8 lbm/cu ft,
$q$ = 988 STB/D,
$B$ = 1.126 RB/STB, and
$h$ = 7 ft.

Data recorded during the buildup test are given in Table 2.8. Plot $p_{ws}$ vs. $(t_p + \Delta t)/\Delta t$ on semilog paper and $(p_{ws} - p_{wf})$ vs. $\Delta t_e$ on log-log paper, and estimate the time at which afterflow ceased distorting the buildup test data.

2.4. Consider the buildup test described in Problem 2.3. Locate the MTR and estimate formation permeability.

2.5. Consider the buildup test described in Problems 2.3 and 2.4. Calculate skin factor, $s$; pressure drop across the altered zone, $(\Delta p)_s$; flow efficiency, $E$; and effective wellbore radius, $r_{wa}$.

2.6. Prove that in a buildup test for a well near a single fault, the technique suggested in the text (extrapolating the late-time line to infinite shut-in time) is the proper method for estimating original reservoir pressure. Comment on the possible errors in original reservoir pressure estimates in these cases: (1) some LTR data were obtained, but final straight line was not established; and (2) no LTR data were obtained.

2.7. Consider the buildup test described in Problems 2.3 and 2.4. Estimate static drainage-area pressure for this well (1) using the $p^*$ method, and (2)

TABLE 2.8 – PRESSURE BUILDUP TEST DATA

| $\Delta t$ (hours) | $p_{ws}$ (psia) | $\Delta t$ (hours) | $p_{ws}$ (psia) |
|---|---|---|---|
| 0 | 709 | 19.7 | 4,198 |
| 1.97 | 3,169 | 24.6 | 4,245 |
| 2.95 | 3,508 | 29.6 | 4,279 |
| 3.94 | 3,672 | 34.5 | 4,306 |
| 4.92 | 3,772 | 39.4 | 4,327 |
| 5.91 | 3,873 | 44.4 | 4,343 |
| 7.88 | 3,963 | 49.3 | 4,356 |
| 9.86 | 4,026 | 59.1 | 4,375 |
| 14.8 | 4,133 | | |

using the modified Muskat method.

2.8. In Example 2.7, explain how we could have applied the modified Muskat method to estimate static drainage-area pressure if we had not had estimates of $k/\phi\mu c_t$ or $r_e$.

2.9. Estimate formation permeability and skin factor from the following data available from a gas well pressure buildup test.

$T$ = 199°F = 659°R,
$h$ = 34 ft,
$\mu_i$ = 0.023 cp,
$S_w$ = 0.33 (water is immobile),
$c_{gi}$ = 0.000315 psi$^{-1}$,
$\phi$ = 0.22,
$z_i$ = 0.87, and
$r_w$ = 0.3 ft.

The well produced 6,068 Mcf/D before the test. A plot of BHP, $p_{ws}$, vs. log $(t_p+\Delta t)/\Delta t$ gave a middle-time line with a slope of 66 psi/cycle. Analysis of the buildup curve showed that static drainage-area pressure, $\bar{p}$, was 3,171 psia. Pressure on the middle-time line at $\Delta t=1$ hour, $p_{1\text{ hr}}$, was 2,745 psia; flowing pressure at shut-in, $p_{wf}$, was 2,486 psia.

2.10. Estimate total mobility, $\lambda_t$, oil, water and gas permeabilities, and skin factor for a well that produced oil, water, and gas simultaneously before a pressure buildup test. Production rates before the test were $q_o=276$ STB/D, $q_w=68$ STB/D, and $q_g=689$ Mcf/D. A plot of $p_{ws}$ vs. log $(t_p+\Delta t)/\Delta t$ showed that the slope $m$ of the middle-time line was 59 psi/cycle. Flowing pressure at shut-in, $p_{wf}$, was 1,581 psia; the pressure on the middle-time straight line at $\Delta t=1$ hour, $p_{1\text{ hr}}$, was 1,744 psia. Plots of $B_o$ vs. $p$ and $R_s$ vs. $p$ showed that $dR_s/dp=0.263$ scf/STB/psi and that $dB_o/dp=0.248\times10^{-3}$ RB/STB/psi. Rock, fluid, and well properties include the following.

$S_o$ = 0.56,
$S_g$ = 0.09,
$S_w$ = 0.35,
$c_w$ = $3.5\times10^{-6}$ psi$^{-1}$,
$c_f$ = $3.5\times10^{-6}$ psi$^{-1}$,
$c_g$ = $0.48\times10^{-3}$ psi$^{-1}$,
$\mu_o$ = 1.1 cp,
$\mu_w$ = 0.6 cp,
$\mu_g$ = 0.026 cp,
$B_o$ = 1.28 RB/STB,
$B_w$ = 1.022 RB/STB,
$B_g$ = 1.122 RB/Mscf,
$R_s$ = 748 scf/STB,
$\phi$ = 0.18,
$r_w$ = 0.3 ft, and
$h$ = 33 ft.

TABLE 2.9 – BUILDUP TEST DATA FOR WELL NEAR FAULT

| $\Delta t$ (hours) | $p_{ws}$ (psi) | $\Delta t$ (hours) | $p_{ws}$ (psi) |
|---|---|---|---|
| 20 | 1,373 | 500 | 2,225 |
| 30 | 1,467 | 800 | 2,360 |
| 40 | 1,533 | 1,000 | 2,434 |
| 50 | 1,585 | 1,500 | 2,545 |
| 100 | 1,752 | 2,000 | 2,616 |
| 200 | 1,940 | | |

2.11. A pressure buildup test was run on an oil well believed to be near a sealing fault in an otherwise infinite-acting reservoir. Estimate the distance to the fault, given the well, rock, and fluid properties below and the buildup data in Table 2.9.

$q$ = 940 STB/D,
$\mu$ = 50 cp,
$\phi$ = 0.2,
$c_t$ = $78\times10^{-6}$ psi$^{-1}$,
$h$ = 195 ft,
$p_i$ = 2,945 psi,
$N_p$ = 84,500 STB,
$B_o$ = 1.11 RB/STB, and
$t_{wbs}$ < 20 hours.

2.12. A well flowed for 10 days at 350 STB/D; it was then shut in for a pressure buildup test. Rock, fluid, and well properties include the following.

$B_o$ = 1.13 RB/STB,
$p_i$ = 3,000 psi,
$\mu$ = 0.5 cp,
$k$ = 25 md,
$s$ = 0,
$h$ = 50 ft,
$c_t$ = $20\times10^{-6}$ psi$^{-1}$,
$\phi$ = 0.16, and
$r_w$ = 0.333 ft.

(a) Determine and plot the pressure distribution in the reservoir for shut-in times of 0, 0.1, 1, and 10 days. (Assume an infinite acting reservoir.)

(b) Calculate the radius of investigation at 0.1, 1, and 10 days. Compare $r_i$ with the depth to which the transient appears to have moved on the plots prepared in Part a.

2.13. In Example 2.6, $\bar{p}$ was determined to be 4,411 psi. Both the Horner plot and the abscissa of the MBH chart used $t_p=13{,}630$ hours. It can be shown that for a well centered in a square drainage area, the time required to reach semisteady state is $t_{pss}=(\phi\mu c_t A/0.000264\ k)(t_{DA})_{pss}$ and that $(t_{DA})_{pss}=0.1$. Show that if $t_{pss}$ is used instead of $t_p$ in both the Horner plot and in the abscissa of the MBH chart, the resulting estimate of $\bar{p}$ is essentially unchanged. Buildup data (from the MTR only) are given in Table 2.10. Other data include:

TABLE 2.10 – MTR DATA FROM BUILDUP TEST

| $\Delta t$ (hours) | $p_{ws}$ (psi) |
|---|---|
| 8 | 4,354 |
| 12 | 4,366 |
| 16 | 4,376 |
| 20 | 4,382 |
| 24 | 4,388 |

$q$ = 250 STB/D,
$B$ = 1.136 RB/STB,
$\mu$ = 0.8 cp,
$h$ = 69 ft,
$\phi$ = 0.039,
$c_t = 17\times10^{-6}$ psi$^{-1}$,
$r_e$ = 1,320 ft, and
$k$ = 7.65 md.

2.14. A well producing only oil and dissolved gas has produced 13,220 STB. To characterize the severe damage believed present, the well is shut in for a buildup test. Well and reservoir data are given below.

$\phi$ = 0.17,
$\mu$ = 0.6 cp,
$c_t = 18\times10^{-6}$ psi$^{-1}$
$r_e$ = 1,320 ft, well centered in square drainage area (160 acres),
$r_w$ = 0.5 ft,
$A_{wb}$ = 0.036 sq ft,
$\rho_o$ = 54.8 lbm/cu ft,
$q$ = 1,135 STB/D (stabilized for several days),
$B$ = 1.214 bbl/STB, and
$h$ = 28 ft.

When the well was shut in for the buildup test, the liquid level rose in the wellbore as pressure increased. Data recorded during the buildup test are given in Table 2.11.

Determine (a) time at which afterflow distortion ceased; (b) time at which boundary effects begin; (c) formation permeability; (d) radius of investigation at beginning and end of MTR; (e) skin factor, $\Delta p_s$, and flow efficiency; (f) $\bar{p}$ using the MBH $p^*$ method; and (g) $\bar{p}$ using the modified Muskat method.

TABLE 2.11 – PRESSURE BUILDUP TEST DATA

| $\Delta t$ (hours) | $p_{ws}$ (psia) | $\Delta t$ (hours) | $p_{ws}$ (psia) |
|---|---|---|---|
| 0 | 2,752 | 10 | 4,272 |
| 0.3 | 3,464 | 12 | 4,280 |
| 0.5 | 3,640 | 14 | 4,287 |
| 1 | 3,852 | 16 | 4,297 |
| 2 | 4,055 | 20 | 4,303 |
| 3 | 4,153 | 24 | 4,308 |
| 4 | 4,207 | 30 | 4,313 |
| 5 | 4,244 | 36 | 4,317 |
| 6 | 4,251 | 42 | 4,320 |
| 8 | 4,263 | 50 | 4,322 |

## References


1. Horner, D.R.: "Pressure Buildup in Wells," *Proc.*, Third World Pet. Cong., The Hague (1951) Sec. II, 503-523; also *Pressure Analysis Methods*, Reprint Series, SPE, Dallas (1967) **9**, 25-43.
2. Cobb, W.M. and Smith, J.T.: "An Investigation of Pressure-Buildup Tests in Bounded Reservoirs," paper SPE 5133 presented at the SPE-AIME 49th Annual Fall Meeting, Houston, Oct. 6-9, 1974. An abridged version appears in *J. Pet. Tech.* (Aug. 1975) 991-996; *Trans.*, AIME, **259**.
3. Miller, C.C., Dyes, A.B., and Hutchinson, C.A. Jr.: "Estimation of Permeability and Reservoir Pressure From Bottom-Hole Pressure Build-Up Characteristics," *Trans.*, AIME (1950) **189**, 91-104.
4. Slider, H.C.: "A Simplified Method of Pressure Buildup Analysis for a Stabilized Well," *J. Pet. Tech.* (Sept. 1971) 1155-1160; *Trans.*, AIME, **271**.
5. Agarwal, R.G.: "A New Method To Account for Producing-Time Effects When Drawdown Type Curves Are Used To Analyze Pressure Buildup and Other Test Data," paper SPE 9289 presented at the SPE 55th Annual Technical Conference and Exhibition, Dallas, Sept. 21-24, 1980.
6. Saidikowski, R.M.: "Numerical Simulations of the Combined Effects of Wellbore Damage and Partial Penetration," paper SPE 8204 presented at the SPE-AIME 54th Annual Technical Conference and Exhibition, Las Vegas, Sept. 23-26, 1979.
7. Russell, D.G. and Truitt, N.E.: "Transient Pressure Behavior in Vertically Fractured Reservoirs," *J. Pet. Tech.* (Oct. 1964) 1159-1170; *Trans.*, AIME, **231**.
8. Matthews, C.S., Brons, F., and Hazebroek, P.: "A Method for Determination of Average Pressure in a Bounded Reservoir," *Trans.*, AIME (1954) **201**, 182-191.
9. Larson, V.C.: "Understanding the Muskat Method of Analyzing Pressure Buildup Curves," *J. Cdn. Pet. Tech.* (Fall 1963) **2**, 136-141.
10. Matthews, C.S. and Russell, D.G.: *Pressure Buildup and Flow Tests in Wells*, Monograph Series, SPE, Dallas (1967) **1**.
11. Pinson, A.E. Jr.: "Concerning the Value of Producing Time Used in Average Pressure Determinations From Pressure Buildup Analysis," *J. Pet. Tech.* (Nov. 1972) 1369-1370.
12. Brons, F. and Miller, W.C.: "A Simple Method for Correcting Spot Pressure Readings," *J. Pet. Tech.* (Aug. 1961) 803-805; *Trans.*, AIME, **222**.
13. Earlougher, R.C. Jr.: *Advances in Well Test Analysis*, Monograph Series, SPE, Dallas (1977) **5**.
14. Gray, K.E.: "Approximating Well-to-Fault Distance From Pressure Buildup Tests," *J. Pet. Tech.* (July 1965) 761-767.
15. Al-Hussainy, R., Ramey, H.J. Jr., and Crawford, P.B.: "The Flow of Real Gases Through Porous Media," *J. Pet. Tech.* (May 1966) 624-636; *Trans.*, AIME, **237**.
16. Wattenbarger, R.A., Ramey, H.J. Jr.: "Gas Well Testing With Turbulence, Damage, and Wellbore Storage," *J. Pet. Tech.* (Aug. 1968) 877-887; *Trans.*, AIME, **243**.
17. Perrine, R.L.: "Analysis of Pressure Buildup Curves," *Drill. and Prod. Prac.*, API, Dallas (1956) 482-509.
18. Martin, J.C.: "Simplified Equations of Flow in Gas Drive Reservoirs and the Theoretical Foundation of Multiphase Pressure Buildup Analyses," *J. Pet. Tech.* (Oct. 1959) 309-311; *Trans.*, AIME, **216**.

# Chapter 3
# Flow Tests

## 3.1 Introduction

This chapter discusses flow tests in wells, including constant-rate drawdown tests, continuously declining-rate drawdown tests, and multirate tests in infinite-acting reservoirs. The more general (and more complex) case of multirate tests in bounded reservoirs is discussed in Appendix E.

## 3.2 Pressure Drawdown Tests

A pressure drawdown test is conducted by producing a well, starting ideally with uniform pressure in the reservoir. Rate and pressure are recorded as functions of time.

The objectives of a drawdown test usually include estimates of permeability, skin factor, and, on occasion, reservoir volume. These tests are particularly applicable to (1) new wells, (2) wells that have been shut in sufficiently long to allow the pressure to stabilize, and (3) wells in which loss of revenue incurred in a buildup test would be difficult to accept. Exploratory wells are frequent candidates for lengthy drawdown tests, with a common objective of determining minimum or total volume being drained by the well.

An idealized constant-rate drawdown test in an infinite-acting reservoir is modeled by the logarithmic approximation to the *Ei*-function solution:

$$p_{wf}=p_i+162.6\frac{qB\mu}{kh}\left[\log\left(\frac{1{,}688\,\phi\mu c_t r_w^2}{kt}\right)-0.869\,s\right]. \quad (3.1)$$

Like buildup tests, drawdown tests are more complex than suggested by simple equations such as Eq. 3.1. The usual test has an ETR, an MTR, and an LTR. The ETR usually is dominated by wellbore unloading: the rate at which fluid is removed from the wellbore exceeds the rate at which fluid enters the wellbore until, finally, equilibrium is established. Until that time, the constant flow rate at the sandface required by Eq. 3.1 is not achieved, and the straight-line plot of $p_{wf}$ vs. log $t$ suggested by Eq. 3.1 is not achieved. Duration of wellbore unloading can be estimated by qualitative comparison of a log-log plot of $(p_i-p_{wf})$ vs. $t$ with the solution of Fig. 1.6 or with the empirical equation based on that figure,

$$t_D \geq (60+3.5\,s)\,C_{sD}, \quad (1.43)$$

or the equivalent form,

$$t_{wbs}=\frac{(200{,}000+12{,}000\,s)\,C_s}{kh/\mu}. \quad (3.2)$$

If the effective radius of the zone of altered permeability is unusually large (e.g., in a hydraulically fractured well), the duration of the ETR may depend on the time required for the radius of investigation to exceed the fracture half-length. (More exactly, for an infinite-conductivity vertical fracture with half-length $L_f$, shut-in time must exceed $1{,}260\,\phi\mu c_t L_f^2/k$ or $r_i$ must exceed $1.15\,L_f$.

The MTR begins when the ETR ends (unless boundaries or important heterogeneities are unusually near the well). In the MTR, a plot of $p_{wf}$ vs. log $t$ is a straight line with slope, $m$, given by

$$m=162.6\frac{qB\mu}{kh}. \quad (3.3)$$

Thus, effective formation permeability, $k$, can be estimated from this slope:

$$k=162.6\frac{qB\mu}{mh}. \quad (3.4)$$

After the MTR is identified, skin factor, $s$, can be determined. The usual equation results from solving Eq. 3.1 for $s$. Setting $t=1$ hour, and letting $p_{wf}=p_{1\,hr}$ be the pressure on the MTR line at 1-hour flow time, the result is

$$s=1.151\left[\frac{(p_i-p_{1\,hr})}{m}-\log\left(\frac{k}{\phi\mu c_t r_w^2}\right)+3.23\right]. \quad (3.5)$$

The LTR begins when the radius of investigation reaches a portion of the reservoir influenced by

TABLE 3.1 – CONSTANT-RATE DRAWDOWN TEST DATA

| $t$ (hours) | $p_{wf}$ (psia) | $p_i - p_{wf}$ (psia) | $t$ (hours) | $p_{wf}$ (psia) | $p_i - p_{wf}$ (psia) | $t$ (hours) | $p_{wf}$ (psia) | $p_i - p_{wf}$ (psia) |
|---|---|---|---|---|---|---|---|---|
| 0 | 4,412 | 0 | 14.4 | 3,573 | 839 | 89.1 | 3,515 | 897 |
| 0.12 | 3,812 | 600 | 17.3 | 3,567 | 845 | 107 | 3,509 | 903 |
| 1.94 | 3,699 | 713 | 20.7 | 3,561 | 851 | 128 | 3,503 | 909 |
| 2.79 | 3,653 | 759 | 24.9 | 3,555 | 857 | 154 | 3,497 | 915 |
| 4.01 | 3,636 | 776 | 29.8 | 3,549 | 863 | 185 | 3,490 | 922 |
| 4.82 | 3,616 | 796 | 35.8 | 3,544 | 868 | 222 | 3,481 | 931 |
| 5.78 | 3,607 | 805 | 43.0 | 3,537 | 875 | 266 | 3,472 | 940 |
| 6.94 | 3,600 | 812 | 51.5 | 3,532 | 880 | 319 | 3,460 | 952 |
| 8.32 | 3,593 | 819 | 61.8 | 3,526 | 886 | 383 | 3,446 | 966 |
| 9.99 | 3,586 | 826 | 74.2 | 3,521 | 891 | 460 | 3,429 | 983 |

reservoir boundaries or massive heterogeneities. For a well centered in a square or circular drainage area, this occurs at a time given approximately by

$$t_{\ell t} \simeq \frac{380\,\phi\mu c_t A}{k}, \qquad (3.6)$$

where $A$ is the drainage area of the tested well. For more general drainage-area shapes, $t_{\ell t}$ can be calculated from the number in the column "Use Infinite System Solution With Less Than 1% Error for $t_{DA} <$" in Table 1.2.[1] The dimensionless time $t_{DA}$ is defined as

$$t_{DA} = \frac{0.000264\,kt}{\phi\mu c_t A}.$$

For this more general case, then,

$$t_{\ell t} \simeq \frac{3{,}800\,\phi\mu c_t A t_{DA}}{k}. \qquad (3.7)$$

Thus, the typical constant-rate drawdown test plot has the shape shown in Fig. 3.1. To analyze the typical test, the following steps are suggested.

1. Plot flowing BHP, $p_{wf}$, vs. flowing time, $t$, on semilog paper as shown in Fig. 3.1.
2. Estimate $t_{wbs}$ from qualitative curve matching (with a full-size version of Fig. 1.6); this usually marks the beginning of the MTR (except for fractured wells).
3. Estimate the beginning of the LTR, $t_{\ell t}$, using deviation from a match with Fig. 1.6 to confirm deviation from an apparent semilog straight line. We must be cautious in drawdown test analysis, though. Even *small* rate changes can cause a drawdown curve to bend just as boundaries do (a method of analyzing this possibility is presented later).
4. Determine the slope $m$ of the most probable MTR, and estimate formation permeability from Eq. 3.4.
5. Estimate the skin factor $s$ from Eq. 3.5.

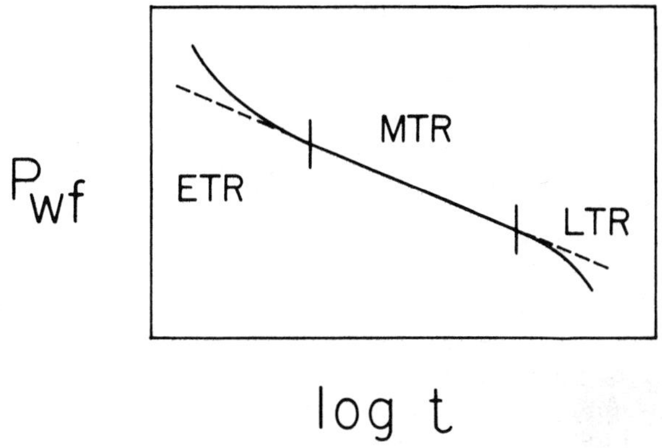


**Fig. 3.1** – Typical constant-rate drawdown test graph.

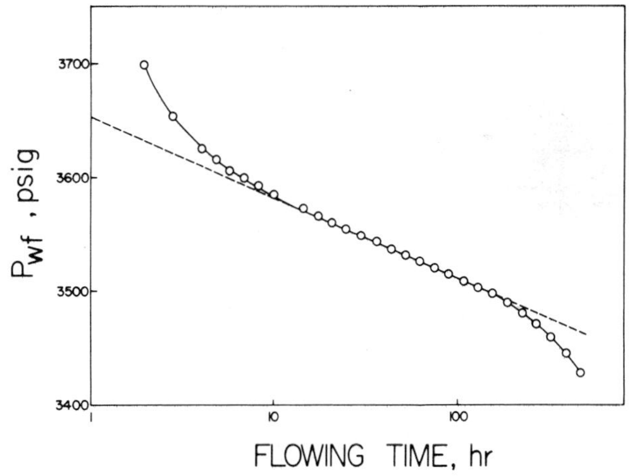


**Fig. 3.2** – Semilog graph of example constant-rate drawdown test.

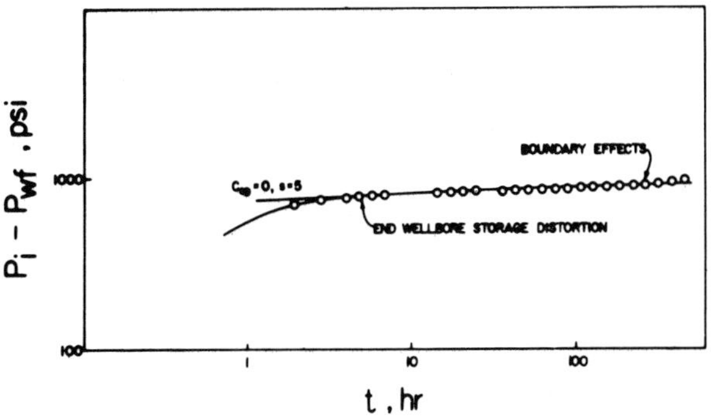


**Fig. 3.3** – Log-log graph of example constant-rate drawdown test.

## *Example 3.1 – Constant-Rate Drawdown Test Analysis*

**Problem.** The data in Table 3.1 were recorded during a constant-rate pressure drawdown test. The wellbore had a falling liquid/gas interface throughout the drawdown test. Other pertinent data include the following.

$q$ = 250 STB/D,
$B$ = 1.136 bbl/STB,
$\mu$ = 0.8 cp,
$r_w$ = 0.198 ft,
$h$ = 69 ft,
$\phi$ = 0.039, and
$c_t$ = $17 \times 10^{-6}$ psi$^{-1}$.

The tubing areas is 0.0218 sq ft; the density of the liquid in the wellbore is 53 lbm/cu ft. Determine the formation permeability and skin factor.

**Solution.** We first plot flowing BHP, $p_{wf}$, vs. time, $t$, on semilog paper and $(p_i - p_{wf})$ vs. $t$ on log-log paper. Then we determine when wellbore effects ceased distorting the curve. From the shape of the semilog graph (Fig. 3.2), this appears to be at about 12 hours; however, we can check this assumption with the log-log graph, Fig. 3.3. For several values of $C_D$ (e.g., $10^3$ to $10^4$), the graph shows a good fit with Fig. 1.6 for $s=5$; wellbore storage distortion ends at $\Delta t = 5$ hours, in approximate agreement with the more sensitive semilog graph.

We have no information about the location of boundaries; therefore, we assume that boundary effects begin when the drawdown curve begins to deviate from the established straight line on the semilog graph at a flowing time of 150 hours. This is confirmed qualitatively on the less sensitive log-log graph by noticeable deviation beginning at $t \cong 260$ hours. The slope of the middle-time line is

$$m = 3{,}652 - 3{,}582$$

$$= 70 \text{ psi/cycle.}$$

Thus, the permeability of the formation is

$$k = 162.6 \frac{qB\mu}{mh}$$

$$= \frac{(162.6)(250)(1.136)(0.8)}{(70)(69)}$$

$$= 7.65 \text{ md.}$$

We now check the radius of investigation at the beginning and end of the apparent middle-time line to ensure that we are sampling a representative portion of the formation.

At the beginning ($t = 12$ hours),

$$\frac{k}{948\,\phi\mu c_t} = \frac{(7.65)}{(948)(0.039)(0.8)(1.7 \times 10^{-5})}$$

$$= 1.521 \times 10^4,$$

and, from Eq. 1.23,

$$r_i = \sqrt{\frac{kt}{948\,\phi\mu c_t}}$$

$$= \sqrt{(1.521 \times 10^4)(12)}$$

$$= 427 \text{ ft.}$$

At the end of the MTR ($t \simeq 150$ hours),

$$r_i = \sqrt{(1.521 \times 10^4)(150)}$$

$$= 1{,}510 \text{ ft.}$$

A substantial amount of formation has been sampled; thus, we can be more confident that the permeability of 7.65 md is representative.

We next calculate the skin factor $s$.

$$s = 1.151\left[\frac{p_i - p_{1\,\text{hr}}}{m} - \log\left(\frac{k}{\phi\mu c_t r_w^2}\right) + 3.23\right]$$

$$= 1.151\left[\frac{4{,}412 - 3{,}652}{70}\right.$$

$$\left. - \log\frac{(1.442 \times 10^7)}{(0.198)^2} + 3.23\right]$$

$$= 6.37.$$

We now can verify more closely the expected end of wellbore storage distortion from Eq. 3.2, using

$$C_s \cong \frac{25.65\,A_{wb}}{\rho}$$

$$= 0.0106 \text{ bbl/psi.}$$

$$t_{wbs} = \frac{(200{,}000 + 12{,}000\,s)\,C_s}{kh/\mu}$$

$$= \frac{[200{,}000 + (12{,}000)(6.37)](0.0106)}{(7.65)(69)10.8}$$

$$= 4.44 \text{ hours.}$$

This closely agrees with the result from the log-log curve fit.

---

Another use of drawdown tests is to estimate reservoir pore volume, $V_p$. This is possible when the radius of investigation reaches all boundaries during a test so that pseudosteady-state flow is achieved. Eqs. 1.12 and 1.13 showed that, in pseudosteady-state flow, flowing BHP, $p_{wf}$, is related linearly to time and that the rate of change in $p_{wf}$ with time is related to the reservoir pore volume. From Eq. 1.13, this relationship is

$$V_p = \frac{-0.234\,qB}{c_t\left(\frac{\partial p_{wf}}{\partial t}\right)},$$

where $\partial p_{wf}/\partial t$ is simply the slope of the straight-line $p_{wf}$ vs. $t$ plot on ordinary Cartesian graph paper.

Even though Eqs. 1.12 and 1.13 were derived for a cylindrical reservoir centered at the wellbore of the tested well, the principles derived from them apply to all closed reservoir shapes. The graph of $p_{wf}$ vs. $t$ is a

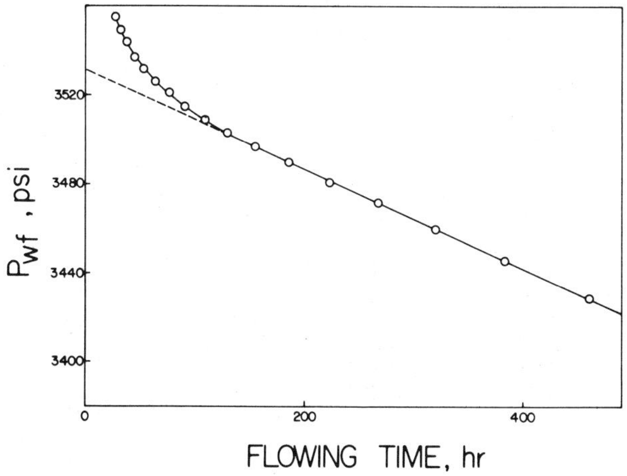


**Fig. 3.4** – Cartesian-coordinate graph of example constant-rate drawdown test.

straight line once pseudosteady-state flow is achieved; the volume of the reservoir can be found from Eq. 1.13. It is important to remember, however, that these equations apply only to closed, or volumetric, reservoirs (i.e., they are not valid if there is water influx or gas-cap expansion). Further, they are limited to reservoirs in which total compressibility, $c_t$, is constant (and, specifically, independent of pressure).

We will illustrate pore-volume estimates with an example.

---

## *Example 3.2 – Estimation of Pore Volume*

**Problem.** Estimate the pore volume of the reservoir with drawdown data reported in Example 3.1.

**Solution.** The first step is to plot $p_{wf}$ vs. $t$ (Fig. 3.4). The slope of this curve is constant for $t > 130$ hours; this slope, $\partial p_{wf}/\partial t$, is

$$\frac{\partial p_{wf}}{\partial t} = \frac{3{,}531 - 3{,}420}{0 - 500}$$

$$= -0.222 \text{ psi/hr}.$$

Thus,

$$V_p = \frac{-0.234\, qB}{c_t\left(\dfrac{\partial p_{wf}}{\partial t}\right)}$$

$$= \frac{(-0.234)(250)(1.136)}{(1.7\times 10^{-5})(-0.222)}$$

$$= 17.61 \times 10^6 \text{ cu ft}$$

$$= 3.14 \times 10^6 \text{ res bbl}.$$

---

The method of permeability determination outlined above applies only to drawdown tests conducted at a *strictly constant rate*. If rate varies during a test – e.g., if rate declines slowly throughout the test – results obtained using those techniques can lead to interpretations that are seriously in error. An analysis method that leads to proper interpretation is available, but it can be used only if the producing rate is changing *slowly* and *smoothly*. Abrupt rate changes will make the drawdown test data impossible to interpret using either the method discussed earlier or this new method.*

Winestock and Colpitts[2] show that when rate is changing slowly and smoothly, the equation modeling the MTR of the drawdown test becomes

$$\frac{p_i - p_{wf}}{q} = 162.6\frac{\mu B}{kh}\left[\log\left(\frac{1{,}688\,\phi\mu c_t r_w^2}{kt}\right) + 0.869\, s\right] + \text{negligible terms}^2. \quad (3.8)$$

The analysis technique is to plot $(p_i - p_{wf})/q$ vs. $t$ on semilog paper;** identify the middle-time straight line; measure the slope $m'$ in psi/STB/D/cycle; calculate $kh$ from

$$kh = 162.6\frac{\mu B}{m'},$$

and, finally, calculate $s$ from

$$s = 1.151\left[\left(\frac{p_i - p_{wf}}{q}\right)_{1\text{ hr}}\frac{1}{m'} - \log\left(\frac{k}{\phi\mu c_t r_w^2}\right) + 3.23\right]. \quad (3.9)$$

In Eq. 3.8, $[(p_i - p_{wf})/q]_{1\text{ hr}}$ is the value of this quantity on the middle-time line or its extrapolation at a flowing time of 1 hour.

We will illustrate use of this method with an example.

---

## *Example 3.3 – Analysis of Drawdown Test With Varying Rate*

**Problem.** The data in Table 3.2 were obtained in a drawdown test in which the rate $q$ was measured as a function of time. Other data include the following

$B$ = 1.136 bbl/STB,
$\mu$ = 0.8 cp,
$h$ = 69 ft,
$\rho$ = 53 lb/cu ft,
$A_{wb}$ = 0.0218 sq ft,
$\phi$ = 0.039,
$c_t$ = $17 \times 10^{-6}$ psi$^{-1}$, and
$r_w$ = 0.198 ft.

Determine formation permeability and skin factor.

**Solution.** We note immediately that conventional drawdown test analysis, using an average rate, would

*Verification of this method is incomplete in cases with severe wellbore storage effects. A nonexhaustive numerical simulation study by this author has shown that the method yields essentially correct permeability and skin factor even in these cases.

**The same analysis technique, but for a different application (analyzing wellbore-storage-dominated data) was suggested earlier by Gladfelter *et al.*[3] and Ramey.[4]

**TABLE 3.2 – VARIABLE-RATE DRAWDOWN TEST DATA**

| $t$ (hours) | $p_{wf}$ (psi) | $q$ (STB/D) | $t$ (hours) | $p_{wf}$ (psi) | $q$ (STB/D) |
|---|---|---|---|---|---|
| 0 | 4,412 | 250 | 8.32 | 3,927 | 147 |
| 0.105 | 4,332 | 180 | 9.99 | 3,928 | 145 |
| 0.151 | 4,302 | 177 | 14.4 | 3,931 | 143 |
| 0.217 | 4,264 | 174 | 20.7 | 3,934 | 140 |
| 0.313 | 4,216 | 172 | 29.8 | 3,937 | 137 |
| 0.450 | 4,160 | 169 | 43.0 | 3,941 | 134 |
| 0.648 | 4,099 | 166 | 61.8 | 3,944 | 132 |
| 0.934 | 4,039 | 163 | 74.2 | 3,946 | 130 |
| 1.34 | 3,987 | 161 | 89.1 | 3,948 | 129 |
| 1.94 | 3,952 | 158 | 107 | 3,950 | 127 |
| 2.79 | 3,933 | 155 | 128 | 3,952 | 126 |
| 4.01 | 3,926 | 152 | 154 | 3,954 | 125 |
| 5.78 | 3,926 | 150 | 185 | 3,956 | 123 |

**TABLE 3.3 – DATA FOR PLOTTING FROM VARIABLE-RATE DRAWDOWN TEST**

| $t$ (hours) | $(p_i - p_{wf})/q$ | $t$ (hours) | $(p_i - p_{wf})/q$ |
|---|---|---|---|
| 0.105 | 0.444 | 8.32 | 3.299 |
| 0.151 | 0.621 | 9.99 | 3.338 |
| 0.217 | 0.851 | 14.4 | 3.364 |
| 0.313 | 1.140 | 20.7 | 3.414 |
| 0.450 | 1.491 | 29.8 | 3.467 |
| 0.648 | 1.886 | 43.0 | 3.515 |
| 0.934 | 2.288 | 61.8 | 3.545 |
| 1.34 | 2.640 | 74.2 | 3.585 |
| 1.94 | 2.911 | 89.1 | 3.597 |
| 2.79 | 3.090 | 107 | 3.638 |
| 4.01 | 3.197 | 128 | 3.651 |
| 5.78 | 3.240 | 154 | 3.664 |
| | | 185 | 3.707 |

be futile. Pressures for flow times greater than about 6 hours are *increasing* even though production continues for another 179 hours and even though the rate decline from this time to the end of the test is only 27 STB/D (from 150 to 123 STB/D).

Thus, we must use the variable-rate analysis technique; the first step is to tabulate $(p_i - p_{wf})/q$, as in Table 3.3. These data are plotted in Fig. 3.5. On the basis of curve shape, wellbore storage appears to end at approximately 6 hours; we will check this assumption with Eq. 3.2 when $k$ and $s$ have been estimated.

There is no deviation from the straight line for $t > 6$ hours; accordingly, we assume the MTR spans the time range 6 hours $< t <$ 185 hours.

From the plot, $m' = 3.616 - 3.328 = 0.288$ psi/STB/D/cycle. Then,

$$k = 162.6 \frac{\mu B}{m' h}$$

$$= \frac{(162.6)(0.8)(1.136)}{(0.288)(69)}$$

$$= 7.44 \text{ md,}$$

and

$$s = 1.151 \left[ \left( \frac{p_i - p_{wf}}{q} \right)_{1\text{ hr}} \left( \frac{1}{m'} \right) - \log \left( \frac{k}{\phi \mu c_t r_w^2} \right) + 3.23 \right]$$

$$= 1.151 \left\{ \frac{3.04}{0.288} - \log \left[ \frac{7.44}{(0.039)(0.8)(17 \times 10^{-6})(0.198)^2} \right] + 3.23 \right\}$$

$$= 6.02.$$

Since $C_s \cong 0.0106$ bbl/psi, as in Example 3.1,

$$t_{wbs} \cong \frac{(200{,}000 + 12{,}000\, s)\, C_s}{kh/\mu}$$

$$= \frac{(200{,}000 + 12{,}000)(6.02)(0.0106)}{(7.44)(69)/0.8}$$

$$= 4.5 \text{ hours.}$$

This qualitatively confirms the choice of wellbore storage distortion end.

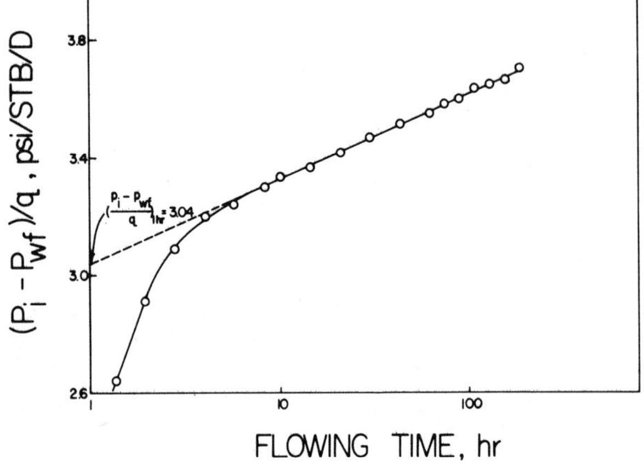


**Fig. 3.5** – Example variable-rate drawdown test.

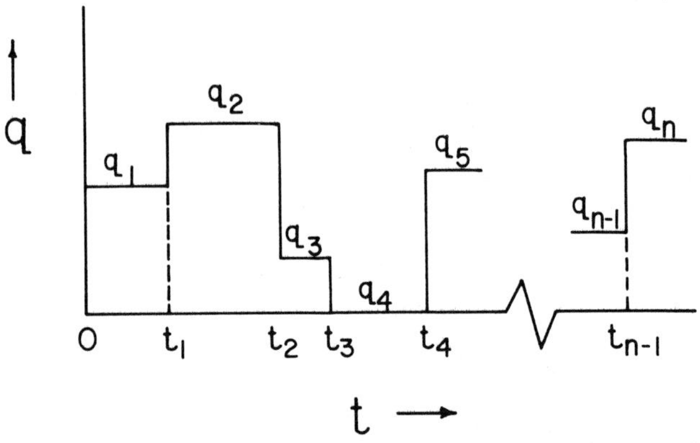


**Fig. 3.6** – Rate history for multirate test.

## 3.3 Multirate Tests

We will develop a general theory for behavior of multirate tests in *infinite-acting* reservoirs for slightly compressible liquids. In Appendix E, we extend this general theory to reservoirs in which boundary effects may become important before the test ends.

Consider a well with $n$ rate changes during its production history, as indicated in Fig. 3.6. Our objective is to determine the wellbore pressure of a well producing with this schedule. We will use superposition of the logarithmic approximation to the *Ei*-function solution; to simplify the algebra, we will write the solution as

$$p_i - p_{wf} = 162.6\frac{qB\mu}{kh}\left[\log\left(\frac{1{,}688\ \phi\mu c_t r_w^2}{kt}\right) - 0.869\,s\right]$$

$$= 162.6\frac{qB\mu}{kh}\left(\log t + \log\frac{k}{\phi\mu c_t r_w^2} - 3.23 + 0.869\,s\right)$$

$$= m' q\left(\log t + \bar{s}\right),$$

where

$$m' = 162.6\,\frac{\mu B}{kh},$$

and

$$\bar{s} = \log\frac{k}{\phi\mu c_t r_w^2} - 3.23 + 0.869\,s.$$

With this nomenclature for $n$ rates and for $t > t_{n-1}$, application of superposition (as in the discussion leading to Eq. 1.27) leads to

$$p_i - p_{wf} = m' q_1\left(\log t + \bar{s}\right) + m'(q_2 - q_1)\cdot\left[\log(t - t_1) + \bar{s}\right] + m'(q_3 - q_2)\cdot\left[\log(t - t_2) + \bar{s}\right] + \ldots + m'(q_n - q_{n-1})\left[\log(t - t_{n-1}) + \bar{s}\right].$$

This can be written more compactly as

$$\frac{p_i - p_{wf}}{q_n} = m' \sum_{j=1}^{n} \frac{(q_j - q_{j-1})}{q_n}\cdot\log(t - t_{j-1}) + m'\bar{s},\ q_n \neq 0 \quad \ldots (3.10)$$

In Eq. 3.10, we define $q_o = 0$ and $t_o = 0$. In terms of more fundamental quantities, Eq. 3.10 becomes

$$\frac{p_i - p_{wf}}{q_n} = m' \sum_{j=1}^{n} \frac{(q_j - q_{j-1})}{q_n}\cdot\log(t - t_{j-1}) + m'\left[\log\left(\frac{k}{\phi\mu c_t r_w^2}\right) - 3.23 + 0.869\,s\right]. \quad \ldots\ldots\ldots\ldots (3.11)$$

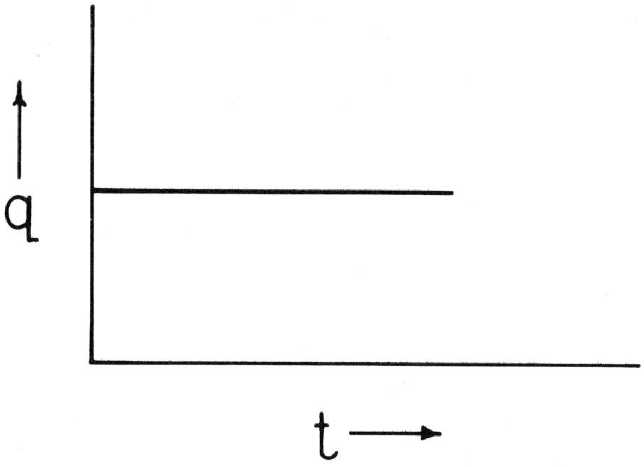


**Fig. 3.7** – Rate history for single-rate drawdown test.

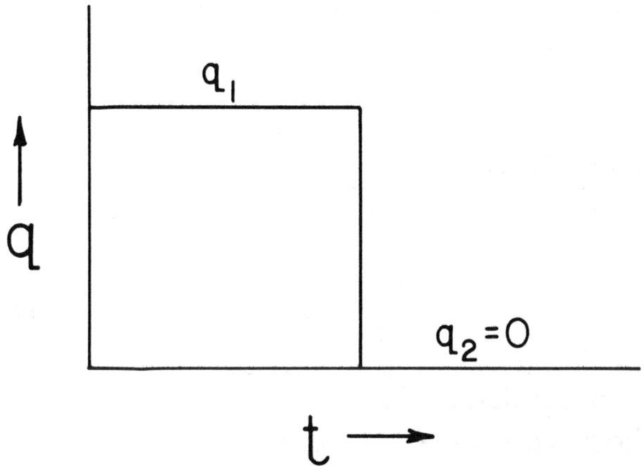


**Fig. 3.8** – Rate history for buildup test following single flow rate.

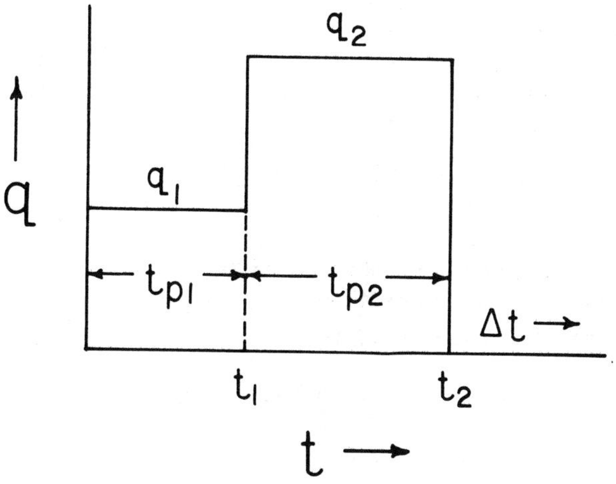


**Fig. 3.9** – Rate history for buildup test following two different flow rates.

For the special case $q_n = 0$ (a pressure buildup test),

$$p_i - p_{ws} = m' q_1 (\log t + \bar{s}) + m' (q_2 - q_1) \cdot [\log(t-t_1) + \bar{s}] + \ldots + m' (q_{n-1} - q_{n-2})[\log (t - t_{n-2}) + \bar{s}] - m' q_{n-1} \cdot [\log (t - t_{n-1}) + \bar{s}]$$

$$= 162.6 \frac{\mu B}{kh} \sum_{j=1}^{n} (q_j - q_{j-1}) \cdot \log (t - t_{j-1}). \quad \ldots\ldots\ldots\ldots\ldots (3.12)$$

Eqs. 3.11 and 3.12 can be used to model several special cases of practical importance, but we must remember that they have an important limitation: the reservoir must be infinite acting for the total time elapsed $t$ since the well began producing at rate $q_1$. Appendix E suggests a more general method of modeling tests in which boundaries have been reached.

**Pressure Drawdown Test**

From Eq. 3.11, for $n = 1$ (Fig. 3.7),

$$\frac{p_i - p_{wf}}{q_1} = 162.6 \frac{\mu B}{kh} \left(\log t + \log \frac{k}{\phi \mu c_t r_w^2} - 3.23 + 0.869\, s\right). \quad \ldots\ldots\ldots\ldots (3.13)$$

**Pressure Buildup Test Preceded by Constant-Rate Production**

From Eq. 3.12 for $n = 2$ (Fig.3.8),

$$p_i - p_{ws} = 162.6 \frac{\mu B}{kh} \left[q_1 \log t - q_1 \log (t - t_1)\right]$$

$$= 162.6 \frac{q_1 \mu B}{kh} \log\left(\frac{t}{t - t_1}\right). \quad \ldots\ldots (3.14)$$

If we let $q_1 = q$, $t - t_1 = \Delta t$, and $t_1 = t_p$, the familiar equation that serves as the basis for the Horner plot results:

$$p_i - p_{ws} = 162.6 \frac{qB\mu}{kh} \log\left(\frac{t_p + \Delta t}{\Delta t}\right).$$

**Pressure Buildup Test Preceded by Two Different Flow Rates**

From Eq. 3.12 (Fig. 3.9),

$$p_i - p_{ws} = m' \left[q_1 \log t + (q_2 - q_1) \log (t - t_1) - q_2 \log (t - t_2)\right].$$

This can be written as

$$p_i - p_{ws} = 162.6 \frac{q_2 B\mu}{kh} \left[\frac{q_1}{q_2} \log\left(\frac{t}{t - t_1}\right) + \log\left(\frac{t - t_1}{t - t_2}\right)\right].$$

Let $t - t_2 = \Delta t$, $t_1 = t_{p1}$, $t_2 = t_{p1} + t_{p2}$, and $t - t_1 = t_{p2} + \Delta t$. Then,

$$p_i - p_{ws} = 162.6 \frac{q_2 B\mu}{kh} \left[\frac{q_1}{q_2} \log\left(\frac{t_{p1} + t_{p2} + \Delta t}{t_{p2} + \Delta t}\right) + \log\left(\frac{t_{p2} + \Delta t}{\Delta t}\right)\right]. \quad \ldots\ldots\ldots\ldots (3.15)$$

Eq. 3.15 has this application: when the producing rate is changed a short time before a buildup test begins, so that there is not sufficient time for Horner's approximation to be valid, we frequently can consider all production before time $t_1$ to have been at rate $q_1$ for time $t_{p1}$ and production just before the test to have been at rate $q_2$ for time $t_{p2}$.

To analyze such a test, we plot

$$p_{ws} \text{ vs. } \left[\frac{q_1}{q_2} \log\left(\frac{t_{p1} + t_{p2} + \Delta t}{t_{p2} + \Delta t}\right) + \log\left(\frac{t_{p2} + \Delta t}{\Delta t}\right)\right].$$

The slope $m$ of this plot is related to formation permeability by the equation

$$m = 162.6 \frac{q_2 B\mu}{kh}.$$

Extrapolation of the plot to $\Delta t = \infty$ gives $p_{ws} = p_i$ because, at $\Delta t = \infty$, the plotting function is zero. *Note that semilog paper is not to be used*; instead, the sum of two logarithms is plotted on an ordinary Cartesian axis.

To calculate skin factor, $s$, note that at the end of the flow period just before shut-in,

$$p_i - p_{ws} = 162.6 \frac{q_2 B\mu}{kh} \left[\frac{q_1}{q_2} \log\left(\frac{t_{p1} + t_{p2}}{t_{p2}}\right) + \log (t_{p2}) + \bar{s}\right],$$

or

$$p_i - p_{wf} = m \left[\frac{q_1}{q_2} \log\left(\frac{t_{p1} + t_{p2}}{t_{p2}}\right) + \log (t_{p2}) + \bar{s}\right],$$

The equation of the MTR line on the buildup test plot is

$$p_i - p_{ws} = m \left[\frac{q_1}{q_2} \log\left(\frac{t_{p1} + t_{p2} + \Delta t}{t_{p2} + \Delta t}\right) + \log\left(\frac{t_{p2} + \Delta t}{\Delta t}\right)\right].$$

Subtracting,

$$p_{ws} - p_{wf} = m \left\{\frac{q_1}{q_2} \log\left[\frac{(t_{p1} + t_{p2})(t_{p2} + \Delta t)}{(t_{p1} + t_{p2} + \Delta t)\,(t_{p2})}\right] + \log\left[\frac{(t_{p2})(\Delta t)}{t_{p2} + \Delta t}\right] + \bar{s}\right\}.$$

Assume $t_{p1} + t_{p2} + \Delta t \simeq t_{p1} + t_{p2}$ and $t_{p2} + \Delta t \simeq t_{p2}$ for small $\Delta t$ (e.g., $\Delta t = 1$ hour). Then,

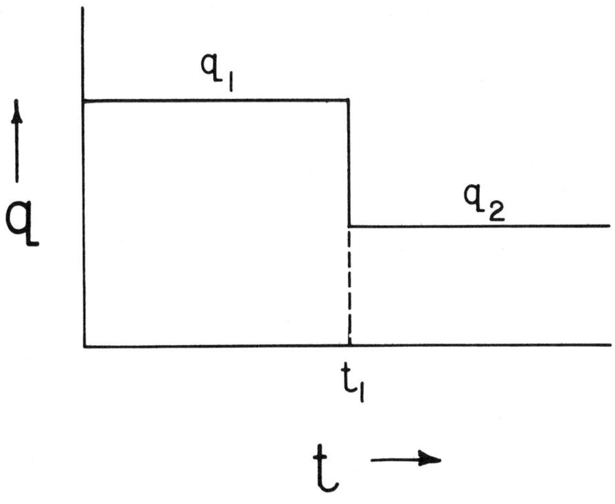


Fig. 3.10 – Rate history for two-rate flow test.

$$p_{ws}-p_{wf}=m(\log \Delta t+\bar{s}).$$

If we choose $\Delta t=1$ hour, $p_{ws}=p_{1\,hr}$ (on the MTR line) and, for $t_{p2} \gg 1$,

$$\bar{s}=\frac{p_{1\,hr}-p_{wf}}{m}$$

$$=\log\frac{k}{\phi\mu c_t r_w^2}-3.23,$$

or

$$s=1.151\left(\frac{p_{1\,hr}-p_{wf}}{m}-\log\frac{k}{\phi\mu c_t r_w^2}+3.23\right),$$

as before.

We also note that duration of wellbore storage distortion is calculated as in the previous analysis for buildup tests.

**Pressure Buildup Test Preceded by $(n-1)$ Different Flow Rates**

From Eq. 3.12,

$$p_i-p_{ws}=162.6\frac{q_{n-1}\mu B}{kh}\left[\frac{q_1}{q_{n-1}}\log\left(\frac{t}{t-t_1}\right)+\frac{q_2}{q_{n-1}}\log\left(\frac{t-t_1}{t-t_2}\right)+\ldots+\frac{q_{n-2}}{q_{n-1}}\cdot\log\left(\frac{t-t_{n-3}}{t-t_{n-2}}\right)+\log\left(\frac{t-t_{n-2}}{t-t_{n-1}}\right)\right]. \quad \ldots\ldots (3.16)$$

Although we introduce no specialized nomenclature for this situation, note that $t-t_{n-1}=\Delta t$ (time elapsed since shut-in) and that $q_{n-1}$ is the production rate just before shut-in.

Applications of Eq. 3.16 in which more than three terms are needed are probably rare; sometimes, though, to satisfy precise legal contracts (e.g., gas deliverability contracts), all significant rates are considered. Improvement in accuracy when using this approach is questionable; further, the fundamental assumption on which Eq. 3.16 is based (that for $t=t_{p1}+t_{p2}+\ldots+t_{p,n-1}+\Delta t$ the reservoir is infinite acting) rarely will be valid for large values of $t$. Nevertheless, when Eq. 3.16 is used to model a buildup test, the following analysis procedure can be used.

1. Write a computer program to calculate the plotting function,

$$\left[\frac{q_1}{q_{n-1}}\log\left(\frac{t}{t-t_1}\right)+\ldots+\log\left(\frac{t-t_{n-2}}{t-t_{n-1}}\right)\right]=X.$$

2. Plot $p_{ws}$ vs. $X$ on ordinary (Cartesian coordinate) graph paper.

3. Determine the slope $m$ of the plot and relate to the formation permeability by the equation

$$m=162.6\frac{q_{n-1}B\mu}{kh}.$$

4. Calculate the skin factor $s$ from the equation

$$s=1.151\left[\frac{(p_{1\,hr}-p_{wf})}{m}-\log\left(\frac{k}{\phi\mu c_t r_w^2}\right)+3.23\right].$$

(The derivation and assumptions implicit in this equation closely parallel those used for a buildup test preceded by two different flow rates.)

5. The original formation pressure $p_i$ is the value of $p_{ws}$ on the MTR line extrapolated to $X=0$.

**Two-Rate Flow Test**

From Eq. 3.11, this test (Fig. 3.10) can be modeled[5] by

$$p_i-p_{wf}=162.6\frac{qB_2\mu}{kh}\left[\frac{q_1}{q_2}\log t+\frac{(q_2-q_1)}{q_2}\cdot\log(t-t_1)+\log\left(\frac{k}{\phi\mu c_t r_w^2}\right)-3.23+0.869\,s\right] \quad \ldots\ldots (3.17)$$

If we rearrange and introduce specialized nomenclature, $t_1=t_{p1}$ and $t-t_{p1}=\Delta t'$, then Eq. 3.17 becomes

$$p_{wf}=p_i-162.6\frac{q_2B\mu}{kh}\left[\log\left(\frac{k}{\phi\mu c_t r_w^2}\right)-3.23+0.869\,s\right]-162.6\frac{q_1B\mu}{kh}\left[\log\left(\frac{t_{p1}+\Delta t'}{\Delta t'}\right)+\frac{q_2}{q_1}\log(\Delta t')\right]. \quad \ldots\ldots (3.18)$$

This type of test can be used when estimates of permeability, skin factor, or reservoir pressure are needed but when the well cannot be shut in because loss of income cannot be tolerated. This test shares a

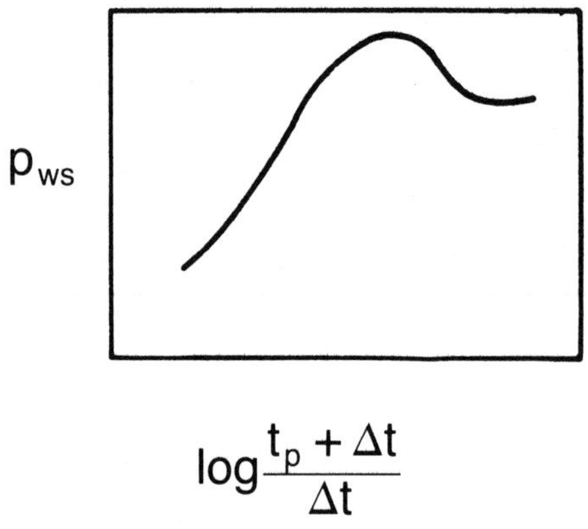


Fig. 3.11 – Buildup test with pressure humping.

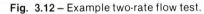

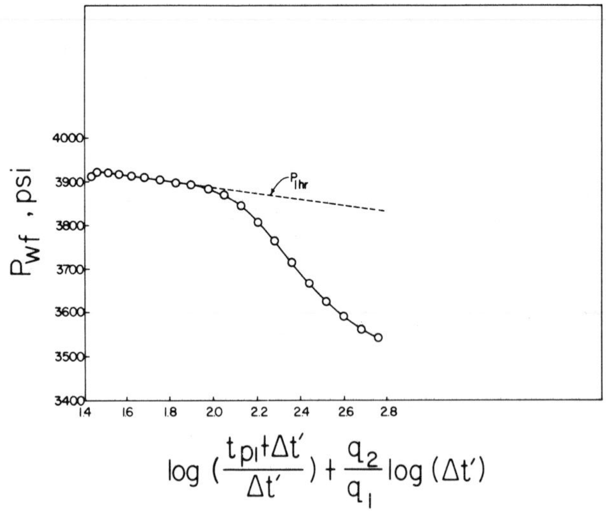


Fig. 3.12 – Example two-rate flow test.

fundamental analysis problem with the conventional drawdown test. The second rate must be kept *strictly constant* or the test interpretation may be substantially in error.

Eq. 3.18 is rigorously correct only when the reservoir is infinite acting for time $(t_{p1}+\Delta t')$ [just as the Horner plot is rigorous only when a reservoir is infinite acting for time $(t_p+\Delta t)$]. Nevertheless, application of the plotting and analysis technique suggested by Eq. 3.18 allows identification of the MTR and determination of formation permeability in a finite-acting reservoir.

The two-rate flow test does not reduce the duration of wellbore storage distortion – the duration of this distortion is essentially the same as in any buildup or drawdown test. However, the test procedure may minimize the effects of phase segregation in the wellbore, an extreme form of which is "pressure humping,"[6] where high-pressure gas trapped in a wellbore in poor communication with a formation may lead to pressures in the wellbore higher than formation pressures (Fig. 3.11). A buildup test with this humping is, at best, difficult to interpret; thus, a test procedure that can minimize these phase-segregation effects can be of value.

The following method of analysis can be used for two-rate flow tests.

1. Plot $p_{wf}$ vs. $\left[\log\left(\frac{t_{p1}+\Delta t'}{\Delta t'}\right)+\frac{q_2}{q_1}\log(\Delta t')\right]$.

2. Determine the slope $m$ from the plot and use it to calculate permeability, $k$, from the relationship

$$k=162.6\frac{q_1 B\mu}{mh}.$$

3. Calculate the skin factor, $s$, from the equation

$$s=1.151\left[\frac{q_1}{(q_1-q_2)}\left(\frac{p_{1\,hr}-p_{wf1}}{m}\right)-\log\left(\frac{k}{\phi\mu c_t r_w^2}\right)+3.23\right]. \quad \text{............ (3.19)}$$

In Eq. 3.19, $p_{1\,hr}$ is the flowing pressure at $\Delta t'=1$ hour on the MTR line and $p_{wf1}$ is the flowing pressure at the time the rate is changed $(\Delta t'=0)$. (Eq. 3.19 was derived by simultaneous solution of Eqs. 3.18 and the drawdown equation for a single rate applied at $t=t_{p1}$, at which time $p_{wf}=p_{wf1}$.)

4. $p_i$ (or, more generally, $p^*$) is obtained by solving for $p_i$ $(p^*)$ from the drawdown equation written to model conditions at the time of the rate change. (It is implied that $s$ and $m$ are known at this point.)

**TABLE 3.4 – TWO-RATE FLOW TEST DATA**

| $\Delta t'$ (hours) | $p_{wf}$ (psi) | $\Delta t'$ (hours) | $p_{wf}$ (psi) |
|---|---|---|---|
| 0 | 3,490 | 8.32 | 3,897 |
| 0.105 | 3,543 | 12.0 | 3,903 |
| 0.151 | 3,564 | 17.3 | 3,908 |
| 0.217 | 3,592 | 24.9 | 3,912 |
| 0.313 | 3,627 | 35.8 | 3,915 |
| 0.450 | 3,669 | 51.5 | 3,918 |
| 0.648 | 3,717 | 74.2 | 3,919 |
| 0.934 | 3,766 | 89.1 | 3,918 |
| 1.344 | 3,810 | 107 | 3,917 |
| 1.936 | 3,846 | 128 | 3,916 |
| 2.788 | 3,868 | 154 | 3,913 |
| 4.01 | 3,882 | 184.7 | 3,910 |
| 5.78 | 3,891 | | |

**TABLE 3.5 – DATA FOR PLOTTING FROM TWO-RATE FLOW TEST**

| $\Delta t'$ (hours) | PF | $p_{wf}$ (psi) | $\Delta t'$ (hours) | PF | $p_{wf}$ (psi) |
|---|---|---|---|---|---|
| 0 | – | 3,490 | 8.32 | 1.826 | 3,897 |
| 0.105 | 2.756 | 3,543 | 12.0 | 1.754 | 3,903 |
| 0.151 | 2.677 | 3,564 | 17.3 | 1.686 | 3,908 |
| 0.217 | 2.599 | 3,592 | 24.9 | 1.623 | 3,912 |
| 0.313 | 2.519 | 3,627 | 35.8 | 1.566 | 3,915 |
| 0.450 | 2.441 | 3,669 | 51.5 | 1.517 | 3,918 |
| 0.648 | 2.362 | 3,717 | 74.2 | 1.478 | 3,919 |
| 0.934 | 2.283 | 3,766 | 89.1 | 1.462 | 3,918 |
| 1.34 | 2.206 | 3,810 | 107 | 1.450 | 3,917 |
| 1.94 | 2.127 | 3,846 | 128 | 1.442 | 3,916 |
| 2.79 | 2.050 | 3,868 | 154 | 1.436 | 3,913 |
| 4.01 | 1.974 | 3,882 | 184.7 | 1.434 | 3,910 |
| 5.78 | 1.899 | 3,891 | | | |

$$p_i = p_{wf1} + m\left[\log\left(\frac{kt_{p1}}{\phi\mu c_t r_w^2}\right) - 3.23 + 0.869\,s\right].$$

$$\ldots\ldots\ldots\ldots\ldots (3.20)$$

---

## *Example 3.4 – Two-Rate Flow Test*

**Problem.** A two-rate flow test was run on a well with properties given below. From these properties and the data in Table 3.4, determine $k$, $s$, and $p^*$.

$q_1$ = 250 STB/D,
$q_2$ = 125 STB/D,
$\mu$ = 0.8 cp,
$B$ = 1.136 RB/STB,
$p_i$ = 4,412 psi,
$c_t$ = $17 \times 10^{-6}$ psi$^{-1}$,
$A_{wb}$ = 0.0218 sq ft,
$r_w$ = 0.198 ft,
$h$ = 69 ft,
$\rho$ = 53 lb/cu ft,
$\phi$ = 0.039, and
$t_{p1}$ = 184.7 hours.

**Solution.**

1. We first tabulate the plotting function (PF),

$$\text{PF} = \left[\log\left(\frac{t_{p1} + \Delta t'}{\Delta t'}\right) + \frac{q_2}{q_1}\log(\Delta t')\right],$$

and plot $p_{wf}$ vs. PF. Note that $t_{p1} = 184.7$ hours and $q_2/q_1 = 125/250 = 0.5$ (Table 3.5). The data are plotted in Fig. 3.12.

2. Next, we determine permeability. Assume that the MTR spans the time range $1.5 < \text{PF} < 1.9$ (50 hours $> \Delta t' > 6$ hours). Then the slope $m = (3{,}927 - 3{,}857)/(1.4 - 2.4) = 70$ psi/cycle and

$$k = 162.6\frac{q_1 B\mu}{mh}$$

$$= \frac{(162.6)(250)(1.136)(0.8)}{(70)(69)}$$

$$= 7.65 \text{ md}.$$

3. We determine skin factor,

$$s = 1.151\left[\frac{q_1}{(q_1 - q_2)}\,\frac{(p_{1\,\text{hr}} - p_{wf1})}{m} - \log\left(\frac{k}{\phi\mu c_t r_w^2}\right) + 3.23\right],$$

at $\Delta t' = 1$ hour, $\text{PF} = \log(184.7 + 1)/1 + 0.5\log(1) = 2.269$, and $p_{1\,\text{hr}} = 3{,}869$ psi (on the MTR line):

$$s = 1.151\left\{\frac{250}{(250 - 125)}\,\frac{(3{,}869 - 3{,}490)}{70} - \log\left[\frac{7.65}{(0.039)(0.8)(17 \times 10^{-6})(0.198)^2}\right] + 3.23\right\}$$

$$= 6.32.$$

4. Determine $p^*$.

$$p^* = p_{wf1} + m\left[\log\left(\frac{kt_{p1}}{\phi\mu c_t r_w^2}\right) - 3.23 + 0.869\,s\right]$$

$$= 3{,}490 + 70\left\{\log\left[(7.65)(184.7)/(0.039)(0.8) \cdot (17 \times 10^{-6})(0.198)^2\right] - 3.23 + (0.869)(6.32)\right\}$$

$$= 4{,}407 \text{ psi}.$$

5. We then check on wellbore-storage duration. For this well, $C_s \cong 25.65\,A_{wb}/\rho = (25.65)\,(0.0218)\,/\,53 = 0.0106$ bbl/psi. Then,

$$t_{wbs} \cong \frac{(200{,}000 + 12{,}000\,s)\,C_s}{kh/\mu}$$

$$= \frac{[200{,}000 + (12{,}000)(6.32)]0.0106}{(7.65)(69)/0.8}$$

$$= 4.4 \text{ hours}.$$

At this time, the plotting function is approximately 1.9; this confirms our choice of the start of the MTR.

### *n*-Rate Flow Test

From Eq. 3.11, an $n$-rate flow test is modeled by

$$\frac{p_i - p_{wf}}{q_n} = 162.6\frac{\mu B}{kh}\left[\sum_{j=1}^{n}\frac{(q_j - q_{j-1})}{q_n} \cdot \log(t_n - t_{j-1})\right] + 162.6\frac{\mu B}{kh} \cdot \left[\log\left(\frac{k}{\phi\mu c_t r_w^2}\right) - 3.23 + 0.869\,s\right].$$

This equation suggests a plot of

$$\frac{p_i - p_{wf}}{q_n} \text{ vs. } \sum_{j=1}^{n}\frac{(q_j - q_{j-1})}{q_n}\log(t_n - t_{j-1}).$$

Permeability is related to the slope $m'$ of such a plot:

$$k = 162.6\frac{\mu B}{m'h}.$$

If we let $b'$ be the value of $(p_i - p_{wf})/q_n$ when the plotting function is zero, then

$$b' = m'\left[\log\left(\frac{k}{\phi\mu c_t r_w^2}\right) - 3.23 + 0.869\,s\right],$$

or

$$s = 1.151\left[\frac{b'}{m'} - \log\left(\frac{k}{\phi\mu c_t r_w^2}\right) + 3.23\right]. \quad \ldots . (3.21)$$

Note that use of Eq. 3.21 and of the proposed

TABLE 3.6 – MULTIRATE FLOW TEST DATA

| $t$ (hours) | $p_{wf}$ (psia) |
|---|---|
| 0 | 3,000 |
| 0.333 | 999 |
| 0.667 | 857 |
| 1.0 | 778.5 |
| 2.0 | 1,378.5 |
| 2.333 | 2,043 |
| 2.667 | 2,067.5 |
| 3.0 | 2,094 |

TABLE 3.7 – DATA FOR PLOTTING MULTIRATE FLOW TEST

| $t$ (hours) | $q_n$ (STB/D) | $p_i-p_{wf}$ (psia) | $\frac{p_i-p_{wf}}{q_n}\left(\frac{\text{psia}}{\text{RB/D}}\right)$ | $\sum_{j=1}^{n}\frac{q_j-q_{j-1}}{q_n}\cdot\log(t_j-t_{j-1})$ |
|---|---|---|---|---|
| 0 | – | – | – | – |
| 0.333 | 478.5 | 2,001 | 4.18 | –0.478 |
| 0.667 | 478.5 | 2,143 | 4.48 | –0.176 |
| 1.0 | 478.5 | 2,221.5 | 4.64 | 0 |
| 2.0 | 319.0 | 1,621.5 | 5.08 | 0.452 |
| 2.333 | 159.5 | 957 | 6.00 | 1.459 |
| 2.667 | 159.5 | 923.5 | 5.79 | 1.232 |
| 3.0 | 159.5 | 906 | 5.68 | 1.130 |

plotting method implies that $p_i$ is known from independent measurements.

Odeh and Jones[7] discussed this analysis technique. They pointed out that it can be applied to the analysis of multirate flow tests commonly run on gas wells and oil wells. In these applications of the technique, it is essential to remember the assumption that the reservoir is infinite acting to the total elapsed time $t$ for all flow rates combined. Further, note that the technique ignores any wellbore storage distortion created by any discrete rate changes.

## *Example 3.6 – Multirate Flow Test Analysis*

**Problem.** Odeh and Jones[7] present data from a 3-hour drawdown test on an oil well; in this test, the rate during the first hour averaged 478.5 STB/D; during the second hour, 319 STB/D; and during the third hour, 159.5 STB/D. Reservoir fluid viscosity is 0.6 cp; initial pressure is 3,000 psia; formation volume factor, $B$, is considered to be 1.0; and the reservoir is assumed to be infinite acting for the entire test. Assume that wellbore storage distortion is minimal at all times during the test. Pressures ($p_{wf}$) at various flow times are given in Table 3.6. From these data, determine the permeability/thickness product of the tested well.

**Solution.** We first prepare the data for plotting – i.e., at each time, we must determine $(p_i-p_{wf})/q_n$ and

$$\sum_{j=1}^{n}\frac{(q_j-q_{j-1})}{q_n}\log(t_n-t_{j-1}).$$

$t=0.333$ hour. Here,

$$q_n=q_1=478.5\text{ STB/D},$$

$$t_n=0.333\text{ hour},$$

$$(p_i-p_{wf})/q_n=(3{,}000-999)/478.5=4.18,\text{ and}$$

$$\sum_{j=1}^{n}\frac{(q_j-q_{j-1})}{q_n}\log(t_n-t_{j-1})$$

$$=\frac{(478.5-0)}{478.5}\log(0.333-0)=-0.478$$

$t=1.0$ hour. Here,

$$q_n=478.5\text{ STB/D},$$

$$t_j=1.0\text{ hour},$$

$$(p_i-p_{wf})/q_n=(3{,}000-778.5)/478.5=4.64,\text{ and}$$

$$\sum_{j=1}^{n}\frac{(q_j-q_{j-1})}{q_n}\log(t_n-t_{j-1})$$

$$=\frac{(478.5-0)}{478.5}\log(1.0-0)=0.$$

$t=3.0$ hours. Here,

$$(p_i-p_{wf})/q_n=(3{,}000-2{,}094)/159.5=5.68,\text{ and}$$

$$\sum_{j=1}^{n}\frac{(q_j-q_{j-1})}{q_n}\log(t_n-t_{j-1})$$

$$=\frac{1}{159.5}[478.5\log(3.0-0)+(319-478.5)\cdot\log(3.0-1.0)+(159.5-319)\cdot\log(3.0-2.0)]$$

$$=1.130.$$

Calculations at these and other times are summarized in Table 3.7.

We next estimate the permeability/thickness product. From Fig. 3.13,

$$m'=\frac{(6.0-4.2)}{[1.459-(-0.452)]}=0.942.$$

Then,

$$kh=162.6\frac{\mu B}{m'}=\frac{(162.6)(0.6)(1.0)}{0.942}$$

$$=104\text{ md-ft}.$$

Odeh and Jones do not state values for $h$, $\phi$, $c_t$, and $r_w$ used to construct these example test data. To illustrate skin-factor calculation, assume that $h=10$ ft and, thus, that $k=10.4$ md. Also assume that $k/\phi\mu c_t r_w^2=4.81\times10^7$. Then, since the graph indicated that $b'=4.63$ [$b'$ is the value of $(p_i-p_{wf})/q_n$ when the plotting function is zero],

$$s=1.151\left[\frac{b'}{m'}-\log\left(\frac{k}{\phi\mu c_t r_w^2}\right)+3.23\right]$$

$$=1.151\left[\frac{4.63}{0.942}-\log\left(4.8\times10^7\right)+3.23\right]$$

$$=0.53.$$

## Exercises

3.1. A constant-rate drawdown test was run in a well with the following characteristics:

$q$ = 500 STB/D (constant),
$\phi$ = 0.2,
$\mu$ = 0.8 cp,
$c_t$ = $10\times10^{-6}$ psi$^{-1}$,
$r_w$ = 0.3 ft,
$h$ = 56 ft,
$B_o$ = 1.2 RB/STB,
$A_{wb}$ = 0.022 sq ft,
$\rho$ = 50 lb/cu ft, and
liquid/gas interface is in well.

From the test data in Table 3.8, estimate formation permeability, skin factor, and area (in acres) drained by the well.

3.2. A drawdown test in which the rate decreased continuously throughout the test was run in a well with the following characteristics.

$\phi$ = 0.2,
$\mu$ = 1.0 cp,
$c_t$ = $10\times10^{-6}$ psi$^{-1}$,
$r_w$ = 0.25 ft,
$h$ = 100 ft,
$B_o$ = 1.3 RB/STB,
$A_{wb}$ = 0.0218 sq ft,
$\rho$ = 55 lb/cu ft, and
liquid/gas interface is in well

From the test data in Table 3.9, estimate formation permeability and skin factor.

3.3. A constant-rate drawdown test was run on the well described in Problem 3.2. Rate was held constant at 600 STB/D. After 96.9 hours, the surface rate was changed abruptly to 300 STB/D. Data for the tests before and after the rate change are given in Table 3.10. From the data obtained with $q=600$ STB/D, estimate $k$, $s$, and area of the reservoir. From the data obtained in the two-rate flow test after $q$ changed to 300 STB/D, confirm the estimates of $k$ and $s$, and calculate the current value of $p^*$.

3.4. For the multirate flow test described in Example 3.6, (a) calculate the value of the plotting function at $t=0.5$, 1.5, and 2.5 hours, and (b) calculate the flowing bottomhole pressure at $t=3.5$ hours, assuming that there is no change in rate for 3 hours $<t<4$ hours.

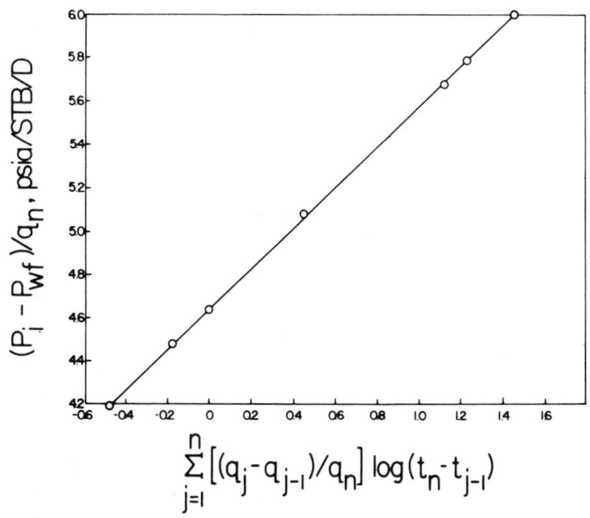


**Fig. 3.13** – Example multirate flow test.

TABLE 3.8 – DATA FOR EXAMPLE CONSTANT-RATE DRAWDOWN TEST

| $t$ (hours) | $p_{wf}$ (psi) | $t$ (hours) | $p_{wf}$ (psi) | $t$ (hours) | $p_{wf}$ (psi) |
|---|---|---|---|---|---|
| 0 | 3,000 | 0.491 | 2,302 | 32.8 | 1,543 |
| 0.0109 | 2,976 | 0.546 | 2,256 | 38.2 | 1,533 |
| 0.0164 | 2,964 | 1.09 | 1,952 | 43.7 | 1,525 |
| 0.0218 | 2,953 | 1.64 | 1,828 | 49.1 | 1,517 |
| 0.0273 | 2,942 | 2.18 | 1,768 | 54.6 | 1,511 |
| 0.0328 | 2,930 | 2.73 | 1,734 | 65.5 | 1,500 |
| 0.0382 | 2,919 | 3.28 | 1,712 | 87.4 | 1,482 |
| 0.0437 | 2,908 | 3.82 | 1,696 | 109.2 | 1,468 |
| 0.0491 | 2,897 | 4.37 | 1,684 | 163.8 | 1,440 |
| 0.0546 | 2,886 | 4.91 | 1,674 | 218.4 | 1,416 |
| 0.109 | 2,785 | 5.46 | 1,665 | 273.0 | 1,393 |
| 0.164 | 2,693 | 6.55 | 1,651 | 327.6 | 1,370 |
| 0.218 | 2,611 | 8.74 | 1,630 | | |
| 0.273 | 2,536 | 10.9 | 1,587 | | |
| 0.328 | 2,469 | 16.4 | 1,568 | | |
| 0.437 | 2,352 | 27.3 | 1,554 | | |

TABLE 3.9 – DATA FOR EXAMPLE VARIABLE-RATE DRAWDOWN TEST

| $t$ (hours) | $p_{wf}$ (psi) | $q$ (STB/D) | $t$ (hours) | $p_{wf}$ (psi) | $q$ (STB/D) |
|---|---|---|---|---|---|
| 0 | 5,000 | 200 | 3.03 | 4,797 | 122 |
| 0.114 | 4,927 | 145 | 3.64 | 4,797 | 121 |
| 0.136 | 4,917 | 143 | 4.37 | 4,798 | 119 |
| 0.164 | 4,905 | 142 | 5.24 | 4,798 | 118 |
| 0.197 | 4,893 | 141 | 6.29 | 4,798 | 117 |
| 0.236 | 4,881 | 140 | 7.54 | 4,799 | 116 |
| 0.283 | 4,868 | 138 | 9.05 | 4,799 | 114 |
| 0.340 | 4,856 | 137 | 10.9 | 4,800 | 113 |
| 0.408 | 4,844 | 136 | 13.0 | 4,801 | 112 |
| 0.490 | 4,833 | 135 | 15.6 | 4,801 | 110 |
| 0.587 | 4,823 | 133 | 18.8 | 4,802 | 109 |
| 0.705 | 4,815 | 132 | 22.5 | 4,803 | 108 |
| 0.846 | 4,809 | 131 | 27.0 | 4,803 | 107 |
| 1.02 | 4,804 | 129 | 32.4 | 4,804 | 105 |
| 1.22 | 4,801 | 128 | 38.9 | 4,805 | 104 |
| 1.46 | 4,799 | 127 | 46.7 | 4,806 | 103 |
| 1.75 | 4,798 | 126 | 56.1 | 4,807 | 102 |
| 2.11 | 4,797 | 124 | 67.3 | 4,807 | 100 |
| 2.53 | 4,797 | 123 | 80.7 | 4,808 | 99 |
| | | | 96.9 | 4,809 | 98 |

TABLE 3.10 – DATA FOR EXAMPLE TWO-RATE FLOW TEST

| t (hours) | $p_{wf}$ at 600 STB/D (psi) | $p_{wf}$ at 300 STB/D (psi) | t (hours) | $p_{wf}$ at 600 STB/D (psi) | $p_{wf}$ at 300 STB/D (psi) |
|---|---|---|---|---|---|
| 0 | 5,000 | 3,833 | 3.03 | 4,012 | 4,324 |
| 0.114 | 4,710 | 3,978 | 3.64 | 4,002 | 4,330 |
| 0.136 | 4,665 | 4,000 | 4.37 | 3,992 | 4,334 |
| 0.164 | 4,616 | 4,025 | 5.24 | 3,982 | 4,339 |
| 0.197 | 4,563 | 4,051 | 6.29 | 3,972 | 4,343 |
| 0.236 | 4,507 | 4,079 | 7.54 | 3,963 | 4,346 |
| 0.283 | 4,449 | 4,108 | 9.05 | 3,953 | 4,350 |
| 0.340 | 4,390 | 4,138 | 10.9 | 3,944 | 4,353 |
| 0.408 | 4,332 | 4,167 | 13.0 | 3,935 | 4,356 |
| 0.490 | 4,277 | 4,194 | 15.6 | 3,926 | 4,359 |
| 0.587 | 4,227 | 4,219 | 18.8 | 3,918 | 4,361 |
| 0.705 | 4,182 | 4,242 | 22.5 | 3,909 | 4,363 |
| 0.846 | 4,144 | 4,261 | 27.0 | 3,900 | 4,364 |
| 1.02 | 4,112 | 4,276 | 32.4 | 3,891 | 4,365 |
| 1.22 | 4,087 | 4,289 | 38.9 | 3,883 | 4,365 |
| 1.46 | 4,067 | 4,299 | 46.7 | 3,874 | 4,364 |
| 1.75 | 4,050 | 4,307 | 56.1 | 3,865 | 4,362 |
| 2.11 | 4,036 | 4,314 | 67.3 | 3,855 | 4,359 |
| 2.53 | 4,024 | 4,319 | 80.7 | 3,845 | 4,354 |
| | | | 96.9 | 3,833 | 4,349 |

## References


1. Earlougher, R.C. Jr.: *Advances in Well Test Analysis*, Monograph Series, SPE, Dallas (1977) **5**.
2. Winestock, A.G. and Colpitts, G.P.: "Advances in Estimating Gas Well Deliverability," *J. Cdn. Pet. Tech.* (July-Sept. 1965) 111-119. Also, *Gas Technology*, Reprint Series, SPE, Dallas (1977) **13**, 122-130.
3. Gladfelter, R.E., Tracy, G.W., and Wilsey, L.E.: "Selecting Wells Which Will Respond to Production-Stimulation Treatment," *Drill. and Prod. Prac.*, API, Dallas (1955) 117-129.
4. Ramey, H.J. Jr.: "Non-Darcy Flow and Wellbore Storage Effects on Pressure Buildup and Drawdown of Gas Wells," *J. Pet. Tech.* (Feb. 1965) 223-233; *Trans.*, AIME, **234**.
5. Russell, D.G.: "Determination of Formation Characteristics From Two-Rate Flow Tests," *J. Pet. Tech.* (Dec. 1963) 1347-1355; *Trans.*, AIME, **228**.
6. Stegemeier, G.L. and Matthews, C.S.: "A Study of Anomalous Pressure Buildup Behavior," *Trans.*, AIME (1958) **213**, 44-50.
7. Odeh, A.S. and Jones, L.G.: "Pressure Drawdown Analysis, Variable-Rate Case," *J. Pet. Tech.* (Aug. 1965) 960-964; *Trans.*, AIME, **234**.

# Chapter 4
# Analysis of Well Tests Using Type Curves

## 4.1 Introduction

This chapter discusses the quantitative use of type curves in well test analysis. The objective of this chapter is limited basically to illustrating how a representative sample of type curves can be used as analysis aids. Other major type curves in use today are discussed in the SPE well testing monograph.[1] However, type curves for specialized situations are appearing frequently in the literature, and even that monograph is not completely current. We hope that the fundamentals of type-curve use presented in this chapter will allow the reader to understand and to apply newer type curves as they appear in the literature.

Specific type curves discussed include (1) Ramey *et al.*'s type curves[2-4] for buildup and constant-rate drawdown tests; (2) McKinley's type curves[5,6] for the same applications; and (3) Gringarten *et al.*'s[7] type curves for vertically fractured wells with uniform flux.

## 4.2 Fundamentals of Type Curves

Many type curves commonly are used to determine formation permeability and to characterize damage and stimulation of the tested well. Further, some are used to determine the beginning of the MTR for a Horner analysis. Most of these curves were generated by simulating constant-rate pressure drawdown (or injection) tests; however, most also can be applied to buildup (or falloff) tests if an equivalent shut-in time[8] is used as the time variable on the graph.

Conventional test analysis techniques (such as the Horner method for buildup tests) share these objectives. However, type curves are advantageous because they may allow test interpretation even when wellbore storage distorts most or all of the test data; in that case, conventional methods fail.

The use of type curves for fractured wells has a further advantage. In a single analytical technique, type curves combine the linear flow that occurs at early times in many fractured reservoirs, the radial flow that may occur later after the radius of investigation has moved beyond the region influenced by the fracture, and the effects of reservoir boundaries that may appear before a true MTR line is established in a pressure transient test on a fractured well.

Fundamentally, a type curve is a preplotted family of pressure drawdown curves. The most fundamental of these curves (Ramey's[2]) is a plot of dimensionless pressure change, $p_D$, vs. dimensionless time change, $t_D$. This curve, reproduced in Fig. 4.1 (identical to Fig. 1.6), has two parameters that distinguish the curves from one another: the skin factor $s$ and a dimensionless wellbore storage constant, $C_{sD}$. For an infinite-acting reservoir, specification of $C_{sD}$ and $s$ *uniquely* determines the value of $p_D$ at a given value of $t_D$. Proof of this follows from application of the techniques discussed in Appendix B. If we put the differential equation describing a flow test in dimensionless form (along with its initial and boundary conditions), then the solution, $p_D$, is determined uniquely by specification of the independent variables (in this case, $t_D$ and $r_D$), of all dimensionless parameters that appear in the equation, and of initial and boundary conditions (in this case, $s$ and $C_{sD}$). Further, in most such solutions, we are interested in wellbore pressures of a tested well; here, dimensionless radius, $r_D = r/r_w$, has a fixed value of unity and thus does not appear as a parameter in the solution.

Thus, type curves are generated by obtaining solutions to the flow equations (e.g., the diffusivity equation) with specified initial and boundary conditions. Some of these solutions are analytical; others are based on finite-difference approximations generated by computer reservoir simulators. For example, Ramey's type curves were generated from analytical solutions to the diffusivity equation, with the initial condition that the reservoir be at uniform pressure before the drawdown test, and with boundary conditions of (1) infinitely large outer drainage radius and (2) constant *surface* withdrawal rate combined with wellbore storage, which results in variable *sandface* withdrawal rate. A skin factor, $s$, is used to characterize wellbore damage or stimulation; as we have seen, this causes an additional pressure drop, $\Delta p_s$, which is proportional to the instantaneous sandface flow rate (which changes with

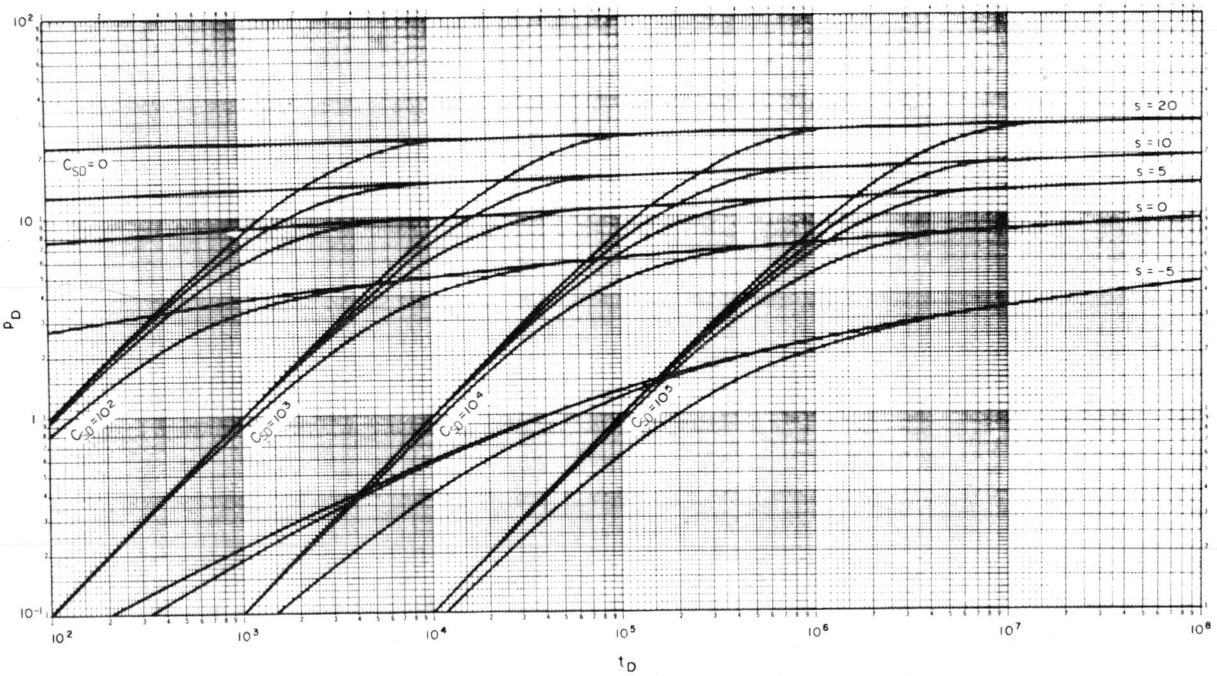


**Fig. 4.1** – Type curves for constant production rate, infinite-acting reservoir (Ramey).

time while wellbore storage is a dominant influence). Dimensionless pressure drawdown at the wellbore, $p_D$, predicted by these solutions thus can be plotted as a function of elapsed time, $t_D$, for fixed values of $C_{sD}$ and $s$. When curves are drawn for the range of $s$ and $C_{sD}$ of greatest practical importance, the type curve results (Fig. 4.1).

To use a type curve to analyze an actual drawdown test, the analyst plots pressure change, $p_i - p_{wf}$ vs. flow time, $t$, on the same size graph paper as the type curve. Then one finds the preplotted curve that most nearly has the same shape as the actual test data plot. When the match is found, $s$, $C_{sD}$, and corresponding values of $[p_D, (p_i - p_{wf})]$ and $(t_D, t)$ will have been established, and $k$ then can be determined. These sentences summarize the *principle* – but the *practice* differs in detail from the principle and is not necessarily as straightforward as this brief discussion implies.

## 4.3 Ramey's Type Curves

Ramey's[2] type curves were generated for the situation of a constant-rate pressure drawdown test in a reservoir with slightly compressible, single-phase liquid flowing; sufficient homogeneity such that the radial diffusivity equation adequately models flow in the reservoir; uniform pressure in the drainage area of the well before production; infinite-acting reservoir (no boundary effects during the flow period of interest for test analysis purposes); constant withdrawal rate at the surface; and wellbore storage and concentrated wellbore damage or stimulation characterized by a skin factor, $s$. This list of assumptions is tedious, but it is also important. When one or more of these assumptions is not valid in a specific case, there is no assurance that use of the type curves can lead to a valid test interpretation. (Some of these limitations can be removed, as we will note later in this chapter. Of major importance is that the curves can be used for buildup tests and for gas well tests.) The result of Ramey's work is shown in Fig. 4.1.

Some important properties of these curves follow.

1. Examination of the analytical solution on which the type curves are based shows that, at earliest times when wellbore unloading is responsible for 100% of the flow in a drawdown test (or afterflow rate equals rate before shut-in in a buildup test), $\Delta p$ is a linear function of $\Delta t$ ($\Delta p$ is pressure change since the test began and $\Delta t$ is time elapsed since the test began). Thus, the log $\Delta p$-log$\Delta t$ curve is also linear with a slope of unity (a 45° line) and the wellbore storage constant $C_s$ can be determined from any point ($\Delta t$, $\Delta p$) on this line (Fig. 4.2) from the relation

$$C_s = \frac{qB}{24}\left(\frac{\Delta t}{\Delta p}\right)_{\text{unit-slope line}} .$$

Note that, in a well with a liquid/gas interface in the wellbore,

$$C_s = \frac{25.65\, A_{wb}}{\rho} \text{ bbl/psi}, \quad \ldots\ldots\ldots\ldots\ldots (4.1)$$

and for a wellbore filled with a single-phase liquid or with gas,

$$C_s = c_{wb} V_{wb} \text{ bbl/psi}, \quad \ldots\ldots\ldots\ldots\ldots (4.2)$$

and

$$C_{sD} = 0.894\, C_s / \phi c_t h r_w^2. \quad \ldots\ldots\ldots\ldots\ldots (4.3)$$

Successful application of Ramey's type curves for quantitative analysis depends significantly on our ability to establish the correct value of $C_{sD}$ to be used for curve matching – type curves for a given value of $s$ and for different values of $C_{sD}$ have very similar shapes, so it is difficult to find the best fit without prior knowledge of $C_{sD}$. Direct calculation of $C_s$,

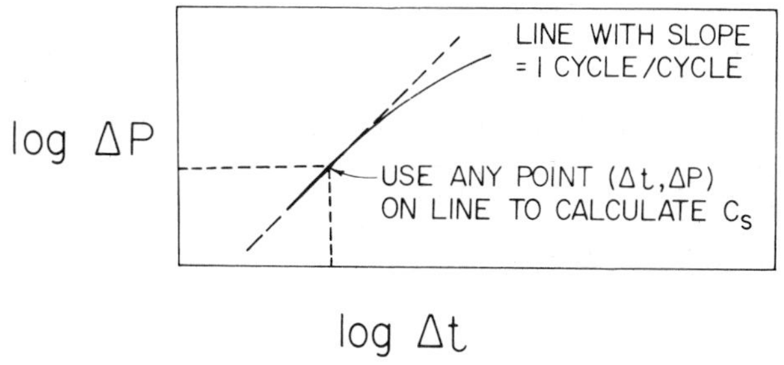


Fig. 4.2 – Use of unit slope line to calculate wellbore storage constant.

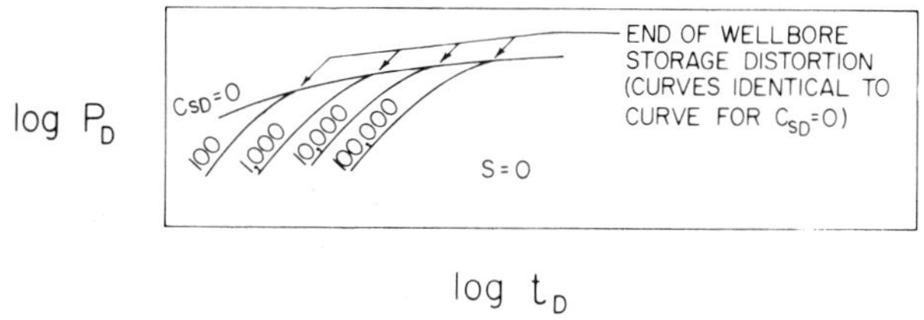


Fig. 4.3 – Use of type curves to determine end of wellbore storage distortion.

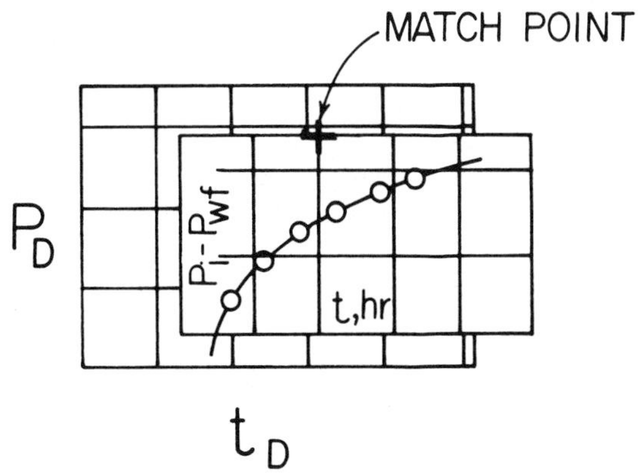


Fig. 4.4 – Horizontal and vertical shifting to find position of fit and match points.

and thus $C_{sD}$, from known values of $A_{wb}$ and $\rho$ or $c_{wb}$ and $V_{wb}$ does not characterize test conditions as well as the value of $C_s$ determined from actual test performance as reflected in the unit-slope lines.[9]

2. Wellbore storage has ceased distorting the pressure transient test data when the type curve for the value of $C_{sD}$ characterizing the test becomes identical to the type curve for $C_{sD}=0$ (Fig. 4.3). (This usually occurs about one and a half to two cycles from the end of the unit-slope line.) Thus, these type curves can be used to determine how much data (if any) can be analyzed by conventional methods such as the Horner plot for buildup tests.

3. The type curves, which were developed for drawdown tests, also can be used for buildup test analysis if an equivalent shut-in time, $\Delta t_e = \Delta t/(1+\Delta t/t_p)$, is used as the time variable. An intuitive proof of this assertion for small $\Delta t$ (where $\Delta t \approx \Delta t_e$) follows.

The equation of the MTR in a drawdown test can be expressed as

$$p_i - p_{wf} = m \log t + C_1 .$$

The equation of the MTR in a buildup test is

$$p_i - p_{ws} = m \log[(t_p+\Delta t)/\Delta t]$$

$$= m \log(t_p+\Delta t) - m \log \Delta t.$$

If $\log(t_p+\Delta t) \simeq \log t_p$ (an adequate assumption for $\Delta t_{max} \le 0.1\, t_p$),

$$p_i - p_{ws} \simeq m \log t_p - m \log \Delta t$$

$$\simeq (p_i - p_{wf}) - m \log \Delta t - C_1 ,$$

or

$$(p_{ws} - p_{wf}) \simeq m \log \Delta t + C_1 .$$

Thus, the equations for MTR's in drawdown and buildup test plots have similar form if we use the analogies

$$(p_i - p_{wf})_{\text{drawdown}} \sim (p_{ws} - p_{wf})_{\text{buildup}},$$

and

$$t_{\text{drawdown}} \sim \Delta t_{\text{buildup}} .$$

If these analogies can be used for the larger values of $\Delta t$ in the MTR, then we would expect intuitively that the approximation used to develop them would be even better for the smaller values of $\Delta t$ in the ETR.

The practical implication of this analysis is this. For use with type curves, we plot actual drawdown test data as $(p_i - p_{wf})$ vs. $t$ and buildup test data as $(p_{ws} - p_{wf})$ vs. $\Delta t$, but we must remember that $\Delta t_e$ must be used instead of $\Delta t$ whenever $\Delta t > 0.1\, t_p$.

4. A log-log plot of $p_D$ vs. $t_D$ differs from a log-log plot of $(p_i - p_{wf})$ vs. $t$ (for a drawdown test) only by a shift in the origin of the coordinate system – i.e., $\log t_D$ differs from $\log t$ by a constant and $\log p_D$ differs from $\log (p_i - p_{wf})$ by another constant. To show this, we note that

$$t_D = \frac{0.000264\, kt}{\phi \mu c_t r_w^2},$$

and

$$p_D = \frac{kh(p_i - p_{wf})}{141.2\, qB\mu} .$$

Thus,

$$\log t_D = \log t + \log \frac{0.000264\, k}{\phi \mu c_t r_w^2},$$

and

$$\log p_D = \log (p_i - p_{wf}) + \log \frac{kh}{141.2\, qB\mu} .$$

The significance of this result is that the plot of an actual drawdown test ($\log t$ vs. $\log \Delta p$) should have a shape identical to that of a plot of $\log t_D$ vs. $\log p_D$, but we have to displace both the horizontal and vertical axes (i.e., shift the origin of the plot) to find the position of best fit (Fig. 4.4).

Once a fit is found by vertical and horizontal shifting, we choose a match point to determine the relationship between actual time and dimensionless time and between actual pressure drawdown and dimensionless pressure for the test being analyzed. Any point on the graph paper will suffice as a match point (i.e., the result is independent of the choice of match point). For the match point chosen, we determine the corresponding values of $(t, t_D)$ and $[(p_i - p_{wf}), p_D]$. Then, from definition of $p_D$ and $t_D$,

$$k = 141.2\frac{qB\mu}{h}\left(\frac{p_D}{p_i - p_{wf}}\right)_{MP}, \quad \ldots\ldots\ldots\ldots (4.4)$$

and

$$\phi c_t = \frac{0.000264\,k}{\mu r_w^2}\left(\frac{t}{t_D}\right)_{MP}. \quad \ldots\ldots\ldots\ldots (4.5)$$

5. Although the type curves were developed from solutions to flow equations for slightly compressible liquids, they also can be used to analyze gas well tests. Transformation of the flow equations to model gas flow in terms of the pseudopressure[10] $\psi(p)$ and comparison of these solutions expressed in terms of dimensionless pseudopressure, $\psi_D$, with solutions $p_D$ for slightly compressible liquids,[11] show that, as a high-order-accuracy approximation for transient flow,

$$\psi_D(t_D, r_D, s', C_{sD}) = p_D(t_D, r_D, s, C_{sD}),$$

where, for gases,

$$t_D = \frac{0.000264\,kt}{\phi \mu_i c_{gi} r_w^2}, \quad \ldots\ldots\ldots\ldots (4.6)$$

$$\psi_D = \frac{kh\,T_{sc}[\psi(p_i) - \psi(p_{wf})]}{50{,}300\,p_{sc} q_g T}, \quad \ldots\ldots\ldots\ldots (4.7)$$

$$s' = s + D|q_g|, \quad \ldots\ldots\ldots\ldots (4.8)$$

and

$$\psi(p) = 2\int_{p_\beta}^{p} \frac{p}{\mu(p) z(p)}\,dp. \quad \ldots\ldots\ldots\ldots (4.9)$$

To use this result as stated would require that we (1) prepare a table or graph of values of $\psi(p)$ vs. $p$ from Eq. 4.9 based on the properties of the specific gas in the well being tested; (2) plot $[\psi(p_i) - \psi(p_{wf})]$ vs. $t$ on log-log paper; and (3) find the best fit just as for a slightly compressible liquid (with the values of $s$ providing the best fit being interpreted for the gas well as $s + D|q_g|$).

Steps 1 and 2 can be simplified in some cases. When $\mu z$ is directly proportional to pressure (for $p > 3{,}000$ psia for some gases), Eq. 4.7 can be replaced by the definition

$$p_D = \frac{kh(p_i - p_{wf})}{141.2\,q_g \mu_i B_{gi}}, \quad \ldots\ldots\ldots\ldots (4.10)$$

where

$$B_{gi} = 5.04\frac{z_i T}{p_i}\ \text{RB/Mscf}. \quad \ldots\ldots\ldots\ldots (4.11)$$

Thus, when $\mu z/p$ = constant, we can plot $(p_i - p_{wf})$ for type-curve use just as for a slightly compressible liquid. Match-point interpretation is

$$k = 141.2\frac{q_g \mu_i B_{gi}}{h}\left(\frac{p_D}{p_i - p_{wf}}\right)_{MP}, \quad \ldots\ldots\ldots\ldots (4.12)$$

$$\phi c_{ti} = \frac{0.000264\,k}{\mu_i r_w^2}\left(\frac{t}{t_D}\right)_{MP}. \quad \ldots\ldots\ldots\ldots (4.13)$$

Note that all gas properties are to be evaluated at original reservoir pressure for a test in an infinite-acting reservoir (or, more generally, at the uniform reservoir pressure preceding the drawdown test or at the current average reservoir pressure for a buildup test).

In some other situations, $\mu z$ is constant (e.g., in many cases for $p < 2{,}000$ psia); as a result, Eq. 4.7 can be replaced by the definition

$$p_D = \frac{kh(p_i^2 - p_{wf}^2)}{1{,}422\,q_g \mu_i z_i T}. \quad \ldots\ldots\ldots\ldots (4.14)$$

Thus, when $\mu z$ = constant, we can plot $(p_i^2 - p_{wf}^2)$ for type-curve use. Match-point interpretation becomes

$$k = 1{,}422\frac{q_g \mu_i z_i T}{h}\left(\frac{p_D}{p_i^2 - p_{wf}^2}\right)_{MP}, \quad \ldots\ldots\ldots\ldots (4.15)$$

$$\phi c_{ti} = \frac{0.000264\,k}{\mu_i r_w^2}\left(\frac{t}{t_D}\right)_{MP} \quad \ldots\ldots\ldots\ldots (4.16)$$

Determination of whether either approximation ($\mu z/p$ = constant or $\mu z$ = constant) is valid must be based on plots or tabulations using the properties of the specific gas in the reservoir being tested.

**Use of Ramey's Type Curves**

The theory of Ramey's type curves leads to the following procedure for using the curves for test analysis. The procedure is given for a slightly compressible liquid; Eqs. 4.6 through 4.16 show the changes necessary when a gas well test is analyzed.

1. Plot $(p_i - p_{wf})$ vs. $t$ (drawdown test) or $(p_{ws} - p_{wf})$ vs. $\Delta t_e = \Delta t/(1 + \Delta t/t_p)$ (buildup test) on log-log paper the same size as Ramey's type curve. Caution: Unless a type curve undistorted in the reproduction process is used, it will not have the same dimensions as commercially available graph paper and finding a fit may be misleading or impossible. The best solution is to use an undistorted type curve; an acceptable alternative is to plot test data on tracing paper, using the grid on the distorted type curve as a plotting aid.

2. If the test has a uniform-slope region (45° line at earliest times), choose any point $[t, (p_i - p_{wf})]$ or $[\Delta t, (p_{ws} - p_{wf})]$ on the unit-slope line and calculate the wellbore storage constant $C_s$:

$$C_s = \frac{qB}{24}\left(\frac{t}{p_i - p_{wf}}\right)_{\text{unit-slope line}}$$

**TABLE 4.1 – CONSTANT-RATE DRAWDOWN TEST DATA**

| t (hours) | $p_{wf}$ (psi) | t (hours) | $p_{wf}$ (psi) | t (hours) | $p_{wf}$ (psi) |
|---|---|---|---|---|---|
| 0.0109 | 2,976 | 0.164 | 2,693 | 2.18 | 1,768 |
| 0.0164 | 2,964 | 0.218 | 2,611 | 2.73 | 1,734 |
| 0.0218 | 2,953 | 0.273 | 2,536 | 3.28 | 1,712 |
| 0.0273 | 2,942 | 0.328 | 2,469 | 3.82 | 1,696 |
| 0.0328 | 2,930 | 0.382 | 2,408 | 4.37 | 1,684 |
| 0.0382 | 2,919 | 0.437 | 2,352 | 4.91 | 1,674 |
| 0.0437 | 2,908 | 0.491 | 2,302 | 5.46 | 1,665 |
| 0.0491 | 2,897 | 0.546 | 2,256 | 6.55 | 1,651 |
| 0.0546 | 2,886 | 1.09 | 1,952 | 8.74 | 1,630 |
| 0.109 | 2,785 | 1.64 | 1,828 | 10.9 | 1,614 |
| | | | | 16.4 | 1,587 |

**TABLE 4.2 – DRAWDOWN DATA TABULATED FOR PLOTTING**

| t (hours) | $p_i-p_{wf}$ (psi) | t (hours) | $p_i-p_{wf}$ (psi) | t (hours) | $p_i-p_{wf}$ (psi) |
|---|---|---|---|---|---|
| 0.0109 | 24 | 0.164 | 307 | 2.18 | 1,232 |
| 0.0164 | 36 | 0.218 | 389 | 2.73 | 1,266 |
| 0.0218 | 47 | 0.273 | 464 | 3.28 | 1,288 |
| 0.0273 | 58 | 0.328 | 531 | 3.82 | 1,304 |
| 0.0328 | 70 | 0.382 | 592 | 4.37 | 1,316 |
| 0.0382 | 81 | 0.437 | 648 | 4.91 | 1,326 |
| 0.0437 | 92 | 0.491 | 698 | 5.46 | 1,335 |
| 0.0491 | 103 | 0.546 | 744 | 6.55 | 1,349 |
| 0.0546 | 114 | 1.09 | 1,048 | 8.74 | 1,370 |
| 0.109 | 215 | 1.64 | 1,172 | 10.9 | 1,386 |
| | | | | 16.4 | 1,413 |

Then calculate the dimensionless wellbore storage constant:

$$C_{sD}=\frac{0.894\,C_s}{\phi c_t h r_w^2}.$$

Note that estimates of $\phi$ and $c_t$ are required at this point – with implications that are discussed later.

If a unit-slope line is not present, $C_s$ and $C_{sD}$ must be calculated from wellbore properties, and inaccuracies may result if these properties do not describe actual test behavior.[9]

3. Using type curves with $C_{sD}$ as calculated in Step 2, find the curve that most nearly fits all the plotted data. This curve will be characterized by some skin factor, $s$; record its value. Interpolation between curves should improve the precision of the analysis, but may prove difficult. Even for fixed $C_{sD}$ from the unit-slope curve, the analyst may experience difficulty in determining that one value of $s$ provides a better fit than another, particularly if all data are distorted by wellbore storage or if the "scatter" or "noise" that characterizes much actual field data is present. If $C_{sD}$ is not known with certainty, the possible ambiguity in finding the best fit is even more pronounced.

4. With the actual test data plot placed in the position of best fit, record corresponding values of $(p_i-p_{wf},\ p_D)$ and $(t,\ t_D)$ from any convenient match point.

5. Calculate $k$ and $\phi c_t$ (Eqs. 4.4 and 4.5):

$$k=141.2\frac{qB\mu}{h}\left(\frac{p_D}{p_i-p_{wf}}\right)_{MP},$$

$$\phi c_t=\frac{0.000264\,k}{\mu r_w^2}\left(\frac{t}{t_D}\right)_{MP}.$$

Eq. 4.5 does not establish $\phi c_t$ based on test performance unless it is possible to establish $C_{sD}$ without assuming values for $\phi c_t$ – it simply reproduces those values assumed in Step 2.

In summary, the procedure outlined in Steps 1 through 5 provides estimates of $k$, $s$, and $C_s$.

---

## *Example 4.1 – Drawdown Test Analysis Using Ramey's Type Curves*

**Problem.** Determine $k$, $s$, and $C_s$ from the data below and in Table 4.1, which were obtained in a pressure drawdown test on an oil well.

$q$ = 500 STB/D,
$\phi$ = 0.2,
$\mu$ = 0.8 cp,
$c_t$ = $10\times10^{-6}$ psi$^{-1}$,
$r_w$ = 0.3 ft,
$h$ = 56 ft,
$B_o$ = 1.2 RB/STB, and
$p_i$ = 3,000 psia.

**Solution.** We must first prepare the data for plotting (Table 4.2). The data are plotted in Fig. 4.5.

From the unit-slope line (on which the data lie for $t\le0.0218$ hour),

$$C_s=\frac{qB}{24}\left[\frac{t}{(p_i-p_{wf})}\right]_{\text{point on unit-slope line}}$$

$$=\frac{(500)(1.2)}{(24)}\times\frac{(0.0218)}{(47)}$$

$$=0.0116 \text{ RB/psi}.$$

Then,

$$C_{sD}=\frac{0.894\,C_s}{\phi c_t h r_w^2}$$

$$=\frac{(0.894)(0.0116)}{(0.2)(1\times10^{-5})(56)(0.3)^2}$$

$$=1.03\times10^3$$

$$\simeq1\times10^3.$$

For $C_{sD}=10^3$, the best-fitting type curve is for $s=5$. A time match point is $t=1$ hour when $t_D=1.93\times10^4$. A pressure match point is $(p_i-p_{wf})=100$ psi, when $p_D=0.85$.

From the match, we also note that wellbore storage distortion ends at $t\simeq5.0$ hours (i.e., the type curve for $C_{sD}=10^3$ becomes identical to the type curve for $C_{sD}=0$).

From the pressure match point,

$$k=141.2\frac{qB\mu}{h}\left(\frac{p_D}{p_i-p_{wf}}\right)_{MP}$$

$$=\frac{(141.2)(500)(1.2)(0.8)}{(56)}\left(\frac{0.85}{100}\right)$$

$$=10.3 \text{ md}.$$

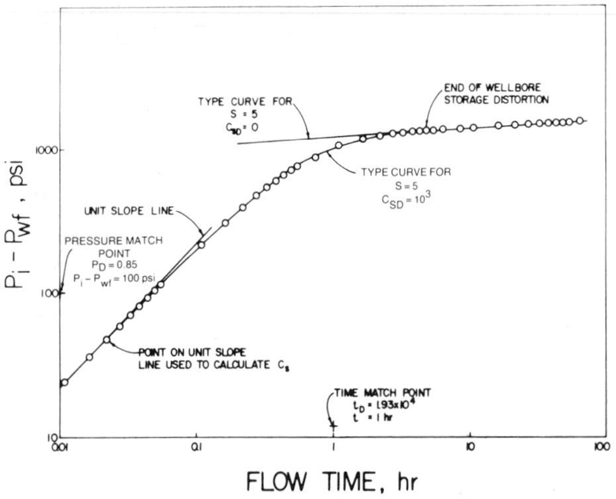


Fig. 4.5 – Drawdown test analysis with Ramey's type curve.

From the time match point,

$$\phi c_t = \frac{0.000264\,k}{\mu r_w^2}\left(\frac{t}{t_D}\right)_{MP}$$

$$= \frac{(0.000264)(10.3)}{(0.8)(0.3)^2}\left(\frac{1}{1.93\times10^4}\right)$$

$$= 1.96\times10^{-6}\ \text{psi}^{-1}.$$

Compare those with values used to determine $C_{sD}$ from $C_s$:

$$\phi c_t = (0.2)(1\times10^{-5})$$

$$= 2\times10^{-6}\ \ldots \text{ values in}$$

$$= \text{values out.}$$

## 4.4 McKinley's Type Curves

McKinley[5] proposed type curves with the primary objective of characterizing damage or stimulation in a drawdown or buildup test in which wellbore storage distorts most or all of the data, thus making this characterization possible with relatively short-term tests.

In constructing his type curves, McKinley observed that the ratio of pressure change, $\Delta p$, to flow rate causing the change, $qB$, is a function of several dimensionless quantities:

$$\frac{\Delta p}{qB} = f\left(\frac{kh\Delta t}{\mu C_s}, \frac{k\Delta t}{\phi\mu c_t r_w^2}, \frac{r_e}{r_w}, \frac{\Delta t}{t_p}\right).$$

Type curves with this many parameters would be difficult, if not impossible, to use. Accordingly, McKinley simplified the problem in the following way.

1. He assumed that the well has produced sufficiently long (essentially to stabilization) that the last group, $\Delta t/t_p$, is not important.

2. He ignored boundary effects except approximately and, thus, ignored $r_e/r_w$ in the basic logic used to construct the type curves.

3. His analysis of simulated buildup and drawdown curves showed that, during the wellbore-storage-dominated portion of a test, the parameter $kh\Delta t/\mu C_s$ was much more important in determining $\Delta p/qB$ than was the parameter $k\Delta t/\phi\mu c_t r_w^2$. Accordingly, he let $k/\phi\mu c_t r_w^2 = 10\times10^6$ md-psi/cp-sq ft (an average value) for all his type curves. It is important to emphasize that even when $k/\phi\mu c_t r_w^2$ varies from this average value by one or two orders of magnitude, the shape of the type curves is not affected significantly. The reason for this approximation was McKinley's judgment that the loss of accuracy is more than offset by the gain in sensitivity in the type curves – i.e., that the shape of each curve is distinctly different at earliest times (Fig. 4.6).

4. To take into account the remaining parameters that *do* have a significant influence on test results, McKinley plotted his type curves as $\Delta t$ (ordinate) vs. $5.615\,C_s\Delta p/qB$ (abscissa), with the single parameter $kh/5.615\,C_s\mu$. A small-scale version of McKinley's curves is shown in Fig. 4.6.

5. Note that the skin factor $s$ does not appear as a parameter in the McKinley curves. Instead, McKinley's curves assess damage or stimulation by noting that the earliest wellbore-storage-distorted data are dominated by the effective near-well transmissibility $(kh/\mu)_{wb}$; thus, a type-curve match of the earliest data in a test should allow calculation of this quantity. Later, after wellbore storage distortion has diminished, the pressure/time behavior is governed by the transmissibility in the formation, $kh/\mu$; this quantity also can be estimated from a type-curve match – but for the later data only.

6. McKinley approximated boundary effects by plotting the simulator-generated type curves for about one-fifth log cycle beyond the end of wellbore storage distortion (where the curve has the same shape as for $C_s = 0$) and then making the curves vertical. This step roughly simulates drainage conditions of 40-acre spacing. Note that this gives the curves early-, middle-, and late-time regions – but remember that the curves were designed to be used primarily to analyze *early-time* data. When the curves are applied to drawdown tests, they *must* be applied to early-time data only; they do not properly simulate boundary effects in drawdown tests.

### Use of McKinley's Type Curves

Before providing a step-by-step procedure for using McKinley's type curves, we note that he actually prepared three different curves: one for the time range 0.01 to 10 minutes; one for 1 to 1,000 minutes; and one for $10^3$ to $10^6$ minutes. The curve for the time range 1 to 1,000 minutes is by far the most useful; accordingly, it is the only one provided with this text. The complete set is provided with Ref. 1.

The steps for using McKinley's curves follow.

1. Plot $\Delta t$ (minutes) as ordinate vs. $\Delta p = p_i - p_{wf}$ (or $p_{ws} - p_{wf}$) as abscissa on 3×5 cycle log-log paper the same size as McKinley's type curve if undistorted type curves are used. Otherwise, use tracing paper for actual test data plotting. The time range on the axis should correspond exactly to one of the type curves (e.g., it should span the time range of

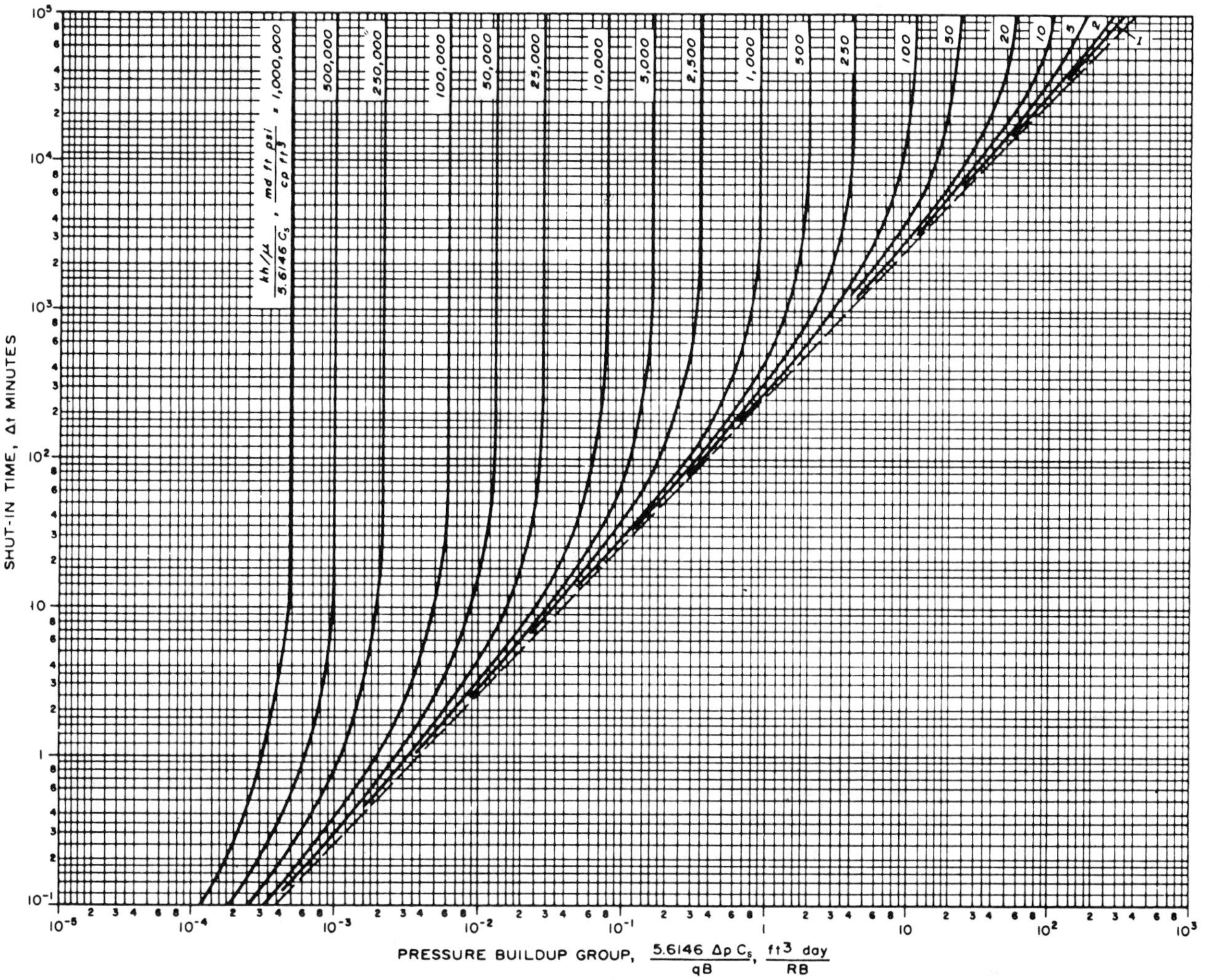


Fig. 4.6 – McKinley's type curve.

1 to 1,000 minutes *or* 0.01 to 10 minutes *or* $10^3$ to $10^6$ minutes).

2. Match the time axis of the test data plot with one of McKinley's. Move the data along the plot horizontally (*no vertical shifting allowed*) until the earliest data fall along one of the type curves.

3. Record the parameter value $(kh/\mu)/5.615\ C_s$ for the correct type curve.

4. Choose a data match point (any $\Delta p$ from the test graph paper and the corresponding value of $5.615\ \Delta pC_s/qB$ from the type curve).

5. Determine the wellbore storage constant $C_s$ from values of $\Delta p = \Delta p_{MP}$ and $5.615\ \Delta pC_s/qB = (5.615\ \Delta pC_s/qB)_{MP}$ at the ma.ch point:

$$C_s = \frac{(5.615\ \Delta pC_s/qB)_{MP}}{\Delta p_{MP}} \times \frac{qB}{5.615}.$$

6. Calculate near-well transmissibility, $(kh/\mu)_{wb}$, from the parameter value recorded in Step 3 and the wellbore storage constant determined in Step 5:

$$(kh/\mu)_{wb} = \left(\frac{kh/\mu}{5.615\ C_s}\right)_{wb} \times 5.615\ C_s.$$

7. If the data trend away from the type curve providing the earliest fit (indicating that formation transmissibility is different from effective near-well transmissibility), shift the data plot horizontally to find another type curve that better fits the later data. A shift to a higher value of $(kh/\mu)/5.615\ C_s$ indicates damage; a shift to a lower value indicates stimulation.

8. Calculate formation transmissibility:

$$(kh/\mu)_f = \left(\frac{kh/\mu}{5.615\ C_s}\right)_{\text{Step 7}} \times (5.615\ C_s)_{\text{Step 5}}.$$

Note that we *do not* find a new pressure match point to redetermine $C_s$; $C_s$ is found once and for all in Step 5. In fact, if only data reflecting formation transmissibility (after wellbore storage distortion has disappeared) are analyzed, error will result using the McKinley method. (However, conventional methods work here, so no problem arises.) The match point must be found with early, wellbore-storage-distorted data (Figs. 4.7 and 4.8).

Flow efficiency also can be estimated fairly directly from the data plotted for use with McKinley's type curves.[6] The definition of flow efficiency, $E$, is

$$E \simeq \frac{p^* - p_{wf} - \Delta p_s}{p^* - p_{wf}} = \frac{\Delta p^* - \Delta p_s}{\Delta p^*}.$$

The quantities $\Delta p^*$ and $\Delta p_s$ can be estimated from

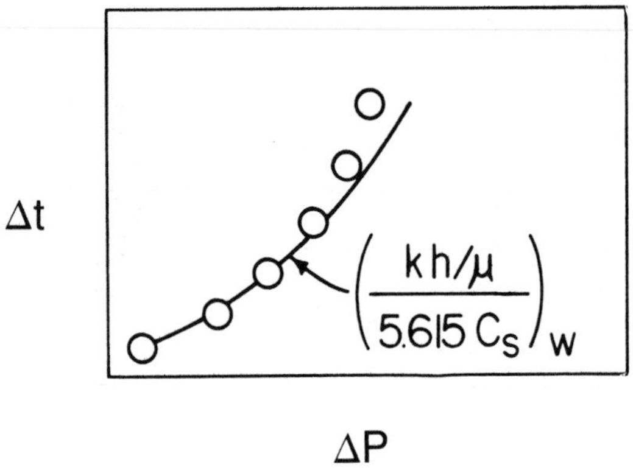


**Fig. 4.7** – Early data fit on McKinley's type curve.

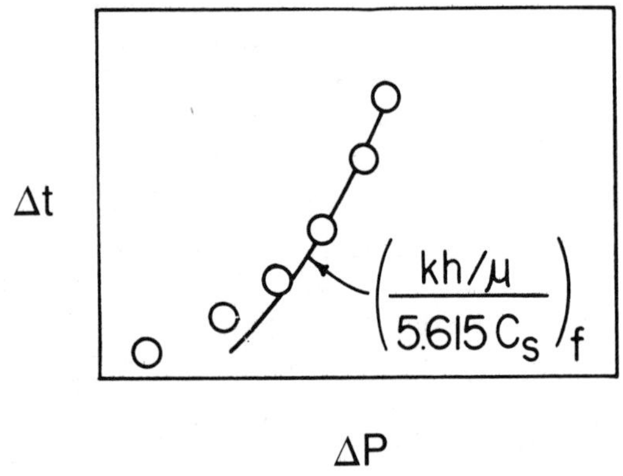


**Fig. 4.8** – Later data fit on McKinley's type curve.

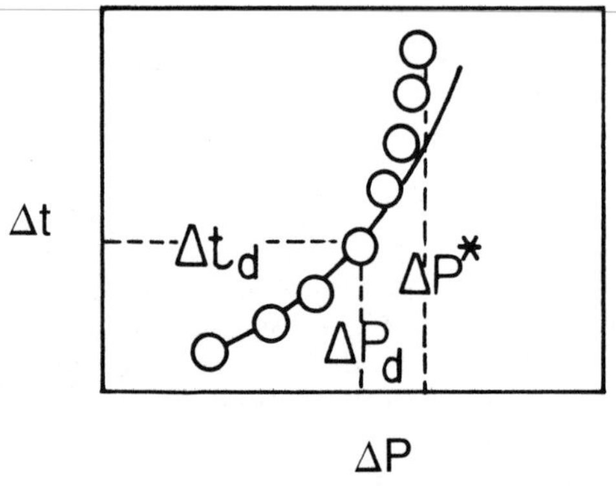


**Fig. 4.9** – Data for flow efficiency calculation from McKinley's type curve.

the McKinley type curves in the following manner (Fig. 4.9):

1. $\Delta p^*$ is the vertical asymptote approached by $\Delta p$ in the McKinley plot.

2. $\Delta p_s$ can be calculated from $\Delta p_d$, the time at which the actual test data depart from the earliest-fitting type curve. Picking a time of departure is subjective, so no great accuracy is assured for this reason alone.

McKinley[6] states that $\Delta p_s$ and $\Delta p_d$ are related by

$$\Delta p_s = \left(1 - \frac{k_{wb}}{k_f}\right)\Delta p_d .$$

3. Thus, $E$ can be calculated:

$$E \simeq \frac{\Delta p^* - \Delta p_s}{\Delta p^*} .$$

## *Example 4.2 – Drawdown Test Analysis Using McKinley's Type Curves*

**Problem.** Estimate near-well and formation permeability and flow efficiency using the data presented in Example 4.1 from a drawdown test on an oil well.

**Solution.** We first prepare those data, most of which lie in the time range 1 minute $< t <$ 1,000 minutes for plotting as $\Delta t$ (minutes) vs. $\Delta p = p_i - p_{wf}$ (Table 4.3). From the data plot (Fig. 4.10) and the match with the best-fitting McKinley curve for the early data,

$$\left(\frac{kh}{\mu}\right)_{wb} \times \frac{1}{5.615\, C_s} = 5{,}000.$$

We also note that the data depart from the best-fitting curve (early time) at $\Delta t_d \simeq 100$ minutes; here, $\Delta p_d \simeq 1{,}180$ psi.

A match point for the early fit is $\Delta p_{MP} = 107$ psi when $5.615\ \Delta\ pC_s/qB = 0.010$. The best fit of the later data is for $(kh/\mu)_f(1/5.615\ C_s) \simeq 10{,}000$. Thus, from the match-point data,

$$C_s = \left(\frac{5.615\ \Delta p C_s}{qB}\right)_{MP}\left(\frac{1}{\Delta p}\right)_{MP} \times \frac{qB}{5.615}$$

$$= (0.010)\left(\frac{1}{107}\right)\left(\frac{500 \times 1.2}{5.615}\right)$$

$$= 0.01 \text{ RB/psi}$$

The near-well transmissibility and apparent permeability, $k_{wL}$, are then

$$\left(\frac{kh}{\mu}\right)_{wb} = \left(\frac{kh/\mu}{5.615\, C_s}\right)_{wb}(5.615\, C_s)$$

$$= (5{,}000)(5.615)(0.010)$$

$$= 281 \text{ md-ft/cp}.$$

Then,

$$k_{wb} = \frac{(281)(0.8)}{(56)} = 4.01 \text{ md}.$$

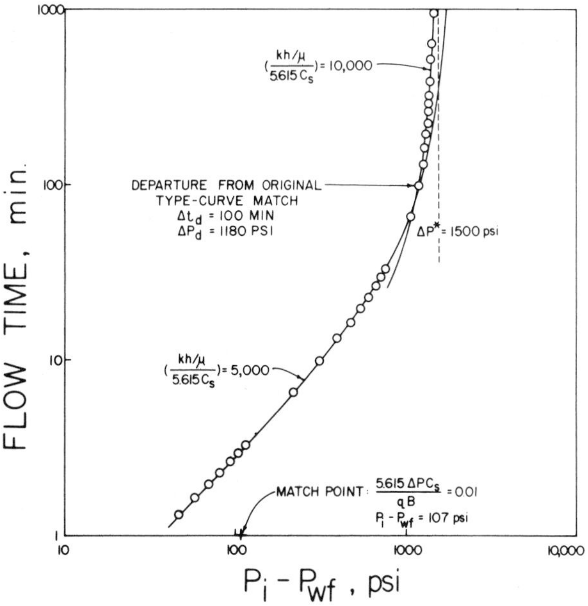


**Fig. 4.10** – Drawdown test analysis with McKinley's type curve.

**TABLE 4.3 – DRAWDOWN DATA FOR McKINLEY CURVE ANALYSIS**

| $\Delta t$ (minutes) | $p_i - p_{wf}$ (psi) | $\Delta t$ (minutes) | $p_i - p_{wf}$ (psi) | $\Delta t$ (minutes) | $p_i - p_{wf}$ (psi) |
|---|---|---|---|---|---|
| 1.31 | 47 | 16.4 | 464 | 197 | 1,288 |
| 1.64 | 58 | 19.7 | 531 | 229 | 1,304 |
| 1.97 | 70 | 22.9 | 592 | 262 | 1,316 |
| 2.29 | 81 | 26.2 | 648 | 295 | 1,326 |
| 2.62 | 92 | 29.5 | 698 | 328 | 1,335 |
| 2.95 | 103 | 32.8 | 744 | 393 | 1,349 |
| 3.28 | 114 | 65.4 | 1,048 | 524 | 1,370 |
| 6.54 | 215 | 98.4 | 1,172 | 654 | 1,386 |
| 9.84 | 307 | 131 | 1,232 | 984 | 1,413 |
| 13.1 | 389 | 164 | 1,266 | | |

The formation transmissibility and permeability are

$$\left(\frac{kh}{\mu}\right)_f = \frac{(kh/\mu \times 5.615\, C_s)_f}{(kh/\mu \times 5.615\, C_s)_{wb}} \times \left(\frac{kh}{\mu}\right)_{wb}$$

$$= \frac{10{,}000}{5{,}000} \times 281$$

$$= 562 \text{ md-ft/cp},$$

and

$$k_f = \left(\frac{kh}{\mu}\right)_f \times \frac{\mu}{h}$$

$$= (562)\left(\frac{0.8}{56}\right)$$

$$= 8.03 \text{ md}.$$

Flow efficiency becomes

$\Delta p^* \simeq 1{,}500$ psi (Fig. 4.10),

$$\Delta p_s = \left(1 - \frac{k_{wb}}{k_f}\right) \Delta p_d$$

$$\Delta p_s = \left(1 - \frac{4.01}{8.03}\right)(1{,}180) = 590 \text{ psi},$$

$$E \simeq \frac{1{,}500 - 590}{1{,}500} = 0.607.$$

## 4.5 Gringarten *et al.*[7] Type Curves for Fractured Wells

Gringarten *et al.*[7] developed type curves for hydraulically fractured wells in which vertical fractures with two equal-length wings were created. The curves discussed in this section assume *uniform flux* into the fracture (same flow rate per unit cross-sectional area of fracture from wellbore to fracture tip). High fracture conductivity is required to achieve uniform flux, but this is not identical to an infinitely conductive fracture (no pressure drop from fracture tip to wellbore), as Gringarten *et al.* demonstrated.[7]

The study was made for finite reservoirs (i.e., boundary effects become important at later times in the test). The reservoir is assumed to be at uniform pressure, $p_i$, initially. The type curve (Fig. 4.11), developed for a constant-rate drawdown test for a slightly compressible liquid, also can be used for buildup tests (for $\Delta t_{max} \leq 0.1\, t_p$) and for gas wells* using the modifications discussed earlier. Wellbore storage effects are ignored.

All the dimensionless variables and parameters considered important are taken into account in Fig. 4.11, which is a log-log plot of $p_D$ vs. $t_D r_w^2/L_f^2$ with parameter $x_e/L_f$. In these parameters, $L_f$ is the fracture half-length and $x_e$ is the distance from the well to the side of the square drainage area in which it is assumed to be centered. Dimensionless pressure has the usual definition,

$$p_D = \frac{kh(p_i - p_{wf})}{141.2\, qB\mu} \text{ (drawdown test)},$$

and

$$\frac{t_D r_w^2}{L_f^2} = \frac{0.000264\, kt}{\phi \mu c_t L_f^2} = t_{DL_f}. \quad \ldots\ldots\ldots\ldots (4.17)$$

Several features of Fig. 4.11 are of interest:

1. The slope of the log-log plot is 1/2 up to $t_{DL_f} \simeq 0.16$ for $x_e/L_f > 1$. This is linear flow. We have shown that, in linear flow,

*This statement may be an oversimplification. Some gas wells exhibit time-dependent non-Darcy flow in the fracture, unlike liquids.[12]

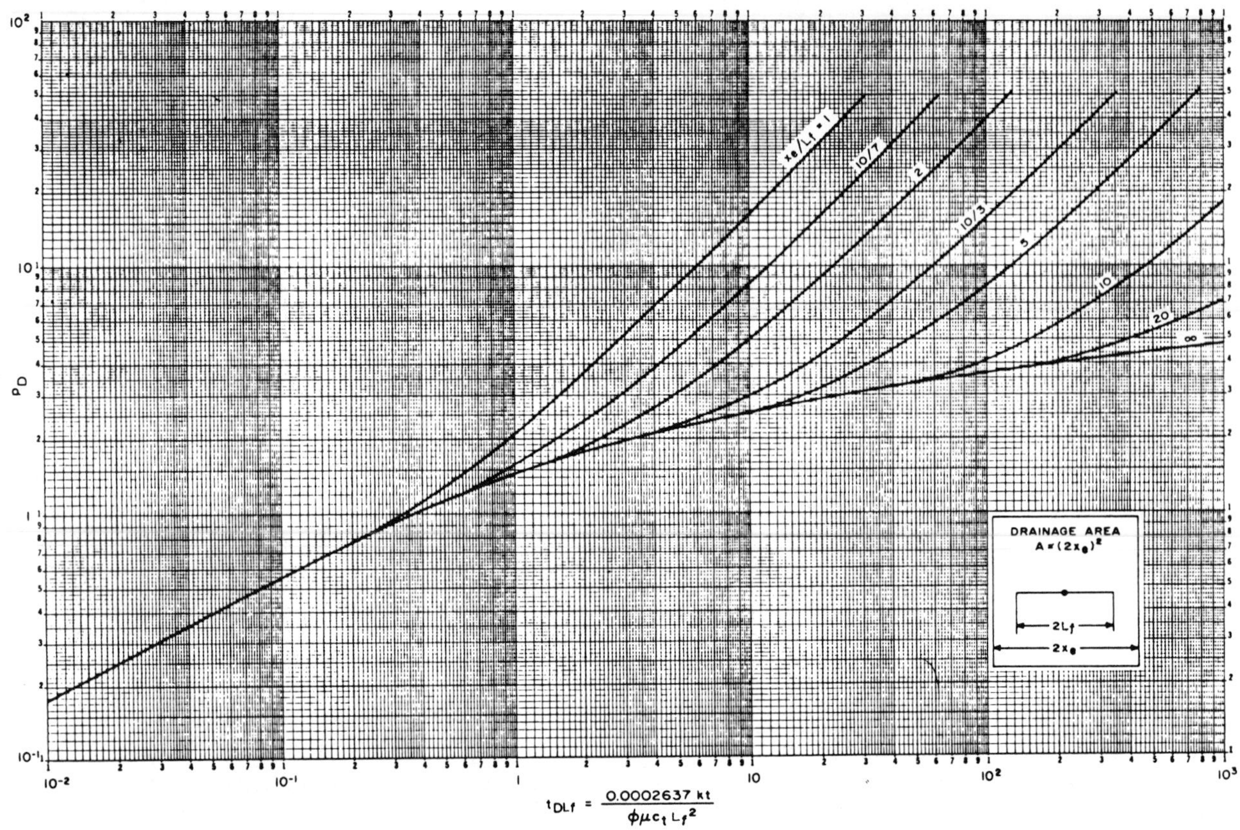


Fig. 4.11 – Gringarten *et al.* type curve for vertically-fractured well centered in closed square, no wellbore storage, uniform flux.

$$p_i - p_{wf} = ct^{1/2},$$

or

$$p_D = c' t_{DL_f}{}^{1/2}.$$

Then,

$$\log p_D = \log c' + 1/2 \log t_{DL_f}.$$

2. Although not apparent on a log-log plot, a semilog plot of the data in Fig. 4.11 ($p_D$ vs. log $t_{DL_f}$) is a straight line, signifying radial flow when, for $x_e/L_f > 5$, $t_{DL_f} \simeq 2$. The straight-line terminates, of course, when boundary effects become important, but a match of actual test data with Fig. 4.11 can show the amount of data in the radial flow region (and, thus, can be analyzed for permeability by the conventional $p_{wf}$ vs. log $t$ or Horner plots). Fig. 4.11 thus combines, in a single graph, the linear flow and radial flow regions (and a region with neither), boundary effects, and the effect of various fracture lengths. *If* fracture conductivity is high and constant throughout the test and *if* wellbore storage has negligible effect on earliest data, then this figure allows a rather complete analysis of a hydraulically fractured well – specifically, estimation of fracture length, $L_f$, and formation permeability, $k$. The method is frequently superior to the nontype-curve methods discussed in Chap. 2.

Steps in use of Fig. 4.11 as a type curve for test analysis include the following.

1. Plot $(p_i - p_{wf})$ (drawdown test) or $(p_{ws} - p_{wf})$ (buildup test) on the ordinate vs. $t$ (drawdown test) or $\Delta t_e$ (buildup test) on the abscissa on a 3×5 cycle log-log paper if an undistorted version of the type curve is available. Otherwise, use tracing paper.

2. Select the best match by sliding the actual test data plot both horizontally and vertically.

3. Note the value of the match points $[(p_D)_{MP}, (p_i - p_{wf})_{MP}]$ and $[(t_{DL_f})_{MP}, t_{MP}]$.

4. Estimate formation permeability from the pressure match point:

$$k = 141.2 \frac{qB\mu}{h} \frac{(p_D)_{MP}}{(p_i - p_{wf})_{MP}}.$$

5. Estimate fracture length from the time match point:

$$L_f = \left[ \frac{0.000264\, kt_{MP}}{\phi\mu c_t (t_{DL_f})_{MP}} \right]^{1/2}. \quad \text{.......... (4.18)}$$

Three useful checks are sometimes possible:

1. If a half-slope (linear flow) region appears on the test data plot, replot data from the region as $p_{wf}$ (or $p_{ws}$) vs. $\sqrt{t}$ (or $\sqrt{\Delta t}$); from the slope $m_L$ and linear flow theory,

$$L_f\sqrt{k} = \frac{4.064\, qB}{hm_L} \sqrt{\frac{\mu}{\phi c_t}},$$

which should agree with the result from the type-curve analysis.

2. If a radial flow region appears (before boundary effects become important – that is, before the data deviate from the $x_e/L_f = \infty$ curve), a plot of $p_{wf}$

**TABLE 4.4 – FRACTURED WELL BUILDUP TEST DATA**

| $\Delta t$ (hours) | $p_{ws}-p_{wf}$ (psi) | $\Delta t$ (hours) | $p_{ws}-p_{wf}$ (psi) |
|---|---|---|---|
| 0 | 0 | 0.75 | 89 |
| 0.0833 | 31 | 0.833 | 100 |
| 0.167 | 43 | 0.917 | 100 |
| 0.250 | 54 | 1.00 | 100 |
| 0.330 | 66 | 1.25 | 114 |
| 0.417 | 66 | 2.00 | 136 |
| 0.500 | 72 | 2.50 | 159 |
| 0.583 | 78 | 4.00 | 181 |
| 0.667 | 83 | 4.75 | 206 |
| | | 6.00 | 218 |

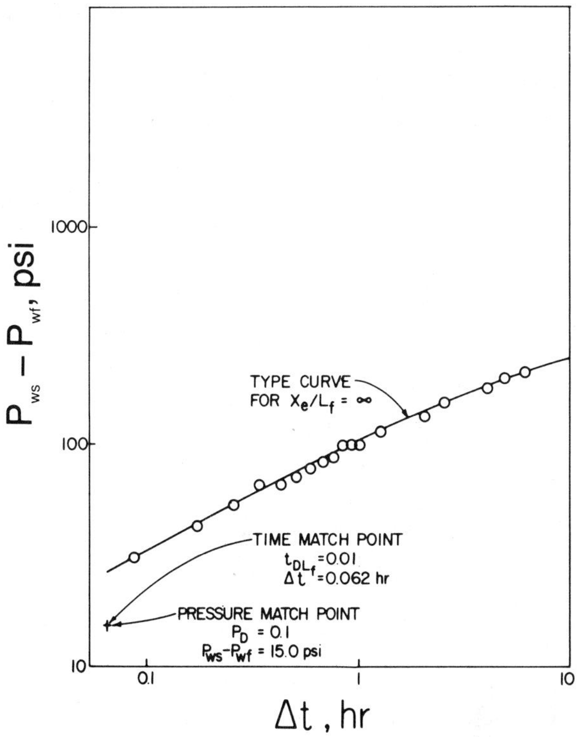


**Fig. 4.12** – Buildup test analysis for vertically-fractured well with Gringarten type curve.

vs. log $t$ [$p_{ws}$ vs. log $\Delta t$ or log $(t_p+\Delta t)/\Delta t$] should show that $k=162.6\, qB\mu/mh$, in agreement with type-curve analysis.

3. If a well proves to be in a finite-acting reservoir, it may be possible to estimate $x_e$ from a matching parameter, $x_e/L_f$, to compare with the known (or assumed) value of $x_e$ to check the quality of the match.

---

## *Example 4.3 – Buildup Test Analysis for a Vertically Fractured Well*

**Problem.** Gringarten *et al.*[13] presented buildup test data for a well believed to be fractured vertically. From these data, presented below and in Table 4.4, estimate fracture length and formation permeability. Producing time, $t_p$, was significantly greater than maximum shut-in time, so that $\Delta t \cong \Delta t_e$.

$$q = 2{,}750 \text{ STB/D},$$
$$\mu = 0.23 \text{ cp},$$
$$B = 1.76 \text{ RB/STB},$$
$$h = 230 \text{ ft},$$
$$\phi = 0.3, \text{ and}$$
$$c_t = 30\times10^{-6} \text{ psi}^{-1}.$$

**Solution.** Fig. 4.12 is a plot of $\Delta p = p_{ws}-p_{wf}$ vs. $\Delta t$. An adequate fit is characterized by the match points ($t=0.062$ hour, $t_{DL_f}=0.01$) and ($\Delta p=15.2$ psi, $p_D=0.1$). From the pressure match point,

$$k=141.2\frac{qB\mu}{h}\frac{(p_D)_{\text{MP}}}{(\Delta p)_{\text{MP}}}$$

$$=\frac{(141.2)(2{,}750)(1.76)(0.23)(0.1)}{(230)(15.2)}$$

$$=4.5 \text{ md}.$$

**TABLE 4.5 – CONSTANT-RATE DRAWDOWN TEST DATA**

| $t$ (hours) | $p_{wf}$ (psi) | $t$ (hours) | $p_{wf}$ (psi) | $t$ (hours) | $p_{wf}$ (psi) | $t$ (hours) | $p_{wf}$ (psi) |
|---|---|---|---|---|---|---|---|
| 0.0109 | 2,976 | 0.218 | 2,611 | 3.28 | 1,712 | 32.8 | 1,543 |
| 0.0164 | 2,964 | 0.273 | 2,536 | 3.82 | 1,696 | 38.2 | 1,533 |
| 0.0218 | 2,953 | 0.328 | 2,469 | 4.37 | 1,684 | 43.7 | 1,525 |
| 0.0273 | 2,942 | 0.382 | 2,408 | 4.91 | 1,674 | 49.1 | 1,517 |
| 0.0328 | 2,930 | 0.437 | 2,352 | 5.46 | 1,665 | 54.6 | 1,511 |
| 0.0382 | 2,919 | 0.491 | 2,302 | 6.55 | 1,651 | 65.5 | 1,500 |
| 0.0437 | 2,908 | 0.546 | 2,256 | 8.74 | 1,630 | 87.4 | 1,482 |
| 0.0491 | 2,897 | 1.09 | 1,952 | 10.9 | 1,614 | 109.2 | 1,468 |
| 0.0546 | 2,886 | 1.64 | 1,828 | 16.4 | 1,587 | 163.8 | 1,440 |
| 0.109 | 2,785 | 2.18 | 1,768 | 21.8 | 1,568 | 218.4 | 1,416 |
| 0.164 | 2,693 | 2.73 | 1,734 | 27.3 | 1,554 | 273.0 | 1,393 |
| | | | | | | 327.6 | 1,370 |

TABLE 4.6 – BUILDUP TEST DATA

| $\Delta t$ (hours) | $p_{ws}$ (psi) | $\Delta t$ (hours) | $p_{ws}$ (psi) | $\Delta t$ (hours) | $p_{ws}$ (psi) | $\Delta t$ (hours) | $p_{ws}$ (psi) |
|---|---|---|---|---|---|---|---|
| 0.0 | 1,370 | 0.546 | 2,114 | 5.46 | 2,703 | 43.7 | 2,828 |
| 0.109 | 1,586 | 1.09 | 2,418 | 6.55 | 2,717 | 49.1 | 2,833 |
| 0.164 | 1,677 | 1.64 | 2,542 | 8.74 | 2,737 | 54.6 | 2,837 |
| 0.218 | 1,760 | 2.18 | 2,602 | 10.9 | 2,752 | 65.5 | 2,844 |
| 0.273 | 1,834 | 2.73 | 2,635 | 16.4 | 2,777 | 87.4 | 2,853 |
| 0.328 | 1,901 | 3.28 | 2,657 | 21.8 | 2,793 | 109.2 | 2,858 |
| 0.382 | 1,963 | 3.82 | 2,673 | 27.3 | 2,805 | 163.8 | 2,863 |
| 0.437 | 2,018 | 4.37 | 2,685 | 32.8 | 2,814 | 218.4 | 2,864 |
| 0.491 | 2,068 | 4.91 | 2,695 | 38.2 | 2,822 | | |

From the time match point,

$$L_f = \left( \frac{0.000264\, k \Delta t_{\mathrm{MP}}}{\phi \mu c_t (t_{DL_f})_{\mathrm{MP}}} \right)^{1/2}$$

$$= \left[ \frac{(0.000264)(4.5)(0.062)}{(0.3)(0.23)(3 \times 10^{-5})(0.01)} \right]^{1/2}$$

$$= 59.7 \text{ ft.}$$

## Exercises

Examples 4.1 and 4.2 were based on a portion of the following data for a drawdown test followed by a pressure buildup test.

$q$ = 500 STB/D (constant),
$\phi$ = 0.2,
$\mu$ = 0.8 cp,
$c_t$ = $10 \times 10^{-6}$ psi$^{-1}$,
$p_i$ = 3,000 psi,
$r_w$ = 0.3 ft,
$h$ = 56 ft,
$B_o$ = 1.2 RB/STB,
$A_{wb}$ = 0.022 sq ft,
$\rho$ = 50 lbm/cu ft,
single-phase liquid,
liquid/gas interface in wellbore, and
well centered in a cylindrical drainage area with
$r_e$ = 1,000 ft.

The drawdown test data are presented in Table 4.5; buildup test data ($t_p$ = 327.6 hours) are given in Table 4.6.

4.1 Using conventional analysis techniques ($p_{wf}$ vs. log $t$ plot in ETR and MTR, $p_{wf}$ vs. $t$ plot in LTR), estimate $k$, $s$, $E$, $t_{wbs}$, $V_p$, $r_e$ (assume cylindrical reservoir), and $r_i$ at the beginning and end of the MTR for the drawdown test.

4.2 Analyze the drawdown test data as completely as possible using Ramey's type curves. Can data in the LTR be analyzed with these curves? Why?

4.3 Analyze the drawdown test data using McKinley's type curves. Estimate $k$, $k_{wb}$, and $E$. Can the data in the LTR be analyzed with these curves? Why?

4.4 Analyze the buildup test using the Horner plotting technique. Estimate (1) $k$, $s$, $E$, $t_{wbs}$, and $r_i$ at the beginning and end of the MTR, (2) $\bar{p}$ from the MBH and modified Muskat techniques, and (3) reservoir pore volume (using $\bar{p}$ before and after the drawdown test).

4.5 Analyze the buildup test as completely as possible using Ramey's type curves. Is there a shut-in time, $\Delta t_{\max}$, beyond which the type-curve technique should not be used? Why?

4.6 Analyze the buildup test using McKinley's type curves. Estimate $k_{wb}$, $k$, and $E$. Is there a shut-in time beyond which the type-curve technique should not be used? Why?

4.7 In the buildup test analyzed in Example 4.3, does a linear flow region appear? If so, analyze the data using the conventional equations for linear flow in a reservoir. Does a radial flow region appear? If so, analyze the data using conventional methods.

4.8 A drawdown test was run in a vertically fractured oil well; the results are given in Table 4.7. Using the Gringarten *et al.* type curve, estimate fracture length and formation permeability. Identify linear flow and radial flow regions and verify the type-curve analysis with conventional analysis of these regions. As part of the conventional analysis of the radial flow region, estimate $r_i$ at the beginning and end of the MTR and estimate fracture length from skin-factor calculation. The test data were as follows.

TABLE 4.7 – FRACTURED WELL DRAWDOWN TEST DATA

| $t$ (hours) | $p_{wf}$ (psi) | $t$ (hours) | $p_{wf}$ (psi) | $t$ (hours) | $p_{wf}$ (psi) |
|---|---|---|---|---|---|
| 0 | 4,000 | 1.5 | 3,932 | 20 | 3,823 |
| 0.15 | 3,982 | 2.0 | 3,922 | 30 | 3,803 |
| 0.2 | 3,978 | 3.0 | 3,907 | 40 | 3,789 |
| 0.3 | 3,975 | 4.0 | 3,896 | 50 | 3,778 |
| 0.4 | 3,969 | 5.0 | 3,886 | 60 | 3,768 |
| 0.5 | 3,965 | 6.0 | 3,879 | 80 | 3,755 |
| 0.6 | 3,960 | 8.0 | 3,866 | 100 | 3,744 |
| 0.8 | 3,957 | 10 | 3,856 | | |
| 1.0 | 3,950 | 15 | 3,837 | | |

$q$ = 200 STB/D (constant),
$B$ = 1.288 RB/STB,
$h$ = 12 ft,
$\phi$ = 0.1,
$\mu$ = 0.5 cp,
$c_t$ = $20 \times 10^{-6}$ psi$^{-1}$, and
wellbore unloading distortion negligible at all times.

## References


1. Earlougher, R.C Jr.: *Advances in Well Test Analysis*, Monograph Series, SPE, Dallas (1977) **5**.
2. Ramey, H.J. Jr.: "Short-Time Well Test Data Interpretation in the Presence of Skin Effect and Wellbore Storage," *J. Pet. Tech.* (Jan. 1970) 97-104; *Trans.*, AIME, **249**.
3. Agarwal, R.G., Al-Hussainy, R., and Ramey, H.J. Jr.: "An Investigation of Wellbore Storage and Skin Effect in Unsteady Liquid Flow: I. Analytical Treatment," *Soc. Pet. Eng. J.* (Sept. 1970) 279-290; *Trans.*, AIME, **249**.
4. Wattenbarger, R.A. and Ramey, H.J. Jr.: "An Investigation of Wellbore Storage and Skin Effect in Unsteady Liquid Flow: II. Finite-Difference Treatment," *Soc. Pet. Eng. J.* (Sept. 1970) 291-297; *Trans.*, AIME, **249**.
5. McKinley, R.M.: "Wellbore Transmissibility From Afterflow-Dominated Pressure Buildup Data," *J. Pet. Tech.* (July 1971) 863-872; *Trans.*, AIME, **251**.
6. McKinley, R.M.: "Estimating Flow Efficiency From Afterflow-Distorted Pressure Buildup Data," *J. Pet. Tech.* (June 1974) 696-697.
7. Gringarten, A.C., Ramey, H.J. Jr., and Raghavan, R.: "Unsteady-State Pressure Distributions Created by a Well With a Single Infinite-Conductivity Vertical Fracture," *Soc. Pet. Eng. J.* (Aug. 1974) 347-360; *Trans.*, AIME, **257**.
8. Agarwal, R.G.: "A New Method To Account for Producing Time Effects When Drawdown Type Curves are Used to Analyze Pressure Buildup and Other Test Data," paper SPE 9289 presented at the SPE 55th Annual Technical Conference and Exhibition, held in Dallas, Sept. 21-24, 1980.
9. Ramey, H.J. Jr.: "Practical Use of Modern Well Test Analysis," paper SPE 5878 presented at the SPE-AIME 51st Annual Technical Conference and Exhibition, New Orleans, Oct. 3-6, 1976.
10. Al-Hussainy, R., Ramey, H.J. Jr., and Crawford, P.B.: "The Flow of Real Gases Through Porous Media," *J. Pet. Tech.* (May 1966) 624-636; *Trans.*, AIME, **237**.
11. Wattenbarger, R.A. and Ramey, H.J. Jr.: "Gas Well Testing With Turbulence, Damage, and Wellbore Storage," *Trans.*, AIME, (1960) **243**, 877-887.
12. Holditch, S.A. and Morse, R.A.: "The Effects of Non-Darcy Flow on the Behavior of Hydraulically Fractured Gas Wells," *J. Pet. Tech.* (Oct. 1976) 1169-1178.
13. Gringarten, A.C., Ramey, H.J. Jr., and Raghavan, R.: "Pressure Analysis for Fractured Wells," paper SPE 4051 presented at the SPE-AIME 47th Annual Fall Meeting, San Antonio, Oct. 8-11, 1972.

# Chapter 5
# Gas Well Testing

## 5.1 Introduction

This chapter discusses deliverability tests of gas wells. The discussion includes basic theory of transient and pseudosteady-state flow of gases, expressed in terms of the pseudopressure $\psi(p)$ and of approximations to the pseudopressure approach that are valid at high and low pressures. This is followed by an examination of flow-after-flow, isochronal, and modified isochronal deliverability tests. The chapter concludes with an introduction to the application of pseudopressure in gas well test analysis.

## 5.2 Basic Theory of Gas Flow in Reservoirs

Investigations[1,2] have shown that gas flow in infinite-acting reservoirs can be expressed by an equation similar to that for flow of slightly compressible liquids if pseudopressure $\psi(p)$ is used instead of pressure:

$$\psi(p_{wf}) = \psi(p_i) + 50{,}300 \frac{p_{sc}}{T_{sc}} \frac{q_g T}{kh} \left[ 1.151 \cdot \log\left( \frac{1{,}688\, \phi \mu_i c_{ti} r_w^2}{kt} \right) - \left(s + D|q_g|\right) \right], \quad \text{(5.1)}$$

where the pseudopressure is defined by the integral

$$\psi(p) = 2\int_{p_\beta}^{p} \frac{p}{\mu z}\, dp. \quad \text{(5.2)}$$

The term $D|q_g|$ reflects a non-Darcy flow pressure loss – i.e., it takes into account the fact that, at high velocities near the producing well (characteristic of large gas production rates), Darcy's law does not predict correctly the relationship between flow rate and pressure drop. As a first approximation, this additional pressure drop can be added to the Darcy's law pressure drop, just as pressure drop across the altered zone is, and $D$ can be considered constant. The absolute value of $q_g$, $|q_g|$, is used so that the term $D|q_g|$ is positive for either production or injection.

For stabilized flow[3] $(r_i \geq r_e)$,

$$\psi(p_{wf}) = \psi(\bar{p}) - 50{,}300 \frac{p_{sc}}{T_{sc}} \frac{q_g T}{kh} \left[ \ln\left( \frac{r_e}{r_w} \right) - 0.75 + s + D|q_g| \right]. \quad \text{(5.3)}$$

Eqs. 5.1 and 5.3 provide the basis for analysis of gas well tests. As noted in Sec. 2.11, for $p > 3{,}000$ psi, these equations assume a simpler form (in terms of pressure, $p$); for $p < 2{,}000$ psi, they assume another simple form (in terms of $p^2$). Thus, we can develop procedures for analyzing gas well tests with equations in terms of $\psi(p)$, $p$, and $p^2$. In most of this chapter, our equations will be written in terms of $p^2$ – not because $p^2$ is more generally applicable or more accurate (the equations in $\psi$ best fit this role), but because the $p^2$ equations illustrate the general method and permit easier comparison with older methods of gas well test analysis that still are used widely.

Before developing the equations, let us generalize Eq. 5.1 to model a drawdown test starting from *any* uniform drainage-area pressure ($\bar{p}$) that may be much lower than initial pressure ($p_i$) after years of production:

$$\psi(p_{wf}) = \psi(\bar{p}) + 50{,}300 \frac{p_{sc}}{T_{sc}} \frac{q_g T}{kh} \left[ 1.151 \cdot \log\left( \frac{1{,}688\, \phi \mu_{\bar{p}} c_{t\bar{p}} r_w^2}{kt} \right) - \left(s + D|q_g|\right) \right], \quad \text{(5.4)}$$

where $p = \bar{p}$ for all $r$ at $t_p = 0$. For $p < 2{,}000$ psia, $\mu z_g \simeq \text{constant} \simeq \mu_{\bar{p}} z_{\bar{p}g}$ for most gases; in this case,

$$\psi(p) = \frac{2}{\mu_{\bar{p}} z_{\bar{p}g}} \left( \frac{p^2}{2} - \frac{p_\beta^2}{2} \right).$$

Substituting into Eq. 5.4,

$$p_{wf}^2 = \bar{p}^2 + 1{,}637 \frac{q_g \mu_{\bar{p}} z_{\bar{p}g} T}{kh} \left[ \log\left( \frac{1{,}688\, \phi \mu_{\bar{p}} c_{t\bar{p}}}{kt_p} \right) \right.$$

$$-\left(\frac{s+D|q_g|}{1.151}\right)\Bigg]. \quad \ldots\ldots\ldots\ldots\ldots\ldots (5.5)$$

For stabilized flow,

$$p_{wf}^2 = \bar{p}^2 - 1{,}422\frac{q_g \mu_{\bar{p}} z_{\bar{p}g} T}{kh}\left[\ln\left(\frac{r_e}{r_w}\right) - 0.75 + s + D|q_g|\right]. \quad \ldots\ldots\ldots\ldots (5.6)$$

Eq. 5.6 is a complete deliverability equation. Given a value of flowing BHP, $p_{wf}$, corresponding to a given pipeline or backpressure, we can estimate the rate $q_g$ at which the well will deliver gas. However, certain parameters must be determined before the equation can be used in this way:

1. The well flowed at rate $q_g$ until $r_i \geq r_e$ (stabilized flow). In this case, note that Eq. 5.6 has the form

$$\bar{p}^2 - p_{wf}^2 = aq_g + bq_g^2, \quad \ldots\ldots\ldots\ldots\ldots (5.7)$$

where

$$a = 1{,}422\frac{\mu_{\bar{p}} z_{\bar{p}g} T}{kh}\left[\ln\left(\frac{r_e}{r_w}\right) - 0.75 + s\right], \quad \ldots\ldots (5.8)$$

and

$$b = 1{,}422\frac{\mu_{\bar{p}} z_{\bar{p}g} T}{kh}D. \quad \ldots\ldots\ldots\ldots\ldots\ldots\ldots\ldots (5.9)$$

The constants $a$ and $b$ can be determined from flow tests for at least two rates in which $q_g$ and the corresponding value of $p_{wf}$ are measured; $\bar{p}$ also must be known.

2. The well flowed for times such that $r_i \leq r_e$ (transient flow). In this case, we will need to estimate *kh, s,* and *D* from transient tests (drawdown or buildup) modeled by Eq. 5.5 (or some adaptation of it using superposition); these parameters then can be combined with known (or assumed) values of $\bar{p}$ and $r_e$ in Eq. 5.6 to provide deliverability estimates.

The gas flow rate $q_g$, used in Eqs. 5.1 through 5.7, should include all substances that are flowing in the vapor phase in the reservoir, with their volumes expressed at standard conditions. These substances include the gas produced as such at the surface, and condensate and liquid water produced at the surface that existed in the vapor phase in the reservoir. Calculation of the vapor equivalent of condensate is discussed in Appendix A of Ref. 4. Craft and Hawkins[5] summarize the calculation of the vapor equivalent of produced *fresh* (nonformation) water.

Most of the remainder of this chapter provides detailed information on testing procedures that lead to estimates of the parameters required to provide deliverability estimates. This discussion is based on recommendations in the ERCB gas well testing manual.[4]

## 5.3 Flow-After-Flow Tests

In this testing method, a well flows at a selected constant rate until pressure stabilizes – i.e., pseudosteady state is reached. The stabilized rate and pressure are recorded; rate is then changed and the

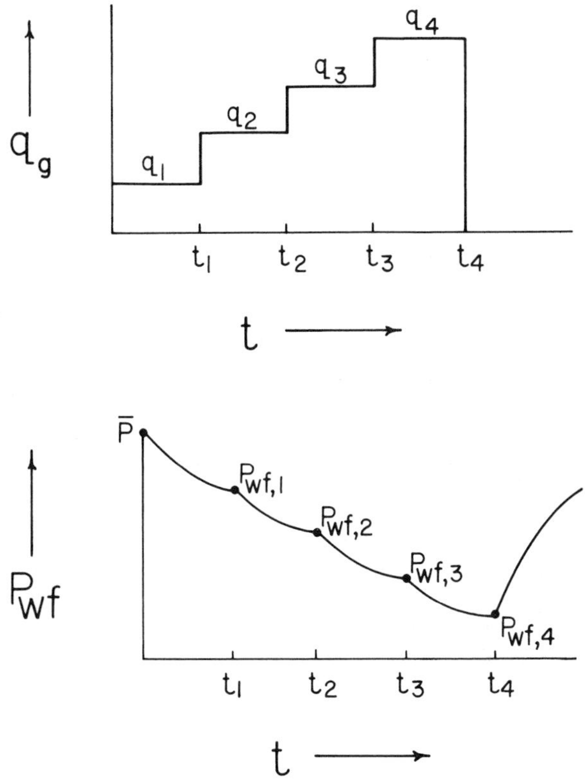


**Fig. 5.1** – Rates and pressures in flow-after-flow test.

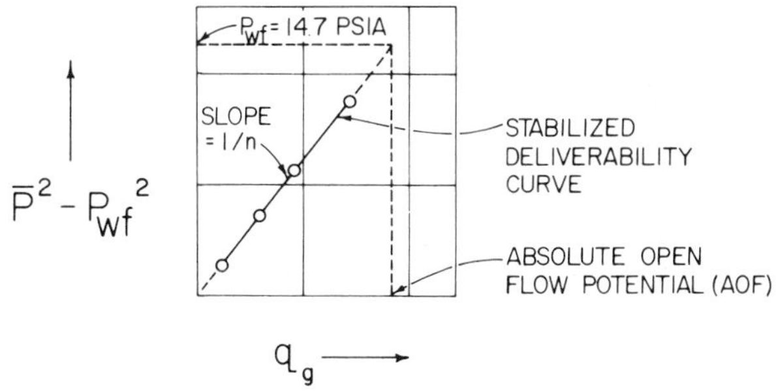


**Fig. 5.2** – Empirical deliverability plot for flow-after-flow test.

TABLE 5.1 – STABILIZED FLOW TEST DATA

| Test | $p_{wf}$ (psia) | $q_g$ (MMscf/D) |
|---|---|---|
| 1 | 403.1 | 4.288 |
| 2 | 394.0 | 9.265 |
| 3 | 378.5 | 15.552 |
| 4 | 362.6 | 20.177 |

TABLE 5.2 – STABILIZED FLOW TEST ANALYSIS

| $p_{wf}$ (psia) | $q_g$ (MMscf/D) | $\bar{p}^2 - p_{wf}{}^2$ (psia$^2$) | $(\bar{p}^2 - p^2_{wf})/q_g$ (psia$^2$/MMscf/D) |
|---|---|---|---|
| 408.2 | 0 | – | – |
| 403.1 | 4.288 | 4,138 | 964.9 |
| 394.0 | 9.265 | 11,391 | 1,229 |
| 378.5 | 15.552 | 23,365 | 1,502 |
| 362.6 | 20.177 | 35,148 | 1,742 |
| 14.7 | AOF | 166,411 | – |

well flows until the pressure stabilizes again at the new rate. The process is repeated for a total of three or four rates.

Rates and pressures in a typical test follow the pattern indicated in Fig. 5.1. Two fundamentally different techniques can be used to analyze these test data.

### Empirical Method

An empirical observation – with a rather tenuous theoretical basis – is that a plot of $\Delta p^2 = \bar{p}^2 - p_{wf}{}^2$ vs. $q_g$ (Fig. 5.2) on log-log paper is approximately a straight line for many wells in which the pseudosteady state is reached at each rate in a flow-after-flow test sequence. The equation of the line in this plot is

$$q_g = C(\bar{p}^2 - p_{wf}{}^2)^n. \quad \ldots\ldots\ldots\ldots\ldots\ldots (5.10)$$

This plot is an empirical correlation of field data. As in any other empirical correlation, there is substantial risk of error in extrapolating the plot a large distance beyond the region in which data were obtained. Unfortunately, such an extrapolation is frequently required. To estimate the absolute open flow potential (AOF) – the theoretical rate at which the well would produce if the flowing pressure $p_{wf}$ were atmospheric – it may be necessary to extrapolate the curve far beyond the range of test data. An AOF determined from such a lengthy extrapolation may be incorrect.

The constants $C$ and $n$ in Eq. 5.10 are not constants at all. They depend on fluid properties that are pressure (and, thus, time) dependent. Accordingly, if this type of deliverability curve is used, periodic retesting of the well will show changes in $C$ and perhaps in $n$.

We must emphasize that deliverability estimates based on this plot assume that pressures were stabilized ($r_i \geq r_e$) during the testing period used to construct the plot. If this is not the case, stabilized deliverability estimates from the curve can be highly misleading.

### Theoretical Method

Eq. 5.7 suggests that we plot $(\bar{p}^2 - p_{wf}{}^2)/q_g$ vs. $q_g$; the result (for *pseudosteady-state flow*) should be a straight line with slope $b$ and intercept $a$. Because this line has a sounder theoretical basis than the log $\Delta p^2$ – log $q_g$ plot, it should be possible to extrapolate it to determine AOF with less error and to correct deliverability estimates for changes in $\mu_{\bar{p}}$, $z_{\bar{p}g}$, etc., more readily.

## *Example 5.1 – Stabilized Flow Test Analysis*

**Problem.** The data in Table 5.1 were reported for a flow-after-flow (or four-point) test in Ref. 6. At each rate, pseudosteady state was reached. Initial (i.e., before the test) shut-in BHP, $\bar{p}$, was determined to be 408.2 psia. Estimate the AOF of the tested well using (1) the empirical plot and (2) the theoretical flow equation. In addition, plot deliverabilities estimated using the theoretical equation on the empirical curve plot.

**Solution.** We prepare a table of data (Table 5.2) to be plotted for both empirical and theoretical analyses.

1. *Empirical Method.* From a plot of $(\bar{p}^2 - p_{wf}{}^2)$ vs. $q_g$ on log-log paper, and extrapolation of this plot to $\bar{p}^2 - p_{wf}{}^2 = 166{,}411$ (where $p_{wf} = 0$ psig or 14.7 psia), AOF $\simeq$ 60 MMscf/D.

The slope of the curve, $1/n$, is

$$1/n = \frac{\log(\bar{p}^2 - p_{wf}{}^2)_2 - \log(\bar{p}^2 - p_{wf}{}^2)_1}{\log q_{g,2} - \log q_{g,1}}$$

$$= \frac{\log\left(\dfrac{10^5}{10^3}\right)}{\log\left(\dfrac{42.5}{1.77}\right)} = 1.449.$$

Thus, $n = 0.690$. Then,

$$C = \frac{q_g}{(\bar{p}^2 - p_{wf}{}^2)^n}$$

$$= \frac{42.5}{(10^5)^{0.690}} = 0.01508.$$

Thus, the empirical deliverability equation is

$$q_g = 0.01508(\bar{p}^2 - p_{wf}{}^2)^{0.690}.$$

These data are plotted in Fig. 5.3.

2. *Theoretical Method.* The theoretical deliverability equation is

$$(\bar{p}^2 - p_{wf}{}^2)/q_g = a + bq_g.$$

Fig. 5.4 is a plot of $(\bar{p}^2 - p_{wf}{}^2)/q_g$ vs. $q_g$ for the test data. Two points on the best straight line through the data are (2.7; 900) and (23.9; 1,900). Thus,

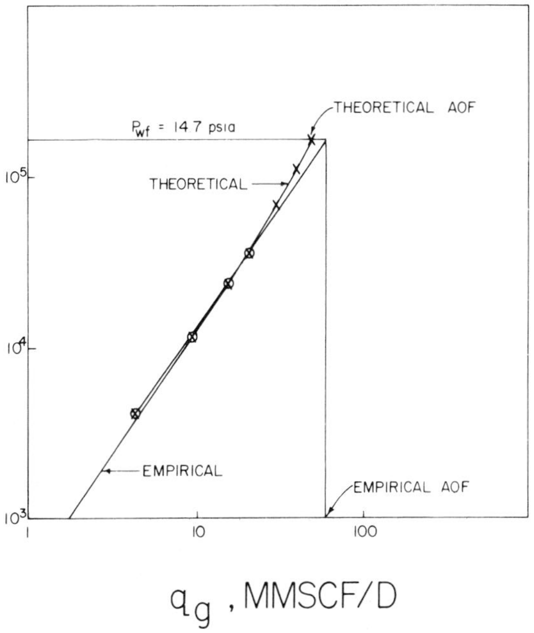


Fig. 5.3 – Stabilized gas well deliverability test.

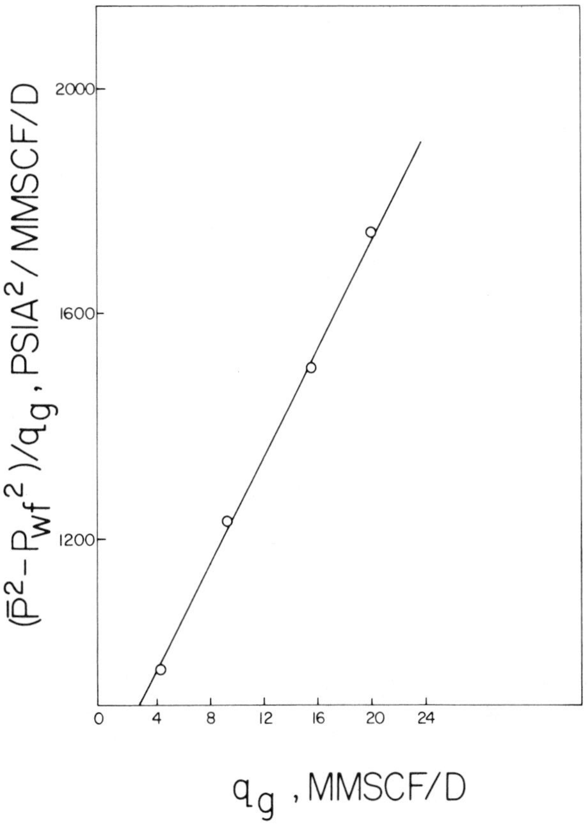


Fig. 5.4 – Stabilized deliverability test, theoretical flow equation, constant determination.

$$900 = a + 2.7\,b,$$

$$1{,}900 = a + 23.9\,b.$$

Solving for $a$ and $b$, we find that $a = 773$ and $b = 47.17$. Thus, the theoretical deliverability equation is

$$47.17\,q_g^{\,2} + 773\,q_g = (\bar{p}^2 - p_{wf}^{\,2}).$$

We can solve this quadratic equation for the AOF:

$$47.17\,q_g^{\,2} + 773\,q_g - 166{,}411 = 0.$$

The solution is

$$q_g = \text{AOF} = 51.8 \text{ MMscf/D}.$$

We also can determine points on the deliverability curve as calculated from the theoretical equation: $\bar{p}^2 - p_{wf}^{\,2} = 47.17\,q_g^{\,2} + 773\,q_g$. See Table 5.3. These results are plotted in Fig. 5.3. The plot is *almost* linear, but there is sufficient curvature to cause a 15.8% error in calculated AOF.

## 5.4 Isochronal Tests

The objective of isochronal testing[7] is to obtain data to establish a stabilized deliverability curve for a gas well without flowing the well for sufficiently long to achieve stabilized conditions ($r_i \geq r_e$) at *each* (or, in some cases, *any*) rate. This procedure is needed for lower-permeability reservoirs, where it frequently is impractical to achieve $r_i = r_e$ during the test. An isochronal test is conducted by flowing a well at a fixed rate, then shutting it in until the pressure builds up to an unchanging (or almost unchanging) value, $\bar{p}$. The well then is flowed at a second rate for the same length of time, followed by another shut-in, etc. If possible, the final flow period should be long enough to achieve stabilized flow. If this is impossible or impractical, it is still possible to predict the stabilized deliverability characteristics (with increased potential for error, of course).

In obtaining data in this testing program, it is essential to record flowing BHP, $p_{wf}$, as a function of time at each flow rate.

Fig. 5.5 illustrates rate and pressures in an isochronal testing sequence. This figure illustrates the following important points about the isochronal testing sequence.

1. Flow periods, excepting the final one, are of equal length [i.e., $t_1 = (t_3 - t_2) = (t_5 - t_4) \leq (t_7 - t_6)$].
2. Shut-in periods have the objective of letting $p \simeq \bar{p}$ rather than the objective of equal length. Thus, in general, $(t_2 - t_1) \neq (t_4 - t_3) \neq (t_6 - t_5)$.
3. A final flow period in which the well stabilizes (i.e., $r_i$ reaches $r_e$ at time $t_7$) is desirable but not essential.

The most general theory of isochronal tests is based on equations using pseudopressure. However, we will once again present the theory in terms of the low-pressure approximations to these equations ($p^2$

**TABLE 5.3 – THEORETICAL DELIVERABILITIES**

| $q_g$ (MMscf/D) | $\bar{p}^2 - p_{wf}^2$ (psia$^2$) |
|---|---|
| 4.288 | 4,182 |
| 9.265 | 11,210 |
| 15.552 | 23,430 |
| 20.177 | 34,800 |
| 30 | 65,640 |
| 40 | 106,400 |
| 49.8 = AOF | 166,600 |

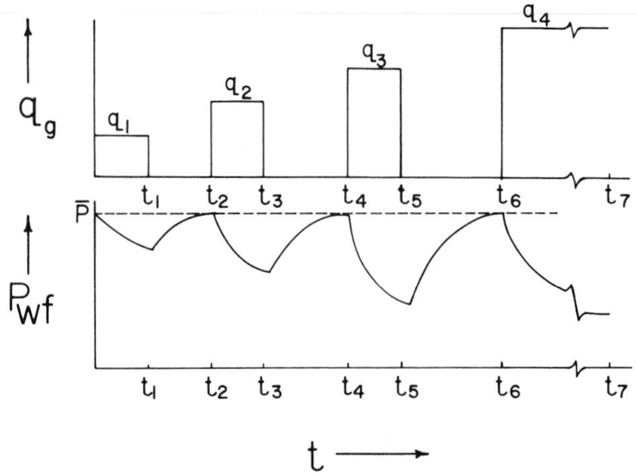


**Fig. 5.5** – Rates and pressures in isochronal test.

equations) because (1) they are somewhat simpler and less abstract than equations in pseudopressure and (2) they allow direct comparison with more conventional analysis methods[7] based on plots of $(\bar{p}^2 - p_{wf}^2)$ vs. $q_g$ on log-log paper.

Eqs. 5.5 and 5.6 provide the basic method for interpreting isochronal tests.

For transient flow $(r_i < r_e)$,

$$p_{wf}^2 = \bar{p}^2 + 1{,}637 \frac{\mu_{\bar{p}} z_{\bar{p}g}}{kh} T q_g \cdot \left[ \log\left( \frac{1{,}688\, \phi \mu_{\bar{p}} c_{t\bar{p}} r_w^2}{kt} \right) - \left( \frac{s + D|q_g|}{1.151} \right) \right]. \quad \text{(5.5)}$$

For stabilized flow $(r_i \geq r_e)$,

$$p_{wf}^2 = \bar{p}^2 - 1{,}422 \frac{\mu_{\bar{p}} z_{\bar{p}g} T q_g}{kh} \left[ \ln\left( \frac{r_e}{r_w} \right) - 0.75 + \left( s + D|q_g| \right) \right]. \quad \text{(5.6)}$$

In addition to the flow equations, an important theoretical consideration in understanding isochronal tests is the radius-of-investigation concept. We observed previously that the radius of investigation achieved at a given time in a flow test is *independent of flow rate* and, thus, should be the same at a given time for each flow rate in an isochronal test. Further, the radius of investigation at a given time can be considered to be proportional to a drainage radius at that time, because it is near (but slightly less than) the point beyond which there has been no appreciable drawdown in reservoir pressure and thus no fluid drainage. Accordingly, at a given time, the same portion of the reservoir is being drained at each rate and, as a good approximation, stabilized flow conditions exist to a point just beyond $r = r_i$. Thus, a deliverability curve can be drawn at each fixed time (hence, the name isochronal) through points $(q_g,\ \bar{p}^2 - p_{wf}^2)$ obtained at that time at several different rates, and a truly stabilized deliverability curve can be drawn when $r_i \geq r_e$.

These assertions can be made more quantitative if we note that for flowing time $t$, at each rate, there corresponds a drainage radius, $r_d' = c r_i$, that is independent of rate. Admittedly anticipating a convenient result, we let $r_d' = 1.585\, r_i$ (but the $p$ vs. log $r$ plot constructed in the solution to Problem 1.2 shows that such an $r_d'$ approximates quite closely the point beyond which no appreciable fluid is being drained). At time $t_1$,

$$p_{wf}^2 = \bar{p}^2 + 1{,}422 \frac{\mu_{\bar{p}} z_{\bar{p}g}}{kh} T q_g \cdot \left[ \frac{1}{2} \ln\left( \frac{1{,}688\, \phi \mu_{\bar{p}} c_{t\bar{p}} r_w^2}{kt_1} \right) - \left( s + D|q_g| \right) \right].$$

Because

$$r_d'^2 = (1.585)^2\, r_i^2 = \frac{(1.585)^2\, kt_1}{948\, \phi \mu_{\bar{p}} c_{t\bar{p}}} = \frac{kt_1}{377\, \phi \mu_{\bar{p}} c_{t\bar{p}}},$$

we may write the transient flow equation as

$$p_{wf}^2 = \bar{p}^2 - 1{,}422 \frac{\mu_{\bar{p}} z_{\bar{p}g} T q_g}{kh} \left[ \ln\left( \frac{r_{d1}'}{r_w} \right) - 0.75 + \left( s + D|q_g| \right) \right].$$

Compare with the stabilized deliverability equation:

$$p_{wf}^2 = \bar{p}^2 - 1{,}422 \frac{\mu_{\bar{p}} z_{\bar{p}b} T q_g}{kh} \left[ \ln\left( \frac{r_e}{r_w} \right) \right.$$

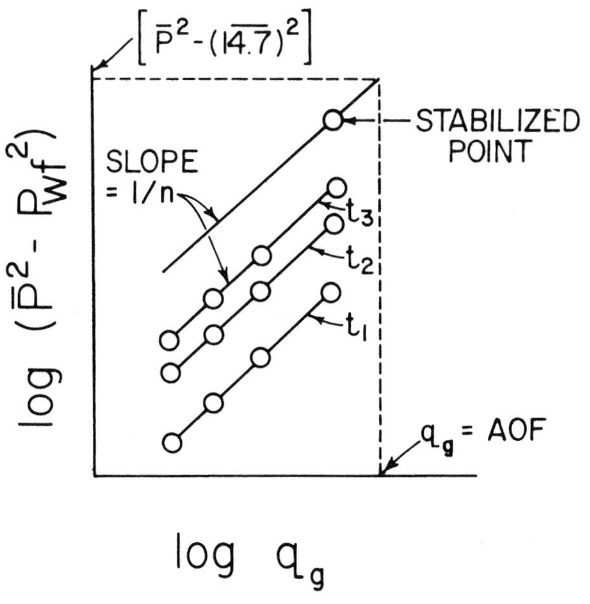


**Fig. 5.6** – Empirical deliverability plot for isochronal test.

**TABLE 5.4 – ISOCHRONAL TEST DATA**

| Test | Duration (hours) | $p_{wf}$ or $p_{ws}$ (psia) | $q_g$ (MMscf/D) |
|---|---|---|---|
| Initial shut-in | 48 | 1,952 | – |
| First flow | 12 | 1,761 | 2.6 |
| First shut-in | 15 | 1,952 | – |
| Second flow | 12 | 1,694 | 3.3 |
| Second shut-in | 17 | 1,952 | – |
| Third flow | 12 | 1,510 | 5.0 |
| Third shut-in | 18 | 1,952 | – |
| Fourth flow | 12 | 1,320 | 6.3 |
| Extended flow (stabilized) | 72 | 1,151 | 6.0 |
| Final shut-in | 100 | 1,952 | – |

**TABLE 5.5 – ISOCHRONAL TEST ANALYSIS**

| $q_g$ (MMscf/D) | $\bar{p}^2 - p_{wf}^2$ (psia²) | $(\bar{p}^2 - p_{wf}^2)/q_g$ (psia²/MMscf/D) |
|---|---|---|
| 2.6 | 709,000 | 273,000 |
| 3.3 | 941,000 | 285,000 |
| 5.0 | 1,530,000 | 306,000 |
| 6.3 | 2,070,000 | 328,200 |

$$-0.75+\left(s+D|q_g|\right)\Big].$$

The equations are identical in form because we have defined a time-dependent drainage radius, $r_d'$, as

$$r_d' = \left(\frac{kt}{377\,\phi\mu_{\bar{p}}c_{t\bar{p}}}\right)^{1/2}.$$

Thus, we conclude that, at each fixed time $t_1$, an analysis of different rates used in an isochronal test should be possible just as it is for a stabilized test – except, of course, the data will not yield a truly stabilized deliverability curve. This is possible only if stabilized data are available or if they can be estimated.

### Analysis of Test Data: One Rate Continued to Stabilization

Experience[7] shows that reasonably satisfactory results can be obtained with the empirical method using the following procedure.

1. The best straight line is drawn through the points $(\bar{p}^2 - p_{wf}^2,\ q_g)$ obtained at a fixed value of time with the different rates used in the isochronal testing program. The data are plotted on log-log paper, just as when analyzing a stabilized deliverability curve.

2. Lines should be drawn for several values of time $t$, and the slope $1/n$ should be established for each isochronal deliverability curve.

3. A line with the slope $1/n$ determined from the nonstabilized, fixed-time curves then is drawn through the single stabilized point, $(q_g,\ \bar{p}^2 - p_{wf}^2)$. This establishes the stabilized deliverability curve. Once the stabilized deliverability curve is determined, AOF is established in the usual way, as indicated in Fig. 5.6.

The theoretical method for analyzing isochronal test data is based on the theoretical equations for stabilized flow and transient flow written in the form below. For stabilized flow,

$$\bar{p}^2 - p_{wf}^2 = aq_g + bq_g^2, \quad \ldots\ldots\ldots\ldots\ldots (5.7)$$

where

$$a = 1{,}422\frac{\mu_{\bar{p}}z_{\bar{p}g}T}{kh}\left[\ln\left(\frac{r_e}{r_w}\right) - 0.75 + s\right], \quad \ldots\ldots (5.8)$$

and

$$b = 1{,}422\frac{\mu_{\bar{p}}z_{\bar{p}g}TD}{kh}. \quad \ldots\ldots\ldots\ldots\ldots (5.9)$$

For transient flow,

$$\bar{p}^2 - p_{wf}^2 = a_t q_g + bq_g^2, \quad \ldots\ldots\ldots\ldots\ldots (5.11)$$

where $b$ has the same meaning as for stabilized flow and where $a_t$, a function of time, is given by

$$a_t = 1{,}422\frac{\mu_{\bar{p}}z_{\bar{p}g}T}{kh}\left[\frac{1}{2}\ln\left(\frac{kt}{1{,}688\,\phi\mu_{\bar{p}}c_{t\bar{p}}r_w^2}\right) + s\right]. \quad \ldots\ldots\ldots\ldots\ldots (5.12)$$

Thus,

$$\bar{p}^2 - p_{wf}^2 = a_t q_g + bq_g^2,\ r_i < r_e, \quad \ldots\ldots\ldots (5.11)$$

and

$$\bar{p}^2 - p_{wf}^2 = aq_g + bq_g^2,\ r_i \geq r_e. \quad \ldots\ldots\ldots (5.7)$$

An analysis method of isochronal tests consistent with the theoretical equations follows.

1. For a *fixed* value of $t$, determine $b$ from a plot of $(\bar{p}^2 - p_{wf}^2)/q_g$ vs. $q_g$.

2. Using the stabilized data point $[q_{gs},\ (\bar{p}^2 - p_{wf}^2)s]$, determine $a$ from

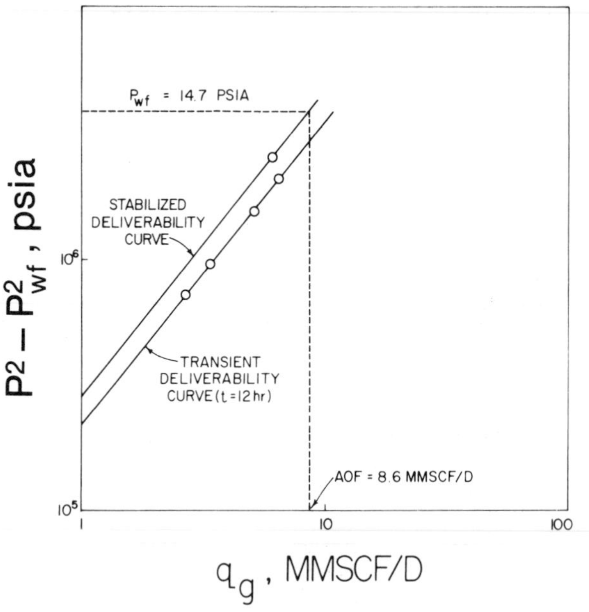


Fig. 5.7 – Isochronal deliverability test analysis.

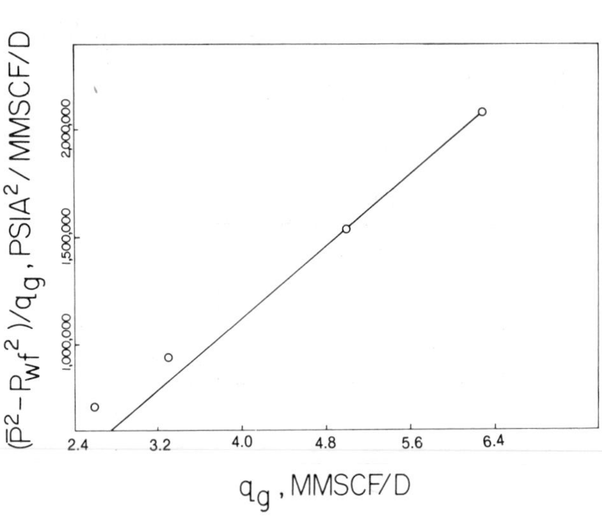


Fig. 5.8 – Isochronal deliverability test analysis, theoretical flow equation, constant determination.

$$a=\left[(\bar{p}^2-p_{wf}{}^2)_s-bq_{gs}{}^2\right]/q_{gs}.$$

3. The stabilized deliverability curve uses the constants determined in Steps 1 and 2:

$$\bar{p}^2-p_{wf}{}^2=aq_g+bq_g{}^2.$$

This equation can be used to calculate the AOF, for example:

$$\text{AOF}=\frac{-a+\sqrt{a^2+4b(\bar{p}^2-14.7^2)}}{2b}.$$

## *Example 5.2 – Isochronal Gas Well Test Analysis*

**Problem.** Determine the stabilized deliverability curve and AOF from the test data in Table 5.4 (adapted from data given in Ref. 4) using (1) the empirical method and (2) the theoretical method. For simplicity, data at only a single flow time are reported.

**Solution.** 1. *Empirical Method.* Data from the 12-hour flow tests are analyzed first to determine the slope of the deliverability curve (Table 5.5). We also note that at stabilized rate $q_g=6.0$ MMscf/D, $\bar{p}^2-p_{wf}{}^2=2{,}485{,}000$ psia$^2$, and that when $q_g$ = AOF, $\bar{p}^2-p_{wf}{}^2=3{,}810{,}000$ psia$^2$.

The data are plotted in Fig. 5.7; from the plot, AOF $\simeq$ 8.6 MMscf/D.

2. *Theoretical Method.* We first determine $b$, using a plot of $(\bar{p}^2-p_{wf}{}^2)/q_g$ vs. $q_g$ from the 12-hour test data. The line drawn (Fig. 5.8) passes through the points with greatest pressure drawdown and, thus, least potential error: [$q_g=5.0$ MMscf/D, $(\bar{p}^2-p_{wf}{}^2)/q_g=306{,}000$ psia$^2$/MMscf/D] and [$q_g=6.3$ MMscf/D, $(\bar{p}^2-p_{wf}{}^2)/q_g=328{,}000$ psia$^2$/MMscf/D]. Thus,

$$306{,}000=a_{12}+5.0\,b,$$

$$328{,}000=a_{12}+6.3\,b.$$

Solving, we find that $b=17{,}080$.

From the stabilized test, $q_g=6.0$ MMscf/D and $\bar{p}^2-p_{wf}{}^2=2{,}485{,}000$ psia$^2$. Thus,

$$a=\frac{(\bar{p}^2-p_{wf}{}^2)_s-bq_{gs}{}^2}{q_{gs}}$$

$$=\frac{2{,}485{,}000-(17{,}080)(6.0)^2}{6.0}$$

$$=311{,}700.$$

Thus, the stabilized deliverability curve is

$$\bar{p}^2-p_{wf}{}^2=311{,}700\,q_g+17{,}080\,q_g{}^2.$$

Solving for AOF, we find that it is equal to 8.38 MMscf/D. This value is quite close to the value established using the empirical method. This is at least partly because the extrapolation of the empirical curve is not far beyond the observed data.

### Analysis of Test Data: No Stabilized Flow Attained

In very low-permeability reservoirs, stabilized flow cannot be achieved in a reasonable length of time. In such cases, a satisfactory determination of the stabilized deliverability curve can be based on use of the theoretical equations for transient flow (Eq. 5.5) and stabilized flow (Eq. 5.6). We have noted that the stabilized equation can be expressed as

$$\bar{p}^2-p_{wf}{}^2=aq_g+bq_g{}^2, \quad\ldots\ldots\ldots\ldots\ldots(5.7)$$

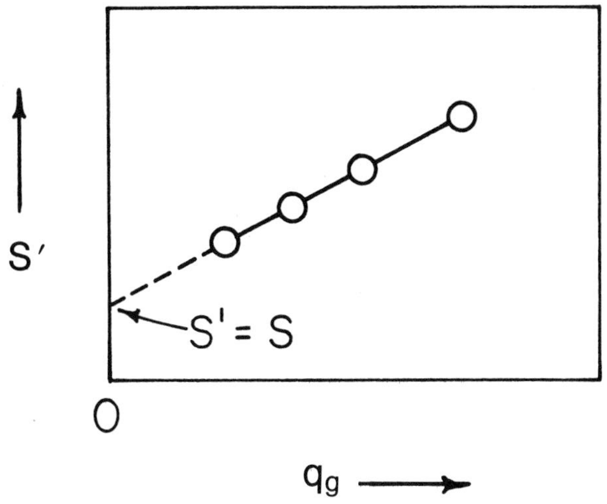


**Fig. 5.9** – Skin factor determination.

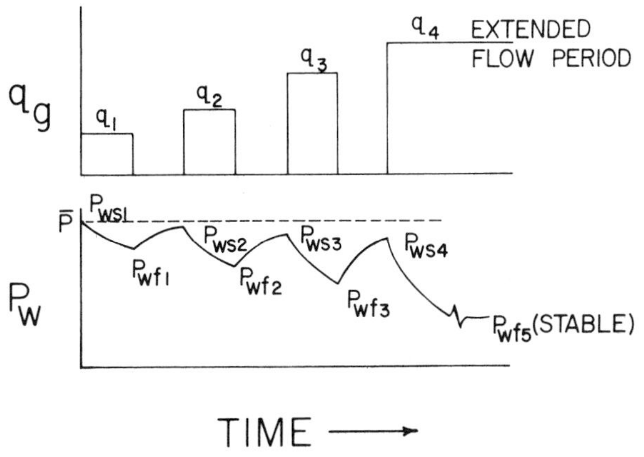


**Fig. 5.10** – Rates and pressures in modified isochronal test.

where

$$a=1{,}422\frac{\mu_{\bar{p}}z_{\bar{p}g}T}{kh}\left[\ln\left(\frac{r_e}{r_w}\right)-0.75+s\right], \quad \ldots\ldots (5.8)$$

and

$$b=1{,}422\frac{\mu_{\bar{p}}z_{\bar{p}g}TD}{kh}. \quad \ldots\ldots (5.9)$$

We also noted that the transient equation has the form

$$\bar{p}^2-p_{wf}^2=a_tq_g+bq_g^2, \quad \ldots\ldots (5.11)$$

where

$$a_t=1{,}422\frac{\mu_{\bar{p}}z_{\bar{p}g}T}{kh}\left[\frac{1}{2}\ ln\left(\frac{kt}{1{,}688\ \phi\mu_{\bar{p}}c_{t\bar{p}}r_w^2}\right)+s\right] \quad \ldots\ldots (5.8)$$

The objective of determining the stabilized flow equation can be achieved if the constants $a$ and $b$ can be determined. We note that the constant $b$ can be determined from the isochronal test data as illustrated [by plotting $(\bar{p}^2-p_{wf}^2)/q_g$ vs. $q_g$ for fixed value of $t$ and determining the slope, $b$, of the resulting best straight line]. Constant $a$, however, is more troublesome: The only satisfactory means of determining $a$ is through knowledge of each term in the defining equation for this quantity. Thus, we need estimates of $kh$, $s$, and $r_e$. (Other quantities in Eq. 5.8 are usually available.)

Since an isochronal test consists of a series of drawdown and buildup tests, $kh$ and $s$ usually can be determined from them. Determination of $kh$ is straightforward in principle; determination of $s$ is less straightforward. Recall that a single test provides only an estimate of $s'=s+D|q_g|$. Accordingly, to determine $s$, we must analyze at least two tests: either drawdown tests run at different rates or buildup tests following drawdown tests at different rates. We can then plot $s'$ vs. $q_g$; extrapolation to $q_g=0$ provides an estimate of true skin factor, $s$ (see Fig. 5.9). The drainage radius $r_e$ must be estimated from expected well spacing (or knowledge of reservoir geometry in a small or irregular reservoir).

The constants $a$ and $b$ determined in this way then can be substituted into the stabilized deliverability equation, Eq. 5.7. If a plot of log $(\bar{p}^2-p_{wf}^2)$ vs. log $q_g$ also is desired, data points to be plotted can be determined from the equation.

## 5.5 Modified Isochronal Tests

The objective of modified isochronal tests is to obtain the same data as in an isochronal test without using the sometimes lengthy shut-in periods required for pressure to stabilize completely before each flow test is run.

In the modified isochronal test (Fig. 5.10), shut-in periods of the same duration as the flow periods are used, and the final shut-in BHP ($p_{ws}$) before the beginning of a new flow period is used as an approximation to $\bar{p}$ in the test analysis procedure. For example, for the first flow period, use $(\bar{p}^2-p_{wf,1}^2)=(p_{ws,1}^2-p_{wf,1}^2)$; for the second flow period, use $(p_{ws,2}^2-p_{wf,2}^2)$. Otherwise, the analysis procedure is the same as for the "true" isochronal test.

Note that the modified isochronal procedure uses *approximations*. Isochronal tests are modeled exactly by rigorous theory (if reservoir and fluid properties cooperate); modified isochronal tests are not. However, modified isochronal tests are used widely because they conserve time and money and because they have proved to be excellent approximations to true isochronal tests.

**TABLE 5.6 – MODIFIED ISOCHRONAL TEST DATA**

| Test | Duration (hours) | $p_{wf}$ or $p_{ws}$ (psia) | $q_g$ (MMscf/D) |
|---|---|---|---|
| Pretest shut-in | 20 | 1,948 | – |
| First flow | 12 | 1,784 | 4.50 |
| First shut-in | 12 | 1,927 | – |
| Second flow | 12 | 1,680 | 5.60 |
| Second shut-in | 12 | 1,911 | – |
| Third flow | 12 | 1,546 | 6.85 |
| Third shut-in | 12 | 1,887 | – |
| Fourth flow | 12 | 1,355 | 8.25 |
| Extended flow (stabilized) | 81 | 1,233 | 8.00 |
| Final shut-in | 120 | 1,948 | – |

**TABLE 5.7 – MODIFIED ISOCHRONAL TEST ANALYSIS**

| $q_g$ (MMscf/D) | $p_{ws}$ (psia) | $p_{wf}$ (psia) | $p_{ws}^2 - p_{wf}^2$ (psia²) | $(p_{ws}^2 - p_{wf}^2)/q_g$ |
|---|---|---|---|---|
| 4.50 | 1,948 | 1,784 | 612,048 | 136,011 |
| 5.60 | 1,927 | 1,680 | 890,929 | 159,094 |
| 6.85 | 1,911 | 1,546 | 1,261,805 | 184,205 |
| 8.25 | 1,887 | 1,355 | 1,724,744 | 209,060 |
| 8.00 (stabilized) | 1,948* | 1,233 | 2,274,415 | – |

*Note that $\bar{p}$, the true current reservoir pressure, is used for the stabilized test analysis.

**TABLE 5.8 – THEORETICAL STABILIZED DELIVERABILITIES**

| $q_g$ (MMscf/D) | $\bar{p}^2 - p_{wf}^2$ (psia²) |
|---|---|
| 4.5 | 972,000 |
| 5.6 | 1,330,000 |
| 6.85 | 1,794,000 |
| 8.0 | 2,274,000 |
| 10.8 | 3,660,000 |

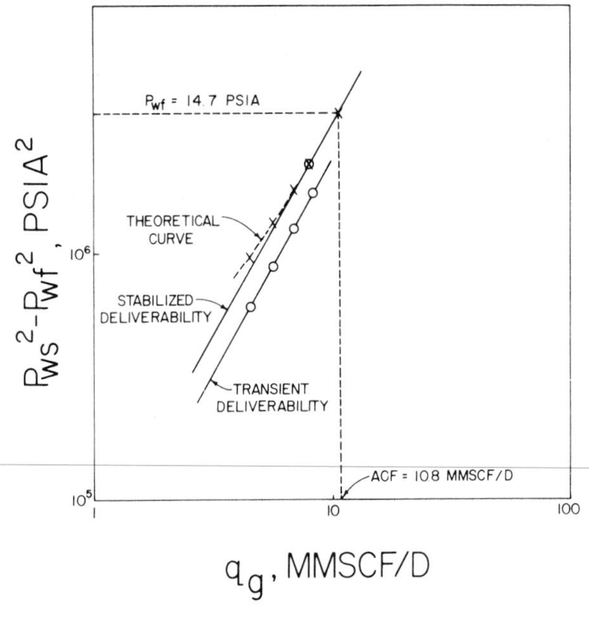


Fig. 5.11 – Modified isochronal test analysis.

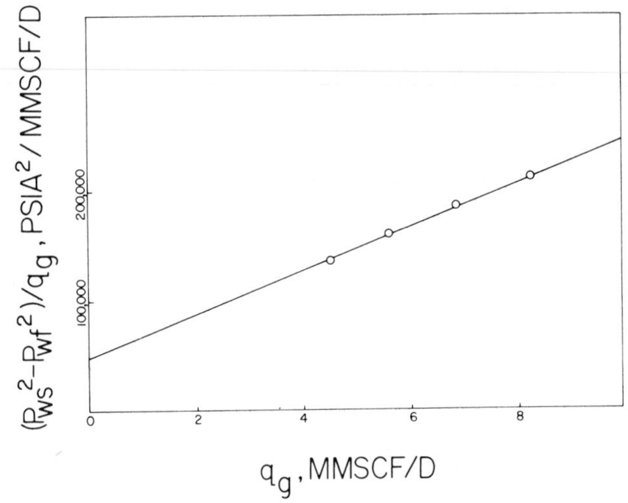


Fig. 5.12 – Modified isochronal test analysis, theoretical flow equation, constant determination.

## *Example 5.3 – Modified Isochronal Test Analysis*

**Problem.** Estimate the AOF from the data in Table 5.6 obtained in a modified isochronal test,[4] using both empirical and theoretical methods.

**Solution.** We first prepare the data for plotting (Table 5.7). Fig. 5.11 shows the data plot for the empirical method. This is a plot of $(p_{ws}^2 - p_{wf}^2)$ vs. $q_g$ on log-log paper. The transient points are used to establish the slope of the curve, and a line with the same slope is drawn through the single stabilized point. The AOF is the value of $q_g$ when $p_{ws}^2 - p_{wf}^2 = \bar{p}^2 - p_{wf}^2 = 1{,}948^2 - 14.7^2 = 3{,}790{,}000$ psia²; this value is AOF = 10.8 MMscf/D.

For the theoretical method, we establish the constant $b$ from the slope of a plot of $(p_{ws}^2 - p_{wf}^2)/q_g$ vs. $q_g$; in this case,

$$b = \frac{243{,}000 - 48{,}000}{10} = 19{,}500,$$

using data from Fig. 5.12. Then,

$$a = \frac{(\bar{p}^2 - p_{wf}^2)_s - bq_{gs}^2}{q_{gs}}$$

$$= \frac{2{,}274{,}415 - (19{,}500)(8.0)^2}{8.0}$$

$$= 128{,}300.$$

Thus, the equation of the stabilized deliverability curve is

$$\bar{p}^2 - p_{wf}^2 = 128{,}300\, q_g + 19{,}500\, q_g^2.$$

Solving this equation for the AOF ($q_g$ when $\bar{p}^2 - p_{wf}^2 = 3{,}790{,}000$), we find that it is equal to 11.0 MMscf/D.

It is also of interest to calculate points on the stabilized deliverability curve and to plot them on the

TABLE 5.9 – GAS PROPERTIES FOR EXAMPLE 5.4

| $p$ (psia) | $\mu_g$ (cp) | $z$ | $p/\mu z$ (psia/cp) |
|---|---|---|---|
| 150 | 0.01238 | 0.9856 | 12,290 |
| 300 | 0.01254 | 0.9717 | 24,620 |
| 450 | 0.01274 | 0.9582 | 36,860 |
| 600 | 0.01303 | 0.9453 | 48,710 |
| 750 | 0.01329 | 0.9332 | 60,470 |
| 900 | 0.01360 | 0.9218 | 71,790 |
| 1,050 | 0.01387 | 0.9112 | 83,080 |
| 1,200 | 0.01428 | 0.9016 | 93,205 |
| 1,350 | 0.01451 | 0.8931 | 104,200 |
| 1,500 | 0.01485 | 0.8857 | 114,000 |
| 1,650 | 0.01520 | 0.8795 | 123,400 |
| 1,800 | 0.01554 | 0.8745 | 132,500 |
| 1,950 | 0.01589 | 0.8708 | 140,900 |
| 2,100 | 0.01630 | 0.8684 | 148,400 |
| 2,250 | 0.01676 | 0.8671 | 154,800 |
| 2,400 | 0.01721 | 0.8671 | 160,800 |
| 2,550 | 0.01767 | 0.8683 | 166,200 |
| 2,700 | 0.01813 | 0.8705 | 171,100 |
| 2,850 | 0.01862 | 0.8738 | 175,200 |
| 3,000 | 0.01911 | 0.8780 | 178,800 |
| 3,150 | 0.01961 | 0.8830 | 181,900 |

TABLE 5.10 – PSEUDOPRESSURE FOR EXAMPLE 5.4

| $p$ (psia) | $\psi(p)$ (psia²/cp) |
|---|---|
| 150 | $1.844\times10^6$ |
| 300 | $7.381\times10^6$ |
| 450 | $1.660\times10^7$ |
| 600 | $2.944\times10^7$ |
| 750 | $4.582\times10^7$ |
| 900 | $6.566\times10^7$ |
| 1,050 | $8.888\times10^7$ |
| 1,200 | $1.154\times10^8$ |
| 1,350 | $1.451\times10^8$ |
| 1,500 | $1.779\times10^8$ |
| 1,650 | $2.135\times10^8$ |
| 1,800 | $2.518\times10^8$ |
| 1,950 | $2.929\times10^8$ |
| 2,100 | $3.363\times10^8$ |
| 2,250 | $3.817\times10^8$ |
| 2,400 | $4.291\times10^8$ |
| 2,550 | $4.781\times10^8$ |
| 2,700 | $5.287\times10^8$ |
| 2,850 | $5.807\times10^8$ |
| 3,000 | $6.338\times10^8$ |
| 3,150 | $6.879\times10^8$ |

empirical data plot. The values are given in Table 5.8. These data are plotted in Fig. 5.11.

## 5.6 Use of Pseudopressure in Gas Well Test Analysis

Accuracy of gas well test analysis can be improved in some cases if the pseudopressure $\psi(p)$ is used instead of approximations written in terms of pressure or pressure squared. In this section, we discuss the calculation of pseudopressure and provide an introduction to direct use of this quantity in gas-well drawdown test analysis. Detailed discussion, including systematic development of working equations and application to drawdown, buildup, and deliverability tests, is provided in Ref. 4.

### Calculation of Pseudopressure

Gas pseudopressure, $\psi(p)$, is defined by the integral

$$\psi(p)=2\int_{p_\beta}^{p}\frac{p}{\mu z}dp, \quad \text{.................... (5.2)}$$

where $p_\beta$ is some arbitrary low base pressure. To evaluate $\psi(p)$ at some value of $p$, we can evaluate the integral in Eq. 5.2 numerically, using values for $\mu$ and $z$ for the *specific* gas under consideration, evaluated at reservoir temperature. An example will illustrate this calculation.

---

### *Example 5.4 – Calculation of Gas Pseudopressure*

**Problem.** Calculate the gas pseudopressure $\psi(p)$ for a reservoir containing 0.7 gravity gas at 200°F as a function of pressure in the range 150 to 3,150 psia. Gas properties as functions of pressure are given in Table 5.9.

**Solution.** We will select $p_\beta=0$ and use the fact that, as $p_\beta\rightarrow0$, $p/\mu z\rightarrow0$. We will use the trapezoidal rule for our numerical integration.

For $p=150$ psia,

$$\psi(150)=2\int_{p_\beta}^{p}\frac{p}{\mu z}dp$$

$$\simeq 2\,\frac{\left[\left(\frac{p}{\mu z}\right)_o+\left(\frac{p}{\mu z}\right)_{150}\right]}{2}(150-0)$$

$$=\frac{2[0+12{,}290]}{2}(150)$$

$$=1.844\times10^6 \text{ psia}^2/\text{cp}.$$

For $p=300$ psia,

$$\psi(300)\simeq1.844\times10^6$$

$$+2\left(\frac{12{,}290+24{,}620}{2}\right)(300-150)$$

$$=7.381\times10^6\text{ psia}^2/\text{cp}.$$

Proceeding in a similar way, we can construct Table 5.10. These results are plotted in Fig. 5.13.

### Drawdown Test Analysis Using Pseudopressure

Transient flow at a constant rate from an infinite-acting gas reservoir is modeled by Eq. 5.1.

$$\psi(p_{wf})=\psi(p_i)+50{,}300\frac{p_{sc}}{T_{sc}}\frac{q_g T}{kh}$$

$$\cdot\left\{1.151\log\left[1{,}688\frac{\phi\mu_i c_{ti} r_w{}^2}{kt}\right.\right.$$

TABLE 5.11 – DRAWDOWN TEST DATA

| $t$ (hours) | $p_{wf}$ (psia) | $\psi(p_{wf})$ (psia²/cp) |
|---|---|---|
| 0 | 3,000 | $6.338\times10^8$ |
| 0.024 | 2,964 | $6.210\times10^8$ |
| 0.096 | 2,920 | $6.055\times10^8$ |
| 0.244 | 2,890 | $5.947\times10^8$ |
| 0.686 | 2,866 | $5.864\times10^8$ |
| 2.015 | 2,848 | $5.801\times10^8$ |
| 6.00 | 2,833 | $5.747\times10^8$ |
| 17.96 | 2,817 | $5.693\times10^8$ |
| 53.82 | 2,802 | $5.640\times10^8$ |
| 161. | 2,786 | $5.585\times10^8$ |
| 281 | 2,777 | $5.553\times10^8$ |
| 401 | 2,771 | $5.532\times10^8$ |
| 521 | 2,766 | $5.517\times10^8$ |
| 641 | 2,763 | $5.505\times10^8$ |
| 761 | 2,760 | $5.494\times10^8$ |
| 881 | 2,757 | $5.485\times10^8$ |

TABLE 5.12 – DRAWDOWN TEST DATA FOR CURVE MATCHING

| $t$ (hours) | $\psi(p_i)-(p_{wf})$ (psia²/cp) |
|---|---|
| 0.024 | $0.128\times10^8$ |
| 0.096 | $0.283\times10^8$ |
| 0.244 | $0.391\times10^8$ |
| 0.686 | $0.474\times10^8$ |
| 2.015 | $0.537\times10^8$ |
| 6.00 | $0.591\times10^8$ |
| 17.96 | $0.645\times10^8$ |
| 53.82 | $0.698\times10^8$ |
| 161 | $0.753\times10^8$ |
| 281 | $0.785\times10^8$ |
| 401 | $0.806\times10^8$ |
| 521 | $0.821\times10^8$ |
| 641 | $0.833\times10^8$ |
| 761 | $0.844\times10^8$ |
| 881 | $0.853\times10^8$ |

$$-\left(s+D\left|q_g\right|\right)\Big]\Big\}. \quad \ldots\ldots\ldots\ldots\ldots (5.1)$$

This equation describes the MTR in a gas well test, just as Eq. 3.1 describes flow of a slightly compressible liquid. In general, of course, drawdown tests in gas wells also have ETR's (usually dominated by wellbore storage distortion) and LTR's (in which boundary effects become important). Analysis of a gas well test using pseudopressure as the dependent variable is illustrated in Example 5.5.

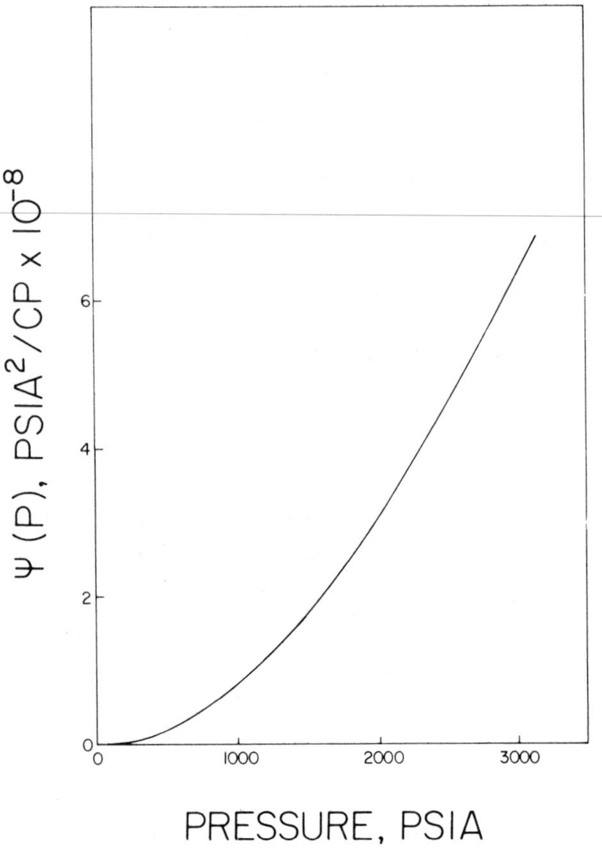


**Fig. 5.13** – Pseudopressure vs. pressure.

## *Example 5.5 – Analysis of Gas Well Drawdown Test Using Pseudopressures*

**Problem.** A constant-rate drawdown test was run on a gas well with the properties given below; results are shown in Table 5.11.

$p_i$ = 3,000 psia,
$\phi$ = 0.19,
$S_{wi}$ = 0.211,
$V_w$ = 286 cu ft,
$h$ = 10 ft,
$T$ = 200°F,
$r_w$ = 0.365 ft,
$\mu_i$ = 0.01911 cp,
$q_g$ = 1,000 Mscf/D,
$\gamma_g$ = 0.7,
$c_{ti}$ = $0.235\times10^{-3}$ psi$^{-1}$,
drainage area = 640 acres (square), and
well centered in drainage area.

This gas is the same as that analyzed in Example 5.4;

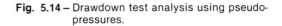

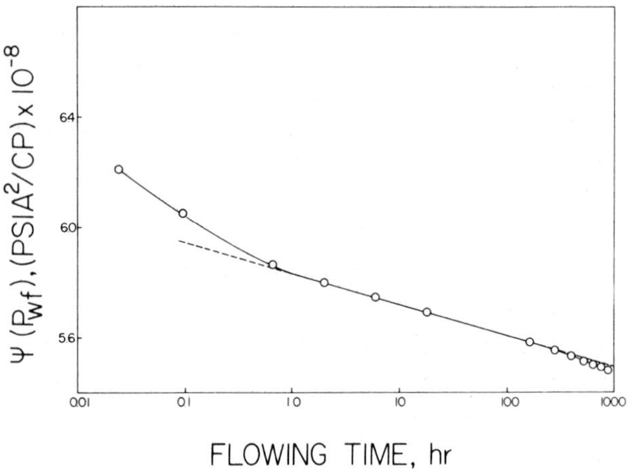


**Fig. 5.14** – Drawdown test analysis using pseudopressures.

TABLE 5.13 – STABILIZED DELIVERABILITY TEST DATA

| Test | Rate (MMscf/D) | $p_{wf}$ (psia) |
|---|---|---|
| initial buildup | – | 3,127 |
| 1 | 3.710 | 3,087 |
| 2 | 5.980 | 3,059 |
| 3 | 8.191 | 3,035 |
| 4 | 14.290 | 2,942 |

TABLE 5.14 – ISOCHRONAL DELIVERABILITY TEST DATA

| $q$ (Mscf/D) | $\bar{p}$ (psia) | $p_{wf}$ (psia) | $t$ (hours) |
|---|---|---|---|
| 983 | 352.4 | 344.7 | 0.5 |
| 977 | | 342.4 | 1.0 |
| 970 | | 339.5 | 2.0 |
| 965 | | 337.6 | 3.0 |
| 2,631 | 352.3 | 329.5 | 0.5 |
| 2,588 | | 322.9 | 1.0 |
| 2,533 | | 315.4 | 2.0 |
| 2,500 | | 310.5 | 3.0 |
| 3,654 | 351.0 | 318.7 | 0.5 |
| 3,565 | | 309.5 | 1.0 |
| 3,453 | | 298.6 | 2.0 |
| 3,390 | | 291.9 | 3.0 |
| 4,782 | 349.5 | 305.5 | 0.5 |
| 4,625 | | 293.6 | 1.0 |
| 4,438 | | 279.6 | 2.0 |
| 4,318 | | 270.5 | 3.0 |

TABLE 5.15 – GAS WELL BUILDUP TEST DATA

| $\Delta t$ (days) | $p_{ws}$ (psia) |
|---|---|
| 0 | 2,430 |
| 0.001883 | 2,542 |
| 0.003392 | 2,600 |
| 0.005738 | 2,650 |
| 0.009959 | 2,692 |
| 0.01903 | 2,726 |
| 0.04287 | 2,756 |
| 0.1144 | 2,785 |
| 0.3289 | 2,814 |
| 0.9724 | 2,843 |
| 2.903 | 2,872 |
| 7.903 | 2,896 |
| 12.90 | 2,907 |
| 17.90 | 2,913 |
| 22.90 | 2,917 |
| 27.90 | 2,920 |
| 32.90 | 2,921 |
| 37.90 | 2,922 |

accordingly, we can determine $\psi(p_{wf})$ for each $p_{wf}$. These values are included in Table 5.11. From these data, determine formation permeability and apparent skin factor.

**Solution.** The first step in the analysis procedure is to plot $\psi(p_{wf})$ vs. log $t$. This plot is shown in Fig. 5.14. The curve shape suggests wellbore storage distortion up to $t \simeq 1$ hour, and boundary effects starting at about 200 hours. A log-log plot of $\Delta\psi = [\psi(p_i) - \psi(p_{wf})]$ is useful to confirm this suspicion. Thus, we tabulate (Table 5.12) and plot $\Delta\psi$ vs. $t$.

Qualitative curve matching of log $\Delta\psi$ vs. log $t$ with Ramey's type curve (not shown) indicates an end to wellbore storage distortion at about 1 hour for skin factors in the range 0 to 5 regardless of $C_{sD}$, confirming the indication on the $\psi(p_{wf})$ vs. log $t$ plot.

The absolute value of the slope of the MTR line is

$$m = \left[(5.944 - 5.497) \times 10^8\right]/4$$

$$= 11.18 \times 10^6 \text{ psia}^2/\text{cp-cycle}.$$

Eq. 5.1 shows that the proper interpretation of this slope is

$$m = 50{,}300 \frac{p_{sc}}{T_{sc}} \frac{q_g T}{kh} (1.151).$$

Thus, for $p_{sc} = 14.7$ psia and $T_{sc} = 520°$R,

$$k = 1{,}637 \frac{q_g T}{mh}$$

$$= \frac{(1{,}637)(1{,}000)(660)}{(11.18 \times 10^6)(10)} = 9.66 \text{ md}.$$

From Eq. 5.1, we see that the apparent skin factor, $s' = s + D|q_g|$, is

$$s' = 1.151\left[\frac{\psi(p_i) - \psi(p_{1\,hr})}{m} - \log\left(\frac{k}{\phi\mu_i c_{ti} r_w^2}\right) + 3.23\right].$$

Thus,

$$s' = 1.151\left\{\frac{6.338 \times 10^8 - 5.833 \times 10^8}{11.18 \times 10^6} - \log\left[\frac{9.66}{(0.19)(0.01911)(2.35 \times 10^{-4})(0.365)^2}\right] + 3.23\right\} = -0.21$$

The well is apparently slightly stimulated.

Radius of investigation at the beginning and end of the MTR is found from Eq. 1.25:

$$r_i = \left(\frac{kt}{948\,\phi\mu_i c_{ti}}\right)^{1/2}$$

$$= \left[\frac{(9.66)(1)}{(948)(0.19)(0.01911)(2.35 \times 10^{-4})}\right]^{1/2}$$

$= 109$ ft at start of MTR ($t \simeq 1$ hour),

and

$$r_i = (109)\left(\frac{200}{1}\right)^{1/2}$$

$= 1{,}550$ ft at end of MTR.

The distance $x_e$ from the well to the edge of the 640-acre square in which it is centered is 2,640 ft; thus, the time at which the observed deviation from the MTR occurs agrees qualitatively with the time at which boundary effects should begin to appear.

## Exercises

5.1. The data in Table 5.13 (from Ref. 6) were obtained on a well believed to be stabilized at each rate. Using equations in $p^2$ (strictly speaking, not applicable in this pressure range), estimate the AOF using (a) the empirical method and (b) the theoretical method.

Also, do the following: (c) plot the theoretical deliverability curve on the same graph paper as the empirical curve; (d) since $p^2$ equations are not accurate at this pressure level, develop and outline a theoretical method based on equations in $p$; and (e) apply equations in $p$ to these data; in particular, calculate the AOF.

5.2. Cullender[7] presented data from an isochronal test and from an earlier, longer test that led to approximate stabilization in 214 hours test time.

In the 214-hour test, the rate was 1,156 Mscf/D, the shut-in pressure was 441.6 psia, and the flowing BHP was 401.4 psia. Using the data in Table 5.14, (a) determine the AOF with both empirical and theoretical methods, and (b) establish plots (on the same graph paper) of the empirical and theoretical stabilized deliverability curves.

5.3. Confirm $\psi(p)$ results stated in Example 5.4 for pressures in the range $450 \leq p \leq 3{,}150$ psia.

5.4. The well discussed in Example 5.5 was produced at 2,000 Mscf/D for 90 days and then shut in for a pressure buildup test. Data obtained in the buildup test are given in Table 5.15. Determine formation permeability and apparent skin factor using an analysis procedure based on equations written in terms of pseudopressure, $\psi(p)$.

## References

1. Al-Hussainy, R., Ramey, H.J. Jr., and Crawford, P.B.: "The Flow of Real Gases Through Porous Media," *J. Pet. Tech.* (May 1966) 624-636; *Trans.*, AIME, **237**.
2. Wattenbarger, R.A. and Ramey, H.J. Jr.: "Gas Well Testing With Turbulence, Damage, and Wellbore Storage," *J. Pet. Tech.* (Aug. 1968) 877-887; *Trans.*, AIME, **243**.
3. Dake, L.P.: *Fundamentals of Reservoir Engineering*, Elsevier Scientific Publishing Co., Amsterdam (1978).
4. *Theory and Practice of the Testing of Gas Wells*, third edition, Pub. ECRB-75-34, Energy Resources and Conservation Board, Calgary, Alta. (1975).
5. Craft, B.C. and Hawkins, M.F. Jr.: *Applied Petroleum Reservoir Engineering*, Prentice-Hall Book Co., Inc., Englewood Cliffs, NJ (1959).
6. *Back Pressure Test for Natural Gas Wells*, Revised edition, Railroad Commission of Texas (1951).
7. Cullender, M.H.: "The Isochronal Performance Method of Determining the Flow Characteristics of Gas Wells," *Trans.*, AIME (1955) **204**, 137-142.

# Chapter 6
# Other Well Tests

## 6.1 Introduction

This concluding chapter surveys four well-testing techniques not yet discussed in the text: interference tests; pulse tests; drillstem tests; and wireline formation tests. These tests and others covered in previous chapters by no means exhaust the subject; however, the comprehensive treatment needed by the practitioner is provided by SPE monographs[1,2] and the Canadian gas well testing manual.[3]

## 6.2 Interference Testing

Interference tests have two major objectives. They are used (1) to determine whether two or more wells are in pressure communication (i.e., in the same reservoir) and (2) when communication exists, to provide estimates of permeability $k$ and porosity/compressibility product, $\phi c_t$, in the vicinity of the tested wells.

An interference test is conducted by producing from or injecting into at least one well (the active well) and by observing the pressure response in at least one other well (the observation well). Fig. 6.1 indicates the typical test program with one active well and one observation well.

As the figure indicates, an active well starts producing from a reservoir at uniform pressure at Time 0. Pressure in an observation well, a distance $r$ away, begins to respond after some time lag (related to the time for the radius of investigation corresponding to the rate change at the active well to reach the observation well). The pressure in the active well begins to decline immediately, of course. The magnitude and timing of the deviation in pressure response at the observation well depends on reservoir rock and fluid properties in the vicinity of the active and observation wells.

Vela and McKinley[4] showed that these properties are values from the area investigated in the test – a rectangle with sides of length $2r_i$ and $2r_i+r$ (see Fig. 6.2). In Fig. 6.2, $r_i$ is the radius of investigation achieved by the active well during the test and $r$ is the distance between active and observation wells. The essential point is that the region investigated is much greater than some small area between wells, as intuition might suggest.

In an infinite-acting, homogeneous, isotropic reservoir, the simple *Ei*-function solution to the diffusivity equation describes the pressure change at the observation well as a function of time:

$$p_i-p_r=-70.6\frac{qB\mu}{kh}Ei\left(-948\frac{\phi\mu c_t r^2}{kt}\right). \quad \ldots (6.1)$$

This is simply a restatement of a familiar result. The pressure drawdown at radius $r$ (i.e., the observation well) resulting from production from the active well at rate $q$, starting from a reservoir initially at uniform pressure $p_i$, is given by the *Ei*-function solution. Eq. 6.1 assumes that the skin factor of the active well does not affect the drawdown at the observation well. Wellbore storage effects also are assumed negligible at both the active and observation wells when Eq. 6.1 is used to model an interference test. Jargon[5] shows that both these assumptions can lead to error in test analysis in some cases.

A convenient analysis technique for interference tests is the use of type curves. Fig. 6.3 is a type curve presented by Earlougher;[1] it is simply the *Ei* function expressed as a function of its usual argument in flow problems, $948\ \phi\mu c_t r^2/kt$. Note that Eq. 6.1 can be expressed completely in terms of dimensionless variables:

$$\frac{p_i-p_r}{\left(141.2\dfrac{qB\mu}{kh}\right)}=-\frac{1}{2}Ei\left|\left(-\frac{1}{4}\right)\cdot\left(\frac{\phi\mu c_t r_w{}^2}{0.000264\,kt}\right)\left(\frac{r}{r_w}\right)^2\right|,$$

or

$$p_D=-\frac{1}{2}Ei\left(\frac{-r_D^2}{4t_D}\right), \quad \ldots (6.2)$$

where

$$p_D=\frac{(p_i-p_r)\,kh}{141.2\,qB\mu},$$

$$r_D=r/r_w,$$

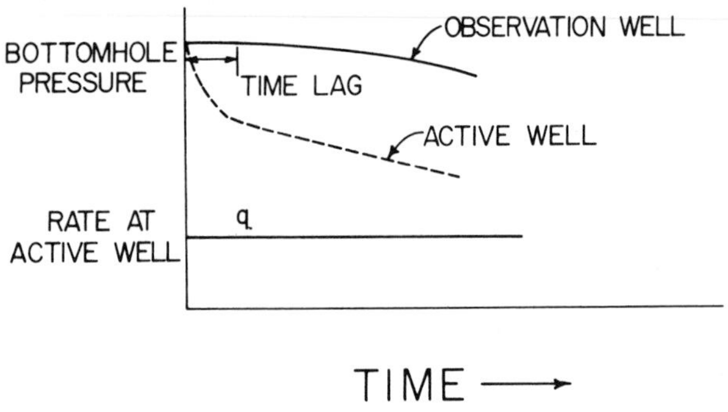


**Fig. 6.1** – Pressure response in interference test.

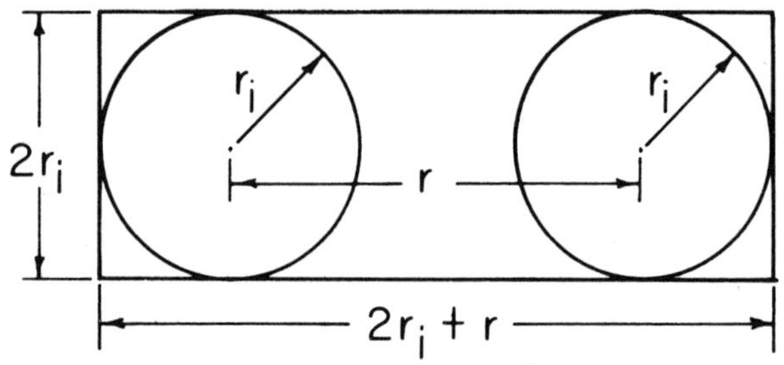


**Fig. 6.2** – Region investigated in interference test.

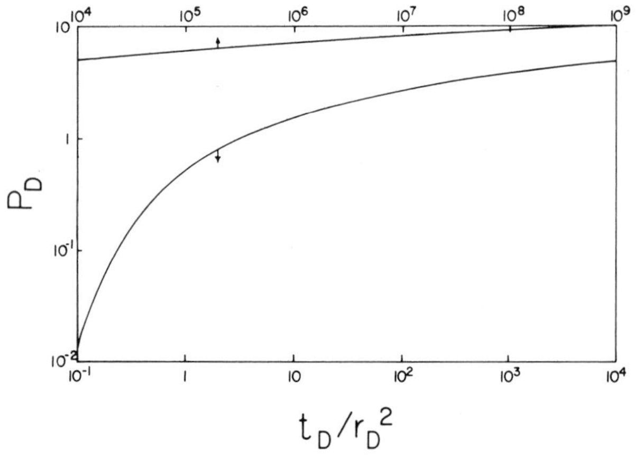


**Fig. 6.3** – Exponential integral solution.

and

$$t_D = \frac{0.000264\,kt}{\phi \mu c_t r_w^2}.$$

Fig. 6.3 can be used in the following way to analyze interference tests.

1. Plot pressure drawdown in an observation well, $\Delta p = p_i - p_r$, vs. elapsed time $t$ on the same size log-log paper as the full-scale, type-curve version of Fig. 6.3 using an undistorted curve (the reader can prepare such a curve easily).
2. Slide the plotted test data over the type curve until a match is found. (Horizontal and vertical sliding both are required.)
3. Record pressure and time match points, $(p_D)_{MP}$, $\Delta p_{MP}$ and $[(t_D/r_D^2)_{MP}, t_{MP}]$.
4. Calculate permeability $k$ in the test region from the pressure match point:

$$k = 141.2 \frac{qB\mu}{h} \frac{(p_D)_{MP}}{(\Delta p)_{MP}}.$$

5. Calculate $\phi c_t$ from the time match point:

$$\phi c_t = \left(\frac{0.000264\,k}{\mu r^2}\right)\left[\frac{t_{MP}}{(t_D/r_D^2)_{MP}}\right].$$

## *Example 6.1 – Interference Test in Water Sand*

**Problem.** An interference test was run in a shallow-water sand. The active well, Well 13, produced 466 STB/D water. Pressure response in shut-in Well 14, which was 99 ft from Well 13, was measured as a function of time elapsed since the drawdown in Well 13 began. Estimated rock and fluid properties include $\mu_w = 1.0$ cp, $B_w = 1.0$ RB/STB, $h = 9$ ft, $r_w = 3$ in., and $\phi = 0.3$. Total compressibility is unknown. Pressure readings in Well 14 were as given in Table 6.1. Estimate formation permeability and total compressibility.

**Solution.** We assume that the aquifer is homogeneous, isotropic, and infinite-acting; we use the *Ei*-function type curves to estimate $k$ and $c_t$. Data to be plotted are presented in Table 6.2. The data fit the *Ei*-function type curve well. A pair of match points are ($\Delta t = 128$ minutes, $t_D/r_D^2 = 10$) and ($\Delta p = 5.1$ psi, $p_D = 1.0$). (See Fig. 6.4.) Thus,

$$k = 141.2 \frac{qB\mu}{h} \frac{(p_D)_{MP}}{(\Delta p)_{MP}}$$

$$= \frac{(141.2)(466)(1.0)(1.0)}{(9.0)} \frac{(1.0)}{(5.1)}$$

$$= 1{,}433 \text{ md},$$

and

$$c_t = \frac{0.000264}{\phi r^2} \frac{k}{\mu} \frac{(t_{MP}/60)}{(t_D/r_D^2)_{MP}}$$

$$= \frac{(0.000264)(1{,}433)(128/60)}{(0.3)(99)^2(1.0)(10)}$$

$$= 2.74 \times 10^{-5} \text{ psi}^{-1}.$$

## 6.3 Pulse Testing

Pulse tests[6] have the same objectives as conventional interference tests – to determine whether well pairs are in pressure communication and to estimate $k$ and $\phi c_t$ in the area of the tested wells. The tests are conducted by sending a coded signal or pulse sequence from an active well (producer or injector) to a shut-in observation well. The pulse sequence is created by producing from (or injecting into) the active well, then shutting it in, and repeating that sequence in a regular pattern. An example is indicated in Fig. 6.5.

The reason for the sequence of pressure pulses is that we readily can determine the effect of an active well on an observation well amid the established trend in reservoir pressure and random perturbations (noise) to that trend. Highly sensitive pressure gauges usually are required to detect these small coded pulses, which may have magnitudes of less than 0.1 psi.

Pulse testing has several advantages over conventional interference tests:

1. It disrupts normal operations much less than interference testing does. It lasts a minimum time, which may range from a few hours to a few days.

2. There are fewer interpretation problems caused by random noise and by trends in reservoir pressure as they affect pressure response at observation wells.

3. Pulse test analysis usually can be based on simple solutions to the flow equations – specifically, superposition of *Ei*-function solutions, which assume infinite-acting, homogeneous reservoirs. In many cases, longer interference tests require that boundaries be taken into account.

Analysis techniques for pulse tests usually are based on simulating the pressure response in an observation well with the familiar *Ei*-function solution to the diffusivity equation, using superposition to model the rate changes in the pulsing sequence. From the simulations of pulse tests, charts relating key characteristics of the tests to reservoir properties have been developed.[7] Before we discuss these charts (Figs. 6.7 through 6.14) and their application, it will be useful to introduce nomenclature used in pulse test analysis, using the system of Earlougher[1] (and his schematic pulse-test rate and pressure-response history).

Fig. 6.6 illustrates the time lag $t_L$ which is the time elapsed between the end of a pulse and the pressure peak caused by the pulse. The radius-of-investigation concept prepares us to expect a time lag. A finite period of time is required for a pulse caused by producing an active well to move to a responding well, and the subsequent transient created by a shut-in period also requires a finite time period to affect pressure response.

The amplitude $\Delta p$ of a pulse can be represented

**TABLE 6.1 – PRESSURE/TIME DATA FROM INTERFERENCE TEST**

| $\Delta t$ (minutes) | $p_{ws}$ (psia) |
|---|---|
| 0 | 148.92 |
| 5 | 148.92 |
| 25 | 144.91 |
| 40 | 143.72 |
| 50 | 143.18 |
| 100 | 141.47 |
| 200 | 139.72 |
| 300 | 138.70 |
| 400 | 137.99 |
| 580 | 137.12 |

**TABLE 6.2 – INTERFERENCE TEST DATA FOR LOG-LOG PLOT**

| $\Delta t$ (minutes) | $\Delta p = p_i - p_{wf}$ (psia) |
|---|---|
| 0 | 0 |
| 5 | 0 |
| 25 | 4.01 |
| 40 | 5.20 |
| 50 | 5.74 |
| 100 | 7.45 |
| 200 | 9.20 |
| 300 | 10.22 |
| 400 | 10.93 |
| 580 | 11.80 |

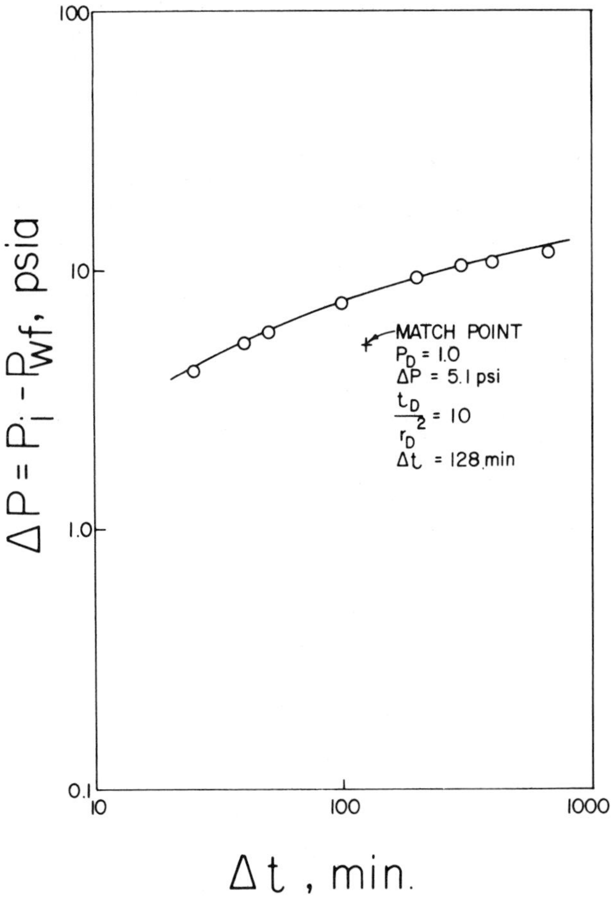


**Fig. 6.4** – Interference test data from water reservoir.

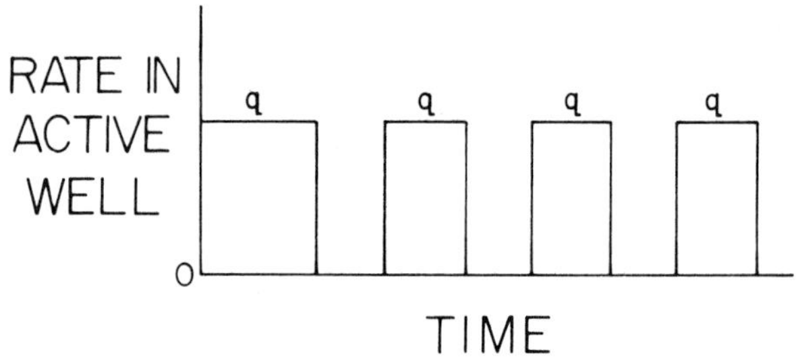


Fig. 6.5 – Typical rate schedule in pulse test.

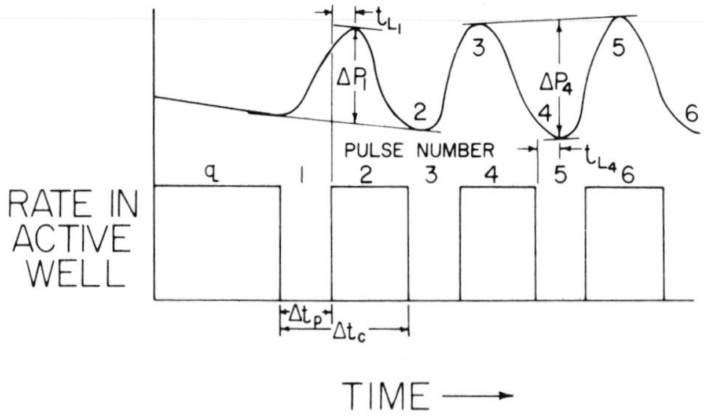


Fig. 6.6 – Pressure response in pulse test.

conveniently as the vertical distance between two adjacent peaks (or valleys) and a line parallel to this through the valley (or peak), as illustrated in Fig. 6.6.

The length of the pulse period and total cycle length (including both shut-in and flow periods) are represented by $\Delta t_p$ and $\Delta t_c$, respectively.

Analysis of simulated pulse tests shows that Pulse 1 (the first odd pulse) and Pulse 2 (the first even pulse) have characteristics that differ from all subsequent pulses; beyond these initial responses, all odd pulses have similar characteristics and all even pulses also have similar characteristics.

We now define dimensionless variables that are used in Figs. 6.7 through 6.14, which were designed for quantitative analysis of pulse tests: dimensionless time lag, $(t_L)_D = 0.000264\ kt_L/\phi\mu c_t r_w^2$; dimensionless distance between active and observation wells, $r_D = r/r_w$; and dimensionless pressure-response amplitude, $\Delta p_D = kh\Delta p/141.2\ qB\mu$ (with sign convention that $\Delta p/q$ is always positive).

Figs. 6.7 through 6.14 provide the correlations to be used in pulse test analysis. Figs. 6.7 and 6.8 are to be used for the first odd (i.e., first) pulse; Figs. 6.9 and 6.10 for the first even (i.e., second) pulse; Figs. 6.11 and 6.12 for all other odd-numbered pulses (third, fifth, etc.); and Figs. 6.13 and 6.14 for all other even-numbered pulses (fourth, sixth, etc.). The figures use the ratios (1) $F'$ = pulse length ($\Delta t_p$) to cycle length ($\Delta t_c$) and (2) time lag ($t_L$) to cycle length ($\Delta t_c$). The figures appropriate for a given pulse number are used to obtain values of $\Delta p_D(t_L/\Delta t_c)^2$ and $[(t_L)_D/r_D^2]$, which are then used to provide estimates of $k$ and $\phi c_t$.

$$k = 141.2\frac{qB\mu\Delta p_D(t_L/\Delta t_c)^2}{h\Delta p(t_L/\Delta t_c)^2},$$

$$\phi c_t = \frac{0.000264\ kt_L}{\mu r^2[(t_L)_D/r_D^2]}.$$

Example 6.2 illustrates how these figures are applied.

---

## *Example 6.2 – Pulse Test Analysis*

**Problem.** A pulse test was run in a reservoir in which the distance between wells, $r$, was 933 ft. Formation fluid viscosity, $\mu$, was 0.8 cp; formation thickness, $h$, was 26 ft; and porosity, $\phi$, was 0.08. In the test following rate stabilization, the active well was shut in for 2 hours, then produced for 2 hours, shut in for 2 hours, etc. Production rate, $q$, was 425 STB/D and formation volume factor, $B$, was 1.26 RB/STB. The amplitude $\Delta p$ of the *fourth* pulse (Fig. 6.15) was 0.629 psi, and the time lag was 0.4 hour. From these data, estimate $k$ and $\phi c_t$.

**Solution.** To analyze the fourth pulse, we use Figs. 6.13 and 6.14. Using Fig. 6.13 first to determine $\Delta p_D(t_L/\Delta t_c)^2$, and thus $k$, we note that

$$F' = \Delta t_p/\Delta t_c = 2/(2+2) = 0.5,$$

$$t_L/\Delta t_c = 0.4/4 = 0.1.$$

Then, from Fig. 6.13,

$$\Delta p_D(t_L/\Delta t_c)^2 = 0.00221,$$

and

$$k = 141.2\frac{qB\mu\ \Delta p_D(t_L/\Delta t_c)^2}{h\Delta p(t_L/\Delta t_c)^2}$$

$$= \frac{(141.2)(425)(1.26)(0.8)(0.00221)}{(26)(0.629)(0.1)^2}$$

$$= 817 \text{ md}.$$

From Fig. 6.14,

$$(t_L)_D/r_D^2 = 0.091.$$

Thus,

$$\phi c_t = \frac{0.000264\ kt_L}{\mu r^2[(t_L)_D/r_D^2]}$$

$$= \frac{(0.000264)(817)(0.4)}{(0.8)(933)^2(0.091)}$$

$$= 1.36 \times 10^{-6}$$

Then,

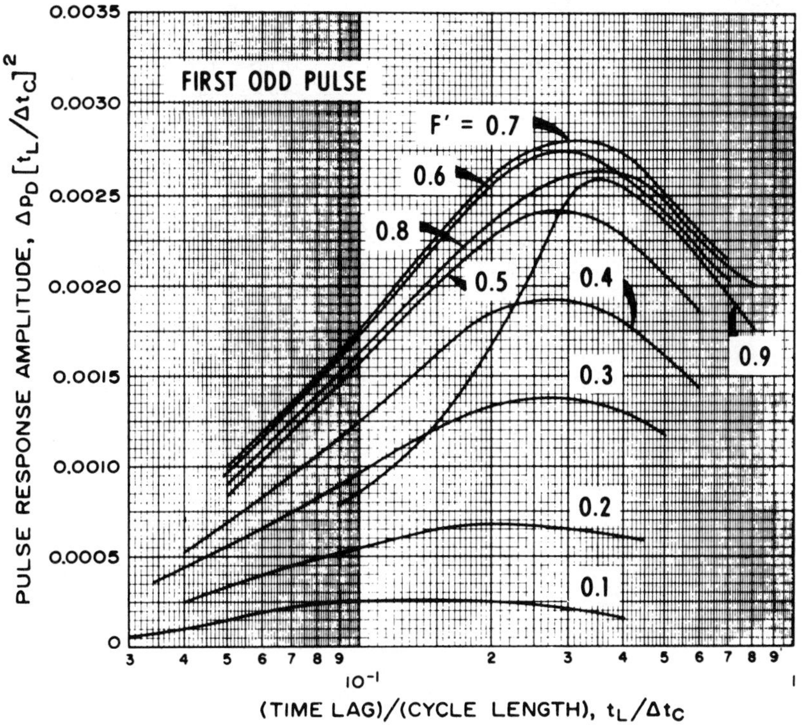


Fig. 6.7 – Relation between time lag and response amplitude for first odd pulse.[1]

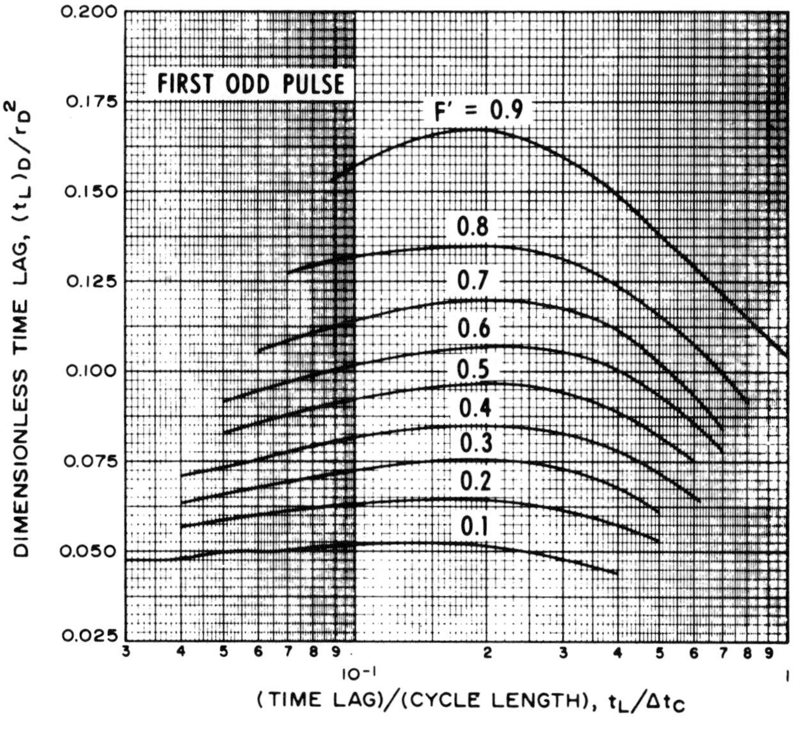


Fig. 6.8 – Relation between time lag and cycle length for first odd pulse.[1]

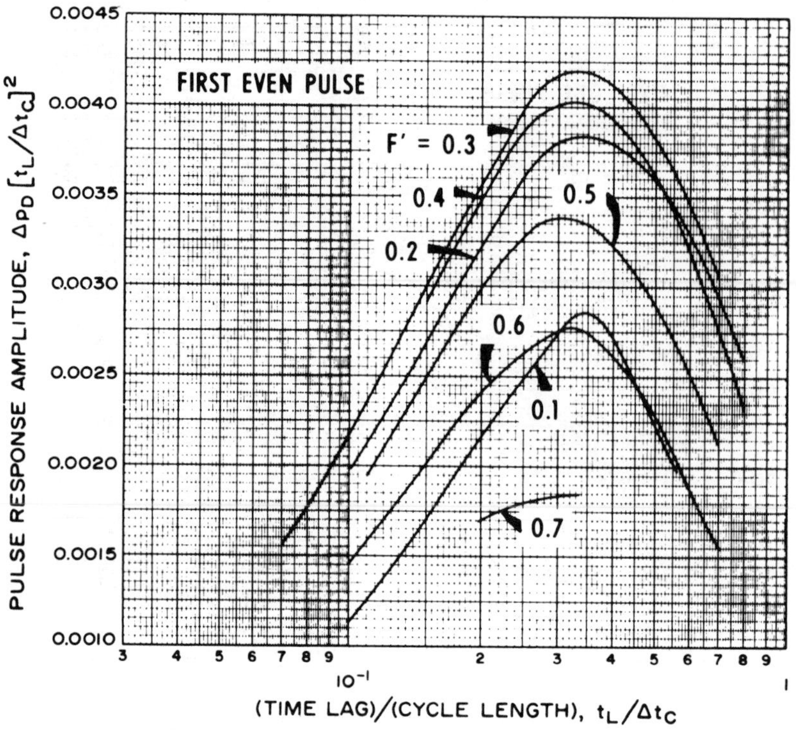


Fig. 6.9 – Relation between time lag and response amplitudes for first even pulse.[1]

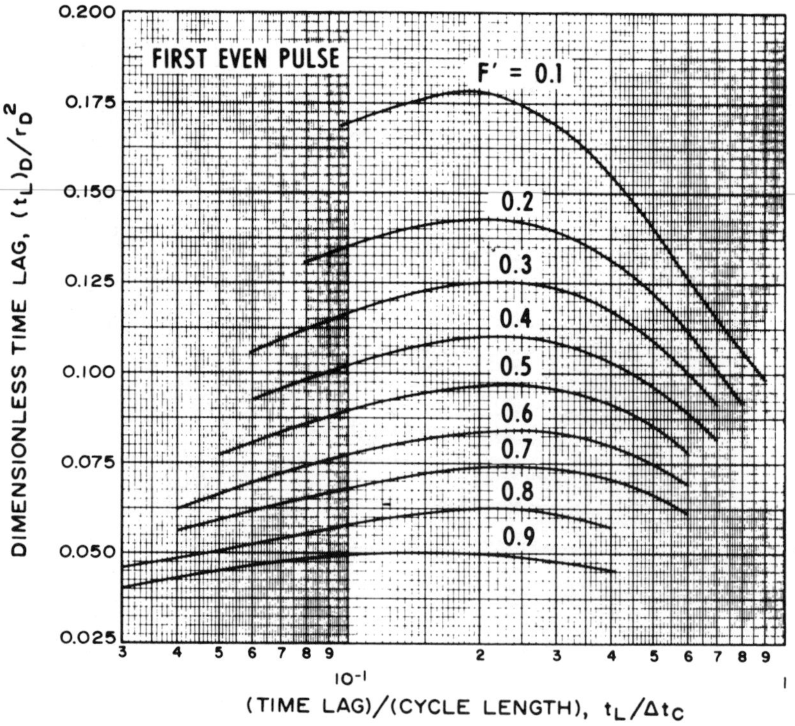


Fig. 6.10 – Relation between time lag and cycle length for first even pulse.[1]

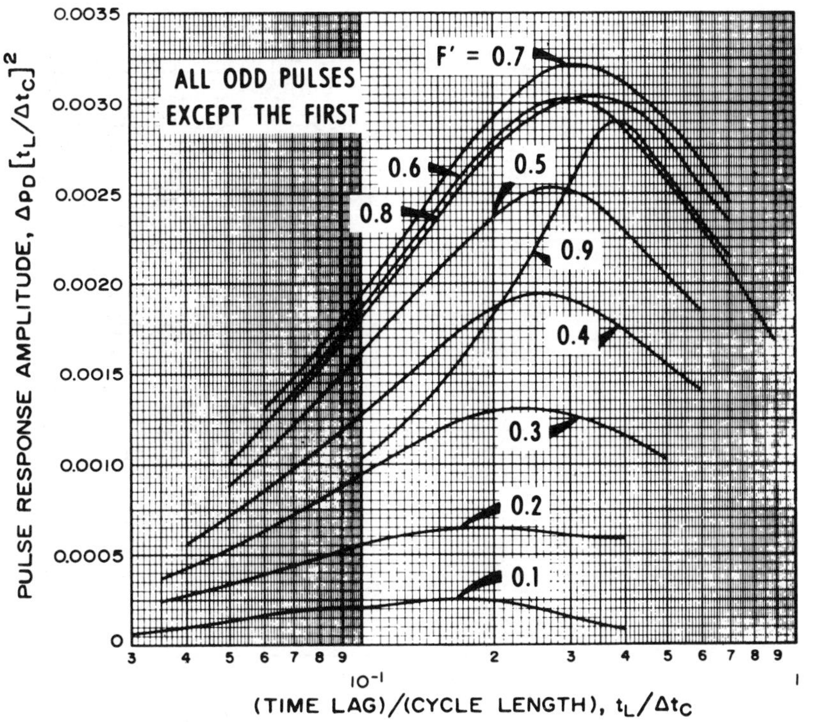


Fig. 6.11 – Relation between time lag and response amplitude for all odd pulses after the first.[1]

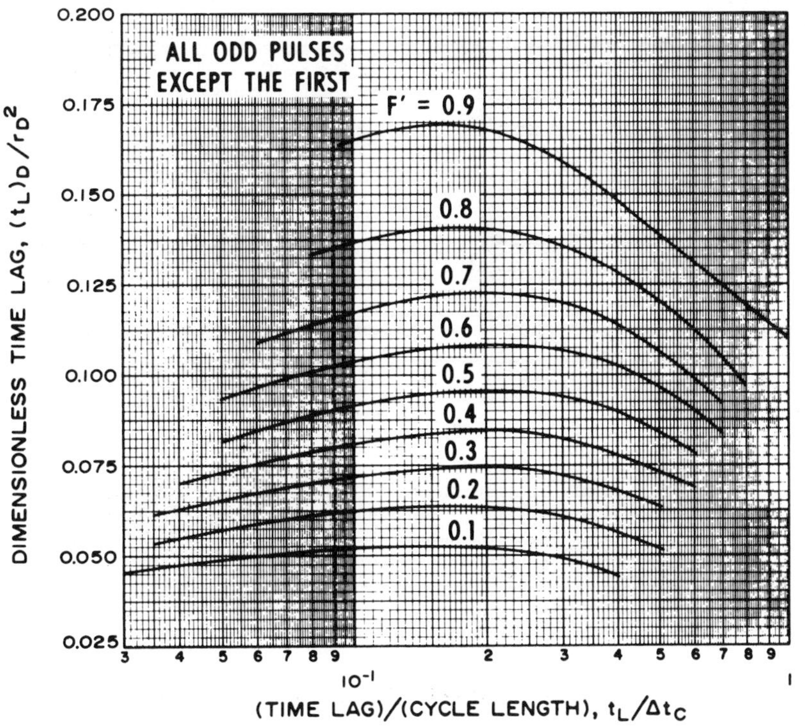


Fig. 6.12 – Relation between time lag and cycle length for all odd pulses after the first.[1]

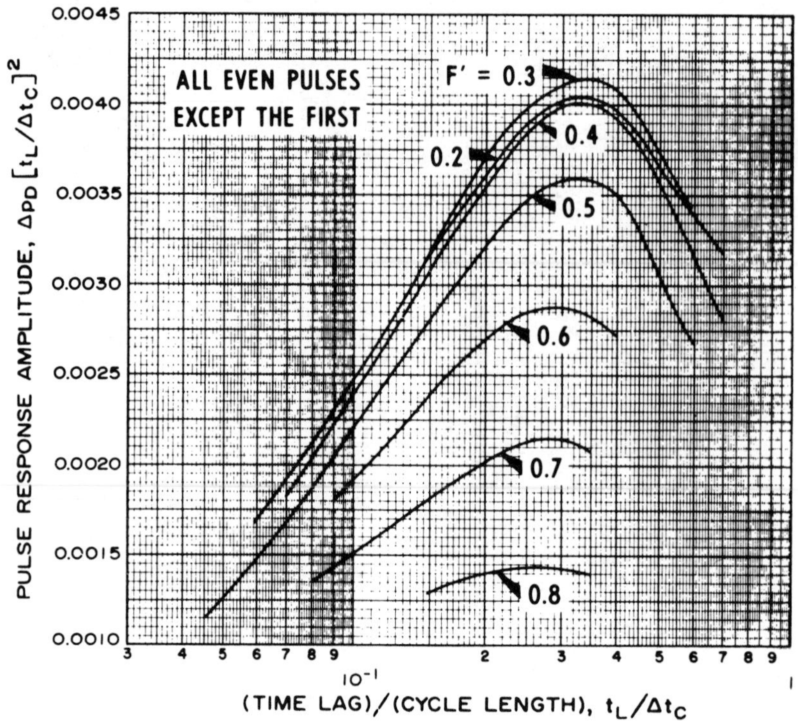


Fig. 6.13 – Relation between time lag and response amplitude for all even pulses after the first.[1]

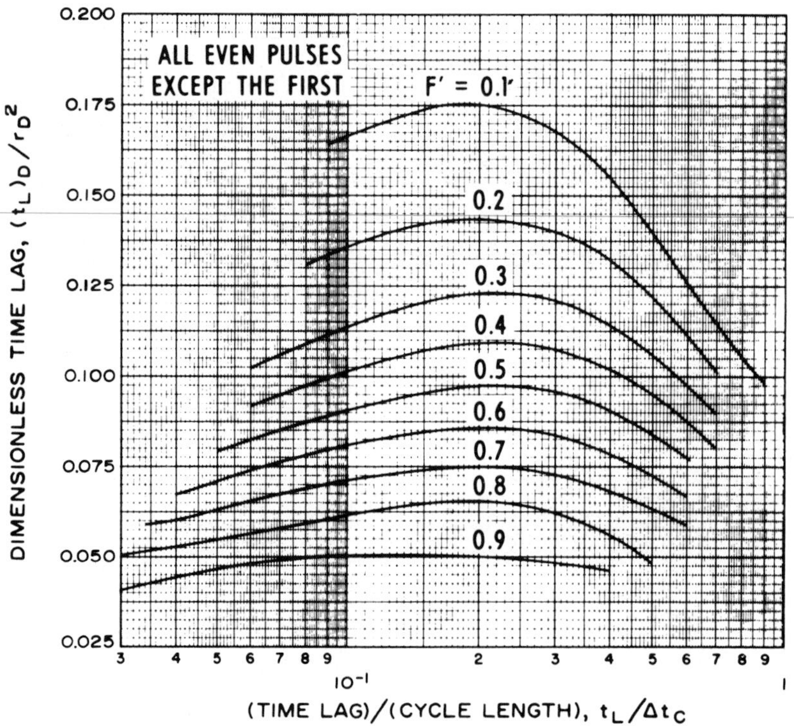


Fig. 6.14 – Relation between time lag and cycle length for all even pulses after the first.[1]

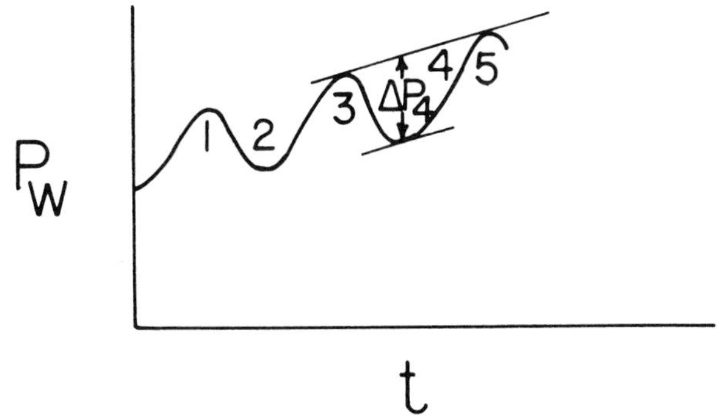


Fig. 6.15 – Schematic of pressure response in pulse test.

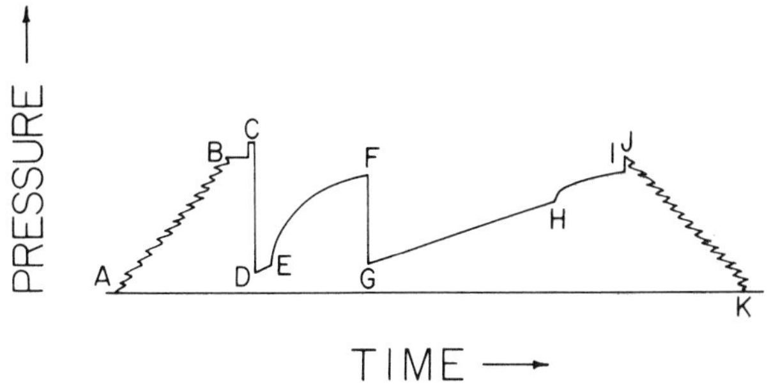


Fig. 6.16 – Schematic of drillstem test pressure chart.

$$c_t = \frac{1.36 \times 10^{-6}}{0.08} = 17 \times 10^{-6} \text{ psi}^{-1}.$$

---

## 6.4 Drillstem Tests

A drillstem test (DST)[8,9] provides a means of estimating formation and fluid properties before completion of a well. Basically, a DST is a temporary completion of a well. The DST tool is an arrangement of packers and valves placed on the end of the drillpipe. This arrangement can be used to isolate a zone of interest and to let it produce into the drillpipe or drillstem. A fluid sample is obtained in the test; thus, the test can tell us the types of fluids the well will produce if it is completed in the tested formation.

With the surface-actuated valves on a DST device, it is possible to have a sequence of flow periods followed by shut-in periods. A pressure recorder on the DST device can record pressures during the flow and shut-in periods. The pressures recorded during the shut-in periods can be particularly valuable for estimating formation characteristics such as permeability/thickness product and skin factor. These data also can be used to determine possible pressure depletion during the test.

To illustrate how a typical DST is performed, we will examine a schematic chart (Fig. 6.16) of pressure vs. time from a test with two flow periods and two shut-in periods.

At Point A, the tool is lowered into the hole. Between Points A and B, the ever-increasing mud-column pressure is recorded; at Point B, the tool is on bottom. When the packers are set, the mud column is compressed and a still higher pressure is recorded at Point C. The tool is opened for an initial flow period, and the pressure drops to Point D as shown. As fluid accumulates in the drillstem above the pressure gauge, the pressure rises. Finally, at Point E, the well is shut in for an initial pressure buildup test. After a suitable shut-in period, the well is reopened for a second final flow period, from Point G to Point H. This final flow period is followed by a final shut-in period (from Point H to Point I). The packers are then released, and the hydrostatic pressure of the mud column is again imposed on the pressure gauge. The testing device is then removed from the hole (Point J to Point K).

The initial flow period is usually brief (5 to 10 minutes); its purpose is to draw down the pressure slightly near the wellbore (perhaps letting any mud-filtrate-invaded zone bleed back to or below static reservoir pressure). The initial shut-in period, often 30 to 60 minutes, is designed to let the pressure build back to true static formation pressure. This initial shut-in pressure on a DST may be the best measurement made of static reservoir pressure.

The second flow period is designed to capture a large sample of formation fluid and to draw down the pressure in the formation to the maximum distance and extent possible within the time that is possible to allow for the DST – frequently 30 minutes to several hours. The second shut-in period is designed to obtain good pressure buildup data so that formation properties can be estimated. In addition, comparison of the final (or extrapolated) pressure from the second shut-in period to the initial shut-in pressure can indicate that pressure depletion has occurred during the DST and that the well thus has been tested in a small, noncommercial reservoir. The desired length of the second shut-in period varies from equal to the second flow period (for high-permeability formations) to twice the length of the second flow period (for low-permeability formations).

Theory much like that used for an ordinary pressure buildup test following production at constant rate is used for analyzing the shut-in periods on a DST. This is true even though the flow rate preceding a shut-in period in a DST usually decreases continuously. Usually, the *average* production rate can be used as a good approximation in buildup test analyses; this average rate of production is determined by dividing the fluid recovery by the length of the flow period.

To analyze the buildup test, we plot $p_{ws}$ vs. log $(t_p + \Delta t)/\Delta t$, where $t_p$ is now the *actual* flowing time at the average rate $q$. The permeability/thickness product is found from the relationship $kh = 162.6 \, qB\mu/m$. Usually, a fluid sample will not yet have been analyzed in the laboratory; accordingly, correlations (Appendix D) relating $\mu$ and $B$ to produced fluid properties must be used.

TABLE 6.3 – DRILLSTEM TEST DATA

| $\Delta t$ (minutes) | $p_{ws}$ (psi) |
|---|---|
| 0 | 350 |
| 5 | 965 |
| 10 | 1,215 |
| 15 | 1,405 |
| 20 | 1,590 |
| 25 | 1,685 |
| 30 | 1,725 |
| 35 | 1,740 |
| 40 | 1,753 |
| 45 | 1,765 |

Static reservoir pressure is found by extrapolating the buildup tests to infinite shut-in time. In a dual shut-in test, we have the opportunity to extrapolate both buildup tests to infinite shut-in time and to compare the estimated static reservoir pressure. If the static reservoir pressure from the final shut-in period is significantly lower than that from the initial shut-in period, it is possible that the reservoir was partially pressure depleted even during the relatively short DST, implying that the formation tested is probably noncommercial.

Skin factor is calculated from the conventional skin-factor equation:

$$s=1.151\left[\frac{(p_{1\,\mathrm{hr}}-p_{wf})}{m}-\log\left(\frac{k}{\phi\mu c_t r_w^2}\right)+3.23\right]. \quad (2.4)$$

There is a further complication in DST analysis. At the time of the test, reservoir rock and fluid properties that appear in Eq. 2.4 may not yet be known accurately. This is particularly true of porosity, $\phi$, and total compressibility, $c_t$. Accordingly, one may be forced to use the best available estimates and to recognize that skin-factor and radius-of-investigation calculations, which also depend on these properties, may be subject to considerable uncertainty. Use of Eq. 2.3 instead of Eq. 2.4 [Eq. 2.3 includes the sometimes important term $\log\,(t_p+\Delta t)/t_p$] to calculate $s$ may be justified in some cases if other data in the equation are known with unusual accuracy.

## 6.5 Wireline Formation Tests

In many areas, hole conditions prohibit use of DST's as temporary wellbore completions. In these areas, and in others where the costs of the required numbers of DST's for complete evaluation are prohibitive, wireline formation tests[10,11] frequently are used in formation evaluation work.

A wireline formation tester is, in effect, a sample chamber of up to several gallons capacity combined with pressure gauges. The test chambers are forced against the borehole wall in a sealing pad, and the formation is perforated by firing a shaped charge. The signal to fire the charge is transmitted on logging cable. Fluid is collected during sampling, and pressure is recorded. Following sample collection, shut-in pressures are recorded as they build up with time.

Flow into the sample chamber is probably approximately spherical (i.e., into a point rather than spread uniformly across an entire productive interval). For this reason, the shut-in test cannot be analyzed as in a DST, although theory based on steady-state spherical flow may explain wireline test buildup pressure satisfactorily in some cases.[11] The device is useful for obtaining samples of formation fluid and estimating initial formation pressure; extensive use of the device for this latter application has been reported in the literature.[11]

## Exercises

6.1 Determine the duration of an interference test required to achieve a pressure drawdown of 25 psig at the observation well for the reservoir described in Example 6.1 if the active well produces 500 STB/D throughout the test. What will be the radius of investigation at this time? If the skin factor in the active well is 2.0, what will be the drawdown in the active well?

6.2 For the pulse test described in Example 6.2, given the results of the test analysis of the fourth pulse, determine the time lag and pressure response for (a) the first odd pulse; (b) the first even pulse; (c) the third pulse; (d) the fifth pulse; and (e) the sixth pulse.

6.3 Johnston-Schlumberger[9] reports data below from a DST:

initial flow period = 5 minutes,
initial shut-in period = 30 minutes,
final flow period = 60 minutes, and
final shut-in period = 45 minutes.

Data obtained in the final shut-in period were as given in Table 6.3. In the initial shut-in period, the pressure reached and remained stable at 1,910 psi. Total fluid recovered in both initial and final flow periods filled 300 ft of 2½-in.-ID drill collars (0.0061 bbl/ft) and 300 ft of 4½-in. drillpipe (0.0142 bbl/ft). The produced fluid was 35°API oil with a measured gas rate of 47 Mscf/D at the surface (assumed solution gas). Formation temperature was measured at 120°F. Porosity is estimated to be 10%; total compressibility is $8.4\times10^{-6}$ psi$^{-1}$; wellbore radius is 4.5 in; and formation thickness is 10 ft.

Estimate formation permeability, skin factor, flow efficiency, and radius of investigation achieved in the test.

6.4 Smolen and Litsey[11] propose that "borehole-corrected," steady-state, spherical flow into the wireline formation tester can be modeled by

$$k=3{,}300\,q\mu/\Delta p,$$

where

$k$ = permeability, md,
$q$ = flow rate, $cm^3/s$ (reservoir conditions),
$\mu$ = fluid viscosity (usually mud filtrate), cp, and
$\Delta p$ = drawdown from formation pressure, psi.

A test showed that formation pressure was 3,850 psi. Pressure was drawn down in the sample chamber to an approximately constant 1,850 psi by withdrawing 10 $cm^3$ of filtrate ($\mu = 0.5$ cp) from the formation in 16 seconds.

Estimate formation permeability from the test data.

Derive an equation for steady-state spherical flow and show that it has the form $k = \text{constant} \times qB\mu/[r(p_i - p_{wf})]$. State the assumptions required for the equation to model a wireline formation tester flow test.

## References


1. Earlougher, R.C. Jr.: *Advances in Well Test Analysis*, Monograph Series, SPE, Dallas (1977) **5**.
2. Matthews, C.S. and Russell, D.G.: *Pressure Buildup and Flow Tests in Wells*, Monograph Series, SPE, Dallas (1967) **1**.
3. *Theory and Practice of the Testing of Gas Wells*, third edition, Pub. ERCB−75-34, Energy Resources and Conservation Board, Calgary, Alta. (1975).
4. Vela, S. and McKinley, R.M.: "How Areal Heterogeneities Affect Pulse-Test Results," *Soc. Pet. Eng. J.* (June 1970) 181-191; *Trans.*, AIME, **249**.
5. Jargon, J.R.: "Effect of Wellbore Storage and Wellbore Damage at the Active Well on Interference Test Analysis," *J. Pet. Tech.* (Aug. 1976) 851-858.
6. Johnson, C.R., Greenkorn, R.A., and Woods, E.G.: "Pulse-Testing: A New Method for Describing Reservoir Flow Properties Between Wells," *J. Pet. Tech.* (Dec. 1966) 1599-1604; *Trans.*, AIME, **237**.
7. Kamal, M. and Brigham, W.E.: "Pulse-Testing Response for Unequal Pulse and Shut-In Periods," *Soc. Pet. Eng. J.* (Oct. 1975) 399-410; *Trans.*, AIME, **259**.
8. Edwards, A.G. and Winn, R.H.: "A Summary of Modern Tools and Techniques Used in Drillstem Testing," Pub. T-4069, Halliburton Co., Duncan, OK (Sept. 1973).
9. "Review of Basic Formation Evaluation," Form J-328, Johnston-Schlumberger, Houston (1976).
10. Schultz, A.L., Bell, W.T., and Urbanosky, H.J.: "Advancements in Uncased-Hole, Wireline Formation-Tester Techniques," *J. Pet. Tech.* (Nov. 1975) 1331-1336.
11. Smolen, J.J. and Litsey, L.R.: "Formation Evaluation Using Wireline Formation Tester Pressure Data," *J. Pet. Tech.* (Jan. 1979) 25-32.

# Appendix A
# Development of Differential Equations for Flow in Porous Media

## Introduction

In this appendix, we develop some of the basic differential equations that describe the flow of fluids in porous media. Results presented include equations for three-dimensional flow of slightly compressible liquids, and for radial flow of slightly compressible liquids, gases, and simultaneous flow of oil, water, and gas. In developing these equations, we start with continuity equations (mass balances); then we introduce flow laws (such as Darcy's law) and appropriate equations of state for the fluid considered.

## Continuity Equation for Three-Dimensional Flow

To develop continuity equations, we use a mass balance on a small element of porous material. The balance has the following form.

(rate of mass flow into element) − (rate of mass flow out of element) = (rate of accumulation of mass within element).

Our element is shown in Fig. A-1. It has dimensions $\Delta x$, $\Delta y$, and $\Delta z$, in the $x$, $y$, and $z$ coordinate system; for convenience, the coordinate system is oriented such that gravitational forces are in the $(-)z$ direction. We denote the components of the volumetric rate of flow per unit cross-sectional area (cubic feet per hour-square feet or feet per hour are typical engineering units) by $u_x$, $u_y$, and $u_z$.

The rate at which mass enters the element in the $x$ direction is $\rho u_x \Delta y \Delta z$ (lbm/cu ft × ft/hr × sq ft = lbm/hr); the rate at which the mass leaves the element in the $x$ direction is $[\rho u_x + \Delta(\rho u_x)]\Delta y \Delta z$. Similar expressions describe rates of mass entering and leaving in the $y$ and $z$ directions. The result of adding these expressions is

(rate of mass flow into element)
− (rate of mass flow out of element)

$$= \rho u_x \Delta y \Delta z + \rho u_y \Delta x \Delta z + \rho u_z \Delta x \Delta y - \left[\rho u_x + \Delta(\rho u_x)\right]\Delta y \Delta z - \left[\rho u_y + \Delta(\rho u_y)\right]\Delta x \Delta z - \left[\rho u_z + \Delta(\rho u_z)\right]\Delta x \Delta y$$

$$= -\Delta(\rho u_x)\Delta y \Delta z - \Delta(\rho u_y)\Delta x \Delta z - \Delta(\rho u_z) \cdot \Delta x \Delta y.$$

To determine the rate at which fluid accumulates within the element, we first note that the mass within the element (of porosity $\phi$) at a given time is $\rho\phi\Delta x \Delta y \Delta z$ (lbm/cu ft × cu ft = lbm). Thus, the rate at which this mass changes over a time interval $\Delta t$ is

$$\frac{\left(\rho\phi|_{t+\Delta t} - \rho\phi|_t\right)}{\Delta t} \Delta x \Delta y \Delta z,$$

where time, $t$, is in hours. Accordingly, the mass balance becomes

$$-\Delta(\rho u_x)\Delta y \Delta z - \Delta(\rho u_y)\Delta x \Delta z - \Delta(\rho u_z)\Delta x \Delta y = \frac{\rho\phi|_{t+\Delta t} - \rho\phi|_t}{\Delta t} \Delta x \Delta y \Delta z.$$

If we divide each term by $\Delta x \Delta y \Delta z$ and take the limit as $\Delta t, \Delta x, \Delta y$, and $\Delta z \rightarrow 0$, the result is

$$\frac{\partial(\rho u_x)}{\partial x} + \frac{\partial(\rho u_y)}{\partial y} + \frac{\partial(\rho u_z)}{\partial z} = -\frac{\partial}{\partial t}(\rho\phi). \quad \text{(A.1)}$$

## Continuity Equation for Radial Flow

From a mass balance similar to that used to develop Eq. A.1, the continuity equation for one-dimensional, radial flow can be shown to be

$$\frac{1}{r}\frac{\partial}{\partial r}(r\rho u_r) = -\frac{\partial}{\partial t}(\phi\rho), \quad \text{(A.2)}$$

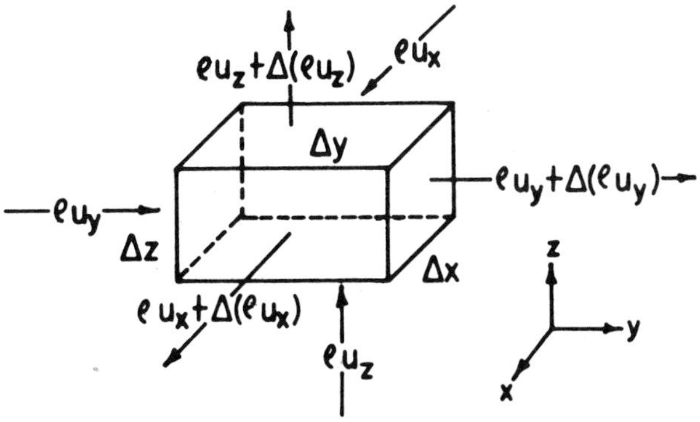


**Fig. A-1** – Element of porous medium used for mass balance.

where $u_r$ is the volumetric flow rate per unit cross-section area in the radial direction. Eq. A.2 is less general than Eq. A.1; in particular, radial flow *only* is assumed [i.e., there is no flow in the $z$ or $\theta$ "direction" in a cylindrical $(r,z,\theta)$ coordinate system].

## Flow Laws

Liquid flow usually is described by Darcy's law. This law, when applied in the coordinate system orientation we have chosen to describe three-dimensional flow, becomes, in field units,

$$u_x = -0.001127 \frac{k_x}{\mu} \frac{\partial p}{\partial x},$$

$$u_y = -0.001127 \frac{k_y}{\mu} \frac{\partial p}{\partial y},$$

$$u_z = -0.001127 \frac{k_z}{\mu} \left( \frac{\partial p}{\partial z} + 0.00694\rho \right), \quad \ldots\ldots \text{(A.3)}$$

The $k_i$ are the permeabilities in direction $i$ and $p$ denotes pressure (psi). In Eq. A.3, we have assumed that the $x$ and $y$ directions are horizontal, so that gravity acts only in the $z$ direction.

After substituting these equations into the continuity equation, the result is

$$\frac{\partial}{\partial x}\left(\frac{k_x \rho}{\mu}\frac{\partial p}{\partial x}\right) + \frac{\partial}{\partial y}\left(\frac{k_y \rho}{\mu}\frac{\partial p}{\partial y}\right)$$

$$+ \frac{\partial}{\partial z}\left[\frac{k_z \rho}{\mu}\left(\frac{\partial p}{\partial z} + 0.00694\rho\right)\right]$$

$$= \frac{1}{0.000264} \frac{\partial}{\partial t}(\phi\rho). \quad \ldots\ldots\ldots\ldots\ldots \text{(A.4)}$$

For radial flow, the result is

$$\frac{1}{r}\frac{\partial}{\partial r}\left(\frac{r\rho k_r}{\mu}\frac{\partial p}{\partial r}\right) = \frac{1}{0.000264}\frac{\partial}{\partial t}(\rho\phi). \quad \ldots\ldots \text{(A.5)}$$

## Single-Phase Flow of Slightly Compressible Fluids

To solve Eqs. A.4 and A.5 analytically, we must introduce additional assumptions. First, we restrict our analysis to slightly compressible liquids – those with constant compressibility, $c$, where $c$ is defined by the equation

$$c = \frac{-1}{V}\frac{dV}{dp} = \frac{1}{\rho}\frac{d\rho}{dp}. \quad \ldots\ldots\ldots\ldots\ldots\ldots \text{(A.6)}$$

For constant compressibility $c$, integration of Eq. A.6 gives

$$\rho = \rho_o e^{c(p-p_o)}, \quad \ldots\ldots\ldots\ldots\ldots\ldots\ldots \text{(A.7)}$$

where $\rho_o$ is the value of $\rho$ at some reference pressure $p_o$. Eq. A.7 describes most single-phase liquids adequately.

If we now introduce Eq. A.7 into Eq. A.4 and assume that (1) $k_x = k_y = k_z = \text{constant}$, (2) gravitational forces are negligible, (3) $\phi = \text{constant}$, and (4) $\mu = \text{constant}$, Eq. A.4 becomes

$$\frac{\partial}{\partial x}\left[e^{c(p-p_o)}\frac{\partial p}{\partial x}\right] + \frac{\partial}{\partial y}\left[e^{c(p-p_o)}\frac{\partial p}{\partial y}\right]$$

$$+ \frac{\partial}{\partial z}\left[e^{c(p-p_o)}\frac{\partial p}{\partial z}\right]$$

$$= \frac{\phi\mu}{0.000264\,k}\frac{\partial}{\partial t}\left[e^{c(p-p_o)}\right].$$

Simplifying,

$$\frac{\partial^2 p}{\partial x^2} + \frac{\partial^2 p}{\partial y^2} + \frac{\partial^2 p}{\partial z^2}$$

$$+ c\left[\left(\frac{\partial p}{\partial x}\right)^2 + \left(\frac{\partial p}{\partial y}\right)^2 + \left(\frac{\partial p}{\partial z}\right)^2\right]$$

$$= \frac{\phi\mu c}{0.000264\,k}\frac{\partial p}{\partial t}.$$

If we further assume that

$$c\left[\left(\frac{\partial p}{\partial x}\right)^2 + \left(\frac{\partial p}{\partial y}\right)^2 + \left(\frac{\partial p}{\partial z}\right)^2\right]$$

is negligible compared with other terms in the equation (which requires either small $c$ or small pressure gradients, or both), then

$$\frac{\partial^2 p}{\partial x^2} + \frac{\partial^2 p}{\partial y^2} + \frac{\partial^2 p}{\partial z^2} = \frac{\phi\mu c}{0.000264\,k}\frac{\partial p}{\partial t}. \quad \ldots\ldots \text{(A.8)}$$

For radial flow, the corresponding equation is

$$\frac{1}{r}\frac{\partial}{\partial r}\left(r\frac{\partial p}{\partial r}\right) = \frac{\phi\mu c}{0.000264\,k}\frac{\partial p}{\partial t}. \quad \ldots\ldots\ldots\ldots \text{(A.9)}$$

Eqs. A.8 and A.9 are diffusivity equations. Analytical solutions to Eq. A.9 are known for several simple boundary conditions; these solutions are used for most well test analysis. We must remember that Eqs. A.8 and A.9 *are not* general; instead, they are based on several important assumptions, including (1) the single-phase liquid flowing has small and constant compressibility; (2) $k$ is constant and the same in all directions (isotropic); (3) $\phi$ is constant; and (4) pressure gradients are small.

## Single-Phase Gas Flow

For gas flow characterized by Darcy's law and for a gas described by the equation of state,

$$\rho = \frac{M}{RT}\frac{p}{z},$$

Eq. A.4 becomes, for constant $\phi$ and $k$ and negligible gravitational forces,

$$\frac{\partial}{\partial x}\left(\frac{p}{\mu z_g}\frac{\partial p}{\partial x}\right)+\frac{\partial}{\partial y}\left(\frac{p}{\mu z_g}\frac{\partial p}{\partial y}\right)+\frac{\partial}{\partial z}\left(\frac{p}{\mu z_g}\frac{\partial p}{\partial z}\right)$$

$$=\frac{\phi}{0.000264\,k}\frac{\partial}{\partial t}\left(\frac{p}{z_g}\right). \quad \text{(A.10)}$$

In this section only, we use $z_g$ to denote the real gas law compressibility factor so that we can continue to use the symbol $z$ for one of the space variables. For gases, $\mu$ and $z_g$ are functions of pressure and cannot be assumed constant except in special cases. To reduce Eq. A.10 to a form similar to Eq. A.8, we define a pseudopressure,[1] $\psi(p)$, as follows:

$$\psi(p)=2\int_{p_\beta}^{p}\frac{p}{\mu z_g}dp, \quad \text{(A.11)}$$

where $p_\beta$ is a low base pressure. Now,

$$\frac{\partial}{\partial t}\left(\frac{p}{z_g}\right)=\frac{d\left(\frac{p}{z_g}\right)}{dp}\frac{\partial p}{\partial t}$$

$$=\frac{c_g p}{z_g}\frac{\partial p}{\partial t},$$

since

$$c_g=\frac{1}{\rho}\frac{d\rho}{dp}=\frac{z_g}{p}\frac{d\left(\frac{p}{z_g}\right)}{dp}.$$

Also note that

$$\frac{\partial \psi}{\partial t}=\frac{\partial \psi}{\partial p}\frac{\partial p}{\partial t}=\frac{2p}{\mu z_g}\frac{\partial p}{\partial t},$$

and

$$\frac{\partial \psi}{\partial x}=\frac{2p}{\mu z_g}\frac{\partial p}{\partial x}.$$

Similar expressions apply for $\partial\psi/\partial y$ and $\partial\psi/\partial z$. Thus, Eq. A.10 becomes

$$\frac{\partial}{\partial x}\left(\frac{\partial \psi}{\partial x}\right)+\frac{\partial}{\partial y}\left(\frac{\partial \psi}{\partial y}\right)+\frac{\partial}{\partial z}\left(\frac{\partial \psi}{\partial z}\right)$$

$$=\frac{\phi\mu c_g}{0.000264\,k}\frac{\partial \psi}{\partial t}. \quad \text{(A.12)}$$

For radial flow, the equivalent of Eq. A.12 is

$$\frac{1}{r}\frac{\partial}{\partial r}\left(r\frac{\partial \psi}{\partial r}\right)=\frac{\phi\mu c_g}{0.000264\,k}\frac{\partial \psi}{\partial t}. \quad \text{(A.13)}$$

Eqs. A.12 and A.13 are similar in form to Eqs. A.8 and A.9, but there is one important difference. The coefficient $\phi\mu c/0.000264\,k$ is constant in Eqs. A.8 and A.9; in Eqs. A.12 and A.13, it is a function of pressure. Despite this complication, many solutions to Eqs. A.8 and A.9 prove to be accurate approximate solutions of Eqs. A.12 and A.13.

## Simultaneous Flow of Oil, Water, and Gas

In this section, we outline a detailed derivation of an equation describing radial, simultaneous flow of oil, gas, and water. More complete discussions and more general equations are given by Matthews and Russell[2] and Martin.[3]

We assume that a porous medium contains oil, water, and gas, and that each phase has saturation-dependent effective permeability ($k_o$, $k_w$, and $k_g$); time-dependent saturation ($S_o$, $S_w$, and $S_g$); pressure-dependent formation volume factor ($B_o$, $B_w$, and $B_g$); and pressure-dependent viscosity ($\mu_o$, $\mu_w$, and $\mu_g$). When gravitational forces and capillary pressures are negligible, the differential equation describing this type of flow is

$$\frac{1}{r}\frac{\partial}{\partial r}\left(r\frac{\partial p}{\partial r}\right)=\frac{\phi c_t}{0.000264\,\lambda_t}\frac{\partial p}{\partial t}, \quad \text{(A.14)}$$

where

$$\lambda_t=\frac{k_o}{\mu_o}+\frac{k_g}{\mu_g}+\frac{k_w}{\mu_w}, \quad \text{(A.15)}$$

$$c_t=S_o c_o+S_w c_w+S_g c_g+c_f, \quad \text{(A.16)}$$

and

$$c_o=-\frac{1}{B_o}\frac{dB_o}{dp}+\frac{B_g}{B_o}\frac{dR_s}{dp}. \quad \text{(A.17)}$$

Note in Eq. A.17 that effective oil compressibility depends not only on the usual change in liquid volume with pressure but also on the change with pressure of the dissolved GOR, $R_s$.

A similar expression can be written for water compressibility:

$$c_w=-\frac{1}{B_w}\frac{dB_w}{dp}+\frac{B_g}{B_w}\frac{dR_{sw}}{dp}. \quad \text{(A.18)}$$

## Exercises

A.1 Derive Eq. A.2; state the assumptions required in this derivation.

A.2 Derive Eq. A.5; state the assumptions required in this derivation.

A.3 Derive Eq. A.9; state the assumptions required in this derivation.

A.4 Derive Eq. A.13; state the assumptions required in this derivation. Compare the assumptions required to derive Eq. A.9 with those required to derive Eq. A.13.

## References

1. Al-Hussainy, R., Ramey, H.J. Jr., and Crawford, P.B.: "The Flow of Real Gases Through Porous Media," *J. Pet. Tech.* (May 1966) 624-636; *Trans.*, AIME, **237**.
2. Matthews, C.S. and Russell, D.G.: *Pressure Buildup and Flow Tests In Wells*, Monograph Series, SPE, Dallas (1967) **1**.
3. Martin, J.C.: "Simplified Equations of Flow in Gas Drive Reservoirs and the Theoretical Foundation of Multiphase Pressure Buildup Analysis," *Trans.*, AIME (1959) **216**, 309-311.

# Appendix B
# Dimensionless Variables

## Introduction

It is convenient and customary to present graphical or tabulated solutions to flow equations, such as Eq. A.9, in terms of dimensionless variables. In this way, it is possible to present compactly solutions for a wide range of parameters $\phi$, $\mu$, $c$, and $k$, and variables $r$, $p$, and $t$.

In this appendix, we show how many of the dimensionless variables that appear in the well-testing literature arise logically and directly in the differential equations (and in their initial and boundary conditions) that describe flow in porous media.

## Radial Flow of a Slightly Compressible Fluid

In this section, we identify the dimensionless variables and parameters required to characterize the solutions to the equations describing radial flow of a slightly compressible liquid in a reservoir. We assume that Eq. A.9 adequately models this flow. Specifically, we analyze the situation in which (1) pressure throughout the reservoir is uniform before production; (2) fluid is produced at a constant rate from a single well of radius $r_w$ centered in the reservoir; and (3) there is no flow across the outer boundary (with radius $r_e$) of the reservoir. Stated mathematically, the differential equation, and initial and boundary conditions are

$$\frac{1}{r}\frac{\partial}{\partial r}\left(r\frac{\partial p}{\partial r}\right)=\frac{\phi\mu c}{0.000264\,k}\frac{\partial p}{\partial t}, \qquad \text{(B.1)}$$

at $t=0$, $p=p_i$ for all $r$,

$$\text{at } r=r_e,\ q=0 \text{ for } t>0, \qquad \text{(B.2)}$$

or

$$\left.\frac{\partial p}{\partial r}\right|_{r_e}=0, \qquad \text{(B.3)}$$

$$\text{at } r=r_w,\ q=\frac{-0.001127\,(2\pi r_w h)}{B}\,\frac{k}{\mu}\left.\frac{\partial p}{\partial r}\right|_{r_w} \quad \text{for } t>0$$

or

$$\left.\frac{\partial p}{\partial r}\right|_{r_w}=-\frac{qB\mu}{0.00708\,khr_w}. \qquad \text{(B.4)}$$

This boundary condition arises from Darcy's law in the form similar to that used in Eq. A.3:

$$u_r=-0.001127\frac{k_r}{\mu}\frac{\partial p}{\partial r}=\frac{qB}{2\pi hr}.$$

Our objective in this analysis is to restate the differential equation, and initial and boundary conditions in dimensionless form so we can determine the dimensionless variables and parameters that characterize this flow situation and that can be used to characterize solutions. These dimensionless parameters and variables are not unique (i.e., more than one choice can be made for each). Further, we want to emphasize that these dimensionless variables are *defined* rather than *derived* quantities. These ideas will become clearer as we proceed.

We *define* a dimensionless radius, $r_D=r/r_w$ (any other convenient reference length, such as $r_e$, could have been used). From the form of the differential equation, we also note that a convenient definition of dimensionless time is $t_D=0.000264\,kt/\phi\mu c r_w^2$.

The initial and boundary conditions suggest that a convenient definition of dimensionless pressure is

$$p_D=\frac{0.00708\,kh\,(p_i-p)}{qB\mu}.$$

With this definition, the boundary condition (Eq. B.4) becomes

$$\frac{-qB\mu}{0.00708\,khr_w}\left.\frac{\partial p_D}{\partial r_D}\right|_{r_{D=1}}=\frac{-qB\mu}{0.00708\,khr_w},$$

or simply

$$\left.\frac{\partial p_D}{\partial r_D}\right|_{r_{D=1}}=1.$$

Expressed in terms of dimensionless variables, the differential equation and its initial and boundary conditions become

$$\frac{1}{r_D}\frac{\partial}{\partial r_D}\left(r_D\frac{\partial p_D}{\partial r_D}\right)=\frac{\partial p_D}{\partial t_D}, \qquad \text{(B.5)}$$

$$p_D = 0 \text{ for all } r_D \text{ at } t_D = 0, \quad \ldots\ldots\ldots\ldots\ldots \text{(B.6)}$$

$$\left.\frac{\partial p_D}{\partial r_D}\right|_{r_D = r_e/r_w = r_{De}} = 0 \text{ for } t_D > 0, \quad \ldots\ldots\ldots\ldots \text{(B.7)}$$

$$\left.\frac{\partial p_D}{\partial r_D}\right|_{r_D = 1} = 1 \text{ for } t_D > 0. \quad \ldots\ldots\ldots\ldots\ldots \text{(B.8)}$$

The implication is that any solution, $p_D$, of Eq. B.5 is a unique function of $r_D$ and $t_D$ for fixed $r_{De}$; no other dimensionless variables appear in either the dimensionless differential equation or in the dimensionless initial or boundary conditions describing this particular problem. Thus, if we wish to present solutions to Eq. B.5, we could do so compactly either by tabulating or by plotting $p_D$ at $r_D = 1$ $(r = r_w)$ as a function of the variable $t_D$ with the parameter $r_{De}$. In fact, such tables and graphs have been prepared and presented; this problem is precisely the well known van Everdingen and Hurst constant terminal-rate problem* for which they present solutions as functions of $t_D$ and $r_{De}$.

We have implied that the groups of parameters and variables that arise from the differential equation and its initial and boundary conditions are dimensionless. We will now provide the proof for the case just considered.

Dimensionless variables are

$$r_D = r/r_w,$$

$$t_D = \frac{0.000264\,kt}{\phi \mu c r_w^2},$$

and

$$p_D = \frac{0.00708\,kh\,(p_i - p)}{qB\mu}.$$

Obviously, $r_D$ is dimensionless. To show that $t_D$ and $p_D$ are dimensionless, we introduce the symbol [ ], which denotes "has units of." Let m denote mass, L, length, and t, time. The quantities that appear in $t_D$ and $p_D$ have the following basic units:

$k \sim [\mathrm{L}^2]$,
$t \sim [\mathrm{t}]$,
$\phi \sim [1]$ (i.e., dimensionless),
$\mu \sim [\mathrm{m/Lt}]$,
$c \sim [\mathrm{Lt}^2/\mathrm{m}]$,
$r_w \sim [\mathrm{L}]$,
$h \sim [\mathrm{L}]$,
$p \sim [\mathrm{m/Lt}^2]$,
$q \sim [\mathrm{L}^3/\mathrm{t}]$, and
$B \sim [1]$.

A comment on the units of pressure, $p$, and compressibility, $c$ (which has the units psi$^{-1}$) may be helpful: Pressure is defined as force per unit area. From Newton's second law, force is mass times acceleration; thus, force has the fundamental units

$$\text{Force} \simeq [\mathrm{mL/t}^2].$$

Thus, pressure has the fundamental units

$$\text{Pressure} \simeq [\mathrm{mL/t^2L^2}] \text{ or } [\mathrm{m/Lt}^2],$$

and compressibility has the fundamental units

$$\text{Compressibility} \simeq [\mathrm{Lt}^2/\mathrm{m}].$$

Then,

$$t_D = \frac{0.000264\,kt}{\phi \mu c r_w^2} \sim \frac{[\mathrm{L}^2][\mathrm{t}]}{[1]\left|\frac{m}{\mathrm{Lt}}\right|\left|\frac{\mathrm{Lt}^2}{m}\right|[\mathrm{L}^2]} \sim [1].$$

Thus, $t_D$ has units of unity, or, more plainly, is dimensionless. Similarly,

$$p_D = \frac{0.00708\,kh\,(p_i - p)}{qB\mu} \sim \frac{[\mathrm{L}^2][\mathrm{L}]\left|\frac{\mathrm{m}}{\mathrm{Lt}^2}\right|}{\left|\frac{\mathrm{L}^3}{\mathrm{t}}\right|[1]\left|\frac{\mathrm{m}}{\mathrm{Lt}}\right|} \sim [1].$$

Thus, $p_D$ also is dimensionless.

Again, we stress the reason for introduction of somewhat unnatural dimensionless quantities. They allow solutions for wide ranges of $k$, $h$, $c$, $t$, $r_e$, $r_w$, $q$, $\mu$, and $B$ to be presented compactly (tables or graphs) as functions of a minimum number of variables and parameters. Such tabulations and graphs are in widespread use in well test analysis.

## Radial Flow With Constant BHP

To illustrate further how the choice of appropriate dimensionless variables depends on the specific differential equation, and initial and boundary conditions, we now determine appropriate dimensionless variables for radial flow of a slightly compressible liquid through a wellbore of radius $r_w$ from a reservoir of radius $r_e$. There is no flow across the outer boundary (at $r = r_e$); initial pressure, $p_i$, is uniform before production; and flowing BHP in the wellbore, $p_{wf}$, is held constant once production begins. The mathematical problem is then

$$\frac{1}{r}\frac{\partial}{\partial r}\left(r\frac{\partial p}{\partial r}\right) = \frac{\phi \mu c}{0.000264\,k}\frac{\partial p}{\partial t}, \quad \ldots\ldots\ldots\ldots \text{(B.9)}$$

$$p = p_i \text{ at } t = 0 \text{ for all } r, \quad \ldots\ldots\ldots\ldots\ldots\ldots \text{(B.10)}$$

$$p = p_{wf} \text{ at } r = r_w \text{ for all } t > 0, \quad \ldots\ldots\ldots\ldots \text{(B.11)}$$

$$\left.\frac{\partial p}{\partial r}\right|_{r_e} = 0 \text{ for all } t > 0. \quad \ldots\ldots\ldots\ldots\ldots\ldots \text{(B.12)}$$

As before, our approach will be to eliminate all parameters and variables with dimensions from the differential equation, and initial and boundary conditions. The appropriate definition of $r_D$ is again $r_D = r/r_w$; for $t_D$, it is

$$t_D = \frac{0.000264\,kt}{\phi \mu c r_w^2}.$$

These definitions will eliminate all parameters from Eq. B.9. The appropriate definition of $p_D$ for this situation is

*See Appendix C.

$$p_D = \frac{p_i - p}{p_i - p_{wf}}.$$

With these definitions, the mathematical statement of the problem becomes.

$$\frac{1}{r_D}\frac{\partial}{\partial r_D}\left(r_D \frac{\partial p_D}{\partial r_D}\right) = \frac{\partial p_D}{\partial t_D}, \quad \ldots\ldots\ldots\ldots (B.13)$$

$$p_D = 0 \text{ at } t_D = 0 \text{ for all } r_D, \quad \ldots\ldots\ldots\ldots (B.14)$$

$$p_D = 1 \text{ at } r_D = 1 \text{ for all } t_D > 0, \quad \ldots\ldots\ldots\ldots (B.15)$$

$$\left.\frac{\partial p_D}{\partial r_D}\right|_{r_D = r_e/r_w = r_{De}} = 0 \text{ for all } t_D > 0. \quad \ldots\ldots\ldots (B.16)$$

Thus, dimensionless pressure $p_D$ is a function of dimensionless time, $t_D$, and dimensionless radius $r_D$ for a given value of the dimensionless parameter $r_{eD}$ since no other variables or parameters appear in the differential equation, or initial or boundary conditions. However, the appropriate definition of dimensionless pressure is different in this constant-pressure case from the constant-rate case – with the appropriate definition dictated by the boundary conditions.

Because this constant-pressure case is of considerable practical importance, we proceed further with our analysis. It is of interest to determine instantaneous production rate and cumulative produced volume for this case. Instantaneous production rate $q$ is

$$q = -\frac{0.001127\,(2\pi r_w h)}{B}\,\frac{k}{\mu}\left.\frac{\partial p}{\partial r}\right|_{r_w}, \quad \ldots\ldots\ldots (B.17)$$

and cumulative production $Q_p$ is

$$Q_p = \frac{1}{24}\int_o^t q\,dt. \quad \ldots\ldots\ldots\ldots\ldots\ldots (B.18)$$

It is convenient to define a dimensionless production rate, $q_D$:

$$q_D = \frac{qB\mu}{0.00708\,kh\,(p_i - p_{wf})}.$$

Then,

$$q_D = \left.\frac{\partial p_D}{\partial r_D}\right|_{r_D=1} \quad \ldots\ldots\ldots\ldots\ldots\ldots (B.19)$$

We also define a dimensionless cumulative produced volume, $Q_{pD}$.

$$Q_{pD} = \int_o^{t_D} q_D\,dt_D = \left[\frac{B\mu}{0.00708\,kh\,(p_i - p_{wf})}\right] \cdot \left(\frac{0.000264\,k}{\phi\mu c r_w^2}\right)\int_o^t q\,dt.$$

Therefore,

$$Q_p = 1.119\,\frac{\phi c h r_w^2\,(p_i - p_{wf})}{B}\,Q_{pD}. \quad \ldots\ldots (B.20)$$

The implication of our analysis is that cumulative production, $Q_p$ (stock-tank barrels), can be determined from dimensionless cumulative production, $Q_{pD}$, which is based on solutions to Eqs. B.13 through B.16. Since $Q_{pD}$ is based on these solutions, it is a function of $t_D$ and $r_{De}$ only (for $r_D = 1$); thus, it appears possible that $Q_{pD}$ could be tabulated (or plotted) as a function of $t_D$ for selected values of $r_{De}$. In fact, this has been done by van Everdingen and Hurst; this problem is their well known constant-pressure case.

The $Q_{pD}$ values also can be interpreted as dimensionless cumulative water influx from a radial aquifer into a reservoir if $r_w$ is interpreted as the reservoir radius and $r_e$ as the aquifer radius.

## Exercise

B.1 Consider the radial flow of a gas expressed in terms of the pseudopressure, $\psi$, by Eq. A.13. Define dimensionless pseudopressure as

$$\psi_D = \frac{kh\,T_{sc}}{50{,}300\,q_g p_{sc} T}(\psi_i - \psi),$$

dimensionless time as

$$t_D = \frac{0.000264\,kt}{\phi\mu_i c_{gi} r_w^2},$$

and

$$r_D = r/r_w,$$

$T$ = reservoir temperature, °R,
$T_{sc}$ = standard-condition temperature, °R,
$p_{sc}$ = standard-condition pressure, psia,
$q_g$ = gas flow rate, Mscf/D,
$\mu_i$ = gas viscosity evaluated at original reservoir pressure, cp, and
$c_{gi}$ = gas compressibility evaluated at original reservoir pressure, psi$^{-1}$.

The cylindrical reservoir is initially at uniform pressure, $p_i$; there is no flow across the outer boundary of radius $r_e$; and flow rate into the wellbore of radius $r_w$ is constant (expressed at standard conditions).

Write the differential equation, and initial and boundary conditions in dimensional form; then rewrite them in dimensionless form.

# Appendix C
# Van Everdingen and Hurst Solutions to Diffusivity Equations

## Introduction

In Appendix B, we showed that solutions to differential equations describing flow in a petroleum reservoir for given initial and boundary conditions can be expressed compactly using dimensionless variables and parameters. In this appendix, we examine four of these solutions that are important in reservoir engineering applications.

## Constant Rate at Inner Boundary, No Flow Across Outer Boundary

This solution of the diffusivity equation models radial flow of a slightly compressible liquid in a homogeneous reservoir of uniform thickness; reservoir at uniform pressure $p_i$ before production; no flow across the outer boundary (at $r=r_e$); and production at constant rate $q$ from the single well (centered in the reservoir) with wellbore radius $r_w$. The solution – pressure as a function of time and radius for fixed values of $r_e$, $r_w$, and rock and fluid properties – is expressed most conveniently in terms of dimensionless variables and parameters:

$$p_D = f(t_D, r_D, r_{eD}), \quad \text{(C.1)}$$

where

$$p_D = \frac{0.00708\,kh(p_i - p)}{qB\mu},$$

$$t_D = \frac{0.000264kt}{\phi\mu c_t r_w^2},$$

$$r_D = r/r_w,$$

and

$$r_{eD} = r_e/r_w.$$

Eq. C.1 states that $p_D$ is a function of the variables $t_D$ and $r_D$ for a fixed value of the parameter $r_{eD}$. The most important solution is that for pressure at the wellbore radius ($r=r_w$ or $r_D=1$):

$$p_D\big|_{r_{D=1}} = f(t_D, r_{eD}).$$

When expressed in terms of dimensionless pressure evaluated at $r_D=1$, Eq. 1.6 shows the functional form of $f(t_D, r_{eD})$ – an infinite series of exponentials and Bessel functions. This series has been evaluated[1] for several values of $r_{eD}$ over a wide range of values of $t_D$. Chatas[2] tabulated these solutions; a modification of Chatas' tabulation is presented in Tables C.1 and C.2.

Some important characteristics of this tabulation include the following.

1. Table C.1 presents values of $p_D$ in the range $t_D < 1{,}000$ for an infinite-acting reservoir. For $t_D < 0.01$, $p_D$ can be approximated by the relation

$$p_D \simeq 2\sqrt{t_D/\pi}. \quad \text{(C.2)}$$

2. Table C.1 is valid for finite reservoirs with $t_D < 0.25\,r_{eD}^2$.

3. For $100 < t_D < 0.25\,r_{eD}^2$, Table C.1 can be extended by use of the equation

$$p_D \simeq 0.5(\ln t_D + 0.80907). \quad \text{(C.3)}$$

(This equation is identical to Eq. 1.10 for $s=0$. It begins to become slightly less accurate at $t_D \simeq 0.0625\,r_{eD}^2$, but there is no simple approximation between this value and $t_D \simeq 0.25\,r_{eD}^2$. Fortunately, this slight loss in accuracy is not of sufficient magnitude to cause problems in most practical applications.)

4. Table C.2 presents $p_D$ as a function of $t_D$ for $1.5 < r_{eD} < 10$.

(a) For values of $t_D$ smaller than the value listed for a given $r_{eD}$, the reservoir is infinite acting, and Table C.1 should be used to determine $p_D$.

(b) For values of $t_D$ larger than the largest value listed for a given $r_{eD}$ (or, more correctly, for $25 < t_D$ and $0.25\,r_{eD}^2 < t_D$), $p_D$ can be calculated from[2]

$$p_D \simeq \frac{2(t_D + 0.25)}{r_{eD}^2 - 1} - \frac{(3r_{eD}^4 - 4r_{eD}^4 \ln r_{eD} - 2r_{eD}^2 - 1)}{4(r_{eD}^2 - 1)^2}. \quad \text{(C.4)}$$

(c) A special case of Eq. C.4 arises when $r_{eD}^2 \gg 1$; then,

**TABLE C.1 – $p_D$ vs. $t_D$ – INFINITE RADIAL SYSTEM, CONSTANT RATE AT INNER BOUNDARY**

| $t_D$ | $p_D$ | $t_D$ | $p_D$ | $t_D$ | $p_D$ |
|---|---|---|---|---|---|
| 0 | 0 | 0.15 | 0.3750 | 60.0 | 2.4758 |
| 0.0005 | 0.0250 | 0.2 | 0.4241 | 70.0 | 2.5501 |
| 0.001 | 0.0352 | 0.3 | 0.5024 | 80.0 | 2.6147 |
| 0.002 | 0.0495 | 0.4 | 0.5645 | 90.0 | 2.6718 |
| 0.003 | 0.0603 | 0.5 | 0.6167 | 100.0 | 2.7233 |
| 0.004 | 0.0694 | 0.6 | 0.6622 | 150.0 | 2.9212 |
| 0.005 | 0.0774 | 0.7 | 0.7024 | 200.0 | 3.0636 |
| 0.006 | 0.0845 | 0.8 | 0.7387 | 250.0 | 3.1726 |
| 0.007 | 0.0911 | 0.9 | 0.7716 | 300.0 | 3.2630 |
| 0.008 | 0.0971 | 1.0 | 0.8019 | 350.0 | 3.3394 |
| 0.009 | 0.1028 | 1.2 | 0.8672 | 400.0 | 3.4057 |
| 0.01 | 0.1081 | 1.4 | 0.9160 | 450.0 | 3.4641 |
| 0.015 | 0.1312 | 2.0 | 1.0195 | 500.0 | 3.5164 |
| 0.02 | 0.1503 | 3.0 | 1.1665 | 550.0 | 3.5643 |
| 0.025 | 0.1669 | 4.0 | 1.2750 | 600.0 | 3.6076 |
| 0.03 | 0.1818 | 5.0 | 1.3625 | 650.0 | 3.6476 |
| 0.04 | 0.2077 | 6.0 | 1.4362 | 700.0 | 3.6842 |
| 0.05 | 0.2301 | 7.0 | 1.4997 | 750.0 | 3.7184 |
| 0.06 | 0.2500 | 8.0 | 1.5557 | 800.0 | 3.7505 |
| 0.07 | 0.2680 | 9.0 | 1.6057 | 850.0 | 3.7805 |
| 0.08 | 0.2845 | 10.0 | 1.6509 | 900.0 | 3.8088 |
| 0.09 | 0.2999 | 15.0 | 1.8294 | 950.0 | 3.8355 |
| 0.1 | 0.3144 | 20.0 | 1.9601 | 1,000.0 | 3.8584 |
| | | 30.0 | 2.1470 | | |
| | | 40.0 | 2.2824 | | |
| | | 50.0 | 2.3884 | | |

Notes: For $t_D < 0.01$, $p_D \cong 2\sqrt{t_D/\pi}$.
For $100 < t_D < 0.25\, r_{eD}^2$, $p_D \cong 0.5\,(\ln t_D + 0.80907)$.

$$p_D \simeq \frac{2t_D}{r_{eD}^2} + \ln r_{eD} - 3/4. \qquad \text{(C.5)}$$

5. The $p_D$ solutions in Tables C.1 and C.2 also apply to a *reservoir* of radius $r_w$ surrounded by an *aquifer* of radius $r_e$ when there is a constant rate of water influx from the aquifer into the reservoir. For this case, values of $r_{eD}$ in the range 1.5 to 10.0, as in Table C.2, are of practical importance. For most well problems, $r_{eD}$ is larger than 10.0, and the approximations given by Eq. C.3 for $100 < t_D < 0.25\, r_{eD}^2$ and Eq. C.5 for $0.25\, r_{eD}^2 < t_D$ are the really useful results of the van Everdingen and Hurst analysis.

6. When analyzing a variable-rate problem with the $p_D$ solution, we use superposition just as we did with the *Ei*-function solution in Sec. 1.5.

## *Example C.1 – Use of $p_D$ Solutions for No-Flow Boundary*

**Problem.** In a large laboratory flow experiment, fluid was produced into a 1-ft-radius perforated cylinder from a sand-packed model with a radius of 10 ft. No fluid flowed across the external radius of the model. Properties of the sandpack and produced fluid include the following.

$k$ = 1 darcy,
$h$ = 0.5 ft,
$p_i$ = 15 psia,
$\mu$ = 2 cp,
$q$ = 1.0 STB/D,
$B$ = 1.0 RB/STB,
$\phi$ = 0.3, and
$c_t$ = $0.11 \times 10^{-3}$ psi$^{-1}$.

Estimate the pressure on the inner boundary of the sandpack at times of 0.001, 0.01, and 0.1 hour.

**Solution.** We first calculate $t_D$ and $r_{eD}$:

$$r_{eD} = 10/1 = 10,$$

$$t_D = \frac{0.000264\,kt}{\phi \mu c_t r_w^2}$$

$$= \frac{(0.000264)(1{,}000)}{(0.3)(2)(0.11\times 10^{-3})(1)}\, t = 4\times 10^3\, t.$$

Then, the following data result.

| $t$ (hour) | $t_D$ | $p_D$ | Source of $p_D$ |
|---|---|---|---|
| 0.001 | 4 | 1.275 | Table C.1 (reservoir infinite acting) |
| 0.01 | 40 | 2.401 | Table C.2 ($r_{eD}$ = 10) |
| 0.1 | 400 | 9.6751 | Eq. C.4 |

Note that for $t_D = 400$, $t_D > 0.25\, r_{eD}^2 = (0.25)(10)^2 = 25$; thus, Eq. C.4 is used to calculate $p_D$.

$$p_D = \frac{2(t_D + 0.25)}{r_{eD}^2 - 1} - \frac{(3r_{eD}^4 - 4r_{eD}^4 \ln r_{eD} - 2r_{eD}^2 - 1)}{4(r_{eD}^2 - 1)^2}$$

$$= 9.6751.$$

A rearrangement of the definition of $p_D$ results in

$$p = p_i - 141.2\frac{qB\mu}{kh} p_D$$

$$= 15 - \frac{(141.2)(1.0)(1.0)(2)}{(1000)(0.5)} p_D$$

$$= 15 - 0.565\, p_D.$$

Thus, we have these values:

| $t$ (hour) | $p$ (psia) |
|---|---|
| 0.001 | 14.28 |
| 0.01 | 13.64 |
| 0.1 | 9.53 |

## Constant Rate at Inner Boundary, Constant Pressure at Outer Boundary

This solution of the diffusivity equation models radial flow of a slightly compressible liquid in a homogeneous reservoir of uniform thickness; reservoir at uniform pressure $p_i$ before production; unchanging pressure, also $p_i$, at the outer boundary (at $r = r_e$); and production at constant rate $q$ from the single well (centered in the reservoir) with wellbore radius $r_w$. The solutions, $p_D$ (evaluated at $r_D = 1$), as a function of $t_D$ for fixed values of $r_{eD}$ in the range $1.5 < r_{eD} < 3{,}000$, are given in Table C.3. The dimensionless variables $p_D$, $t_D$, $r_D$, and $r_{eD}$ have the same definitions as in the previous section.

**TABLE C.2 – $p_D$ vs. $t_D$ – FINITE RADIAL SYSTEM WITH CLOSED EXTERIOR BOUNDARY, CONSTANT RATE AT INNER BOUNDARY**

| $r_{eD}=1.5$ | | $r_{eD}=2.0$ | | $r_{eD}=2.5$ | | $r_{eD}=3.0$ | | $r_{eD}=3.5$ | | $r_{eD}=4.0$ | |
|---|---|---|---|---|---|---|---|---|---|---|---|
| $t_D$ | $p_D$ | $t_D$ | $p_D$ | $t_D$ | $p_D$ | $t_D$ | $p_D$ | $t_D$ | $p_D$ | $t_D$ | $p_D$ |
| 0.06 | 0.251 | 0.22 | 0.443 | 0.40 | 0.565 | 0.52 | 0.627 | 1.0 | 0.802 | 1.5 | 0.927 |
| 0.08 | 0.288 | 0.24 | 0.459 | 0.42 | 0.576 | 0.54 | 0.636 | 1.1 | 0.830 | 1.6 | 0.948 |
| 0.10 | 0.322 | 0.26 | 0.476 | 0.44 | 0.587 | 0.56 | 0.645 | 1.2 | 0.857 | 1.7 | 0.968 |
| 0.12 | 0.355 | 0.28 | 0.492 | 0.46 | 0.598 | 0.60 | 0.662 | 1.3 | 0.882 | 1.8 | 0.988 |
| 0.14 | 0.387 | 0.30 | 0.507 | 0.48 | 0.608 | 0.65 | 0.683 | 1.4 | 0.906 | 1.9 | 1.007 |
| 0.16 | 0.420 | 0.32 | 0.522 | 0.50 | 0.618 | 0.70 | 0.703 | 1.5 | 0.929 | 2.0 | 1.025 |
| 0.18 | 0.452 | 0.34 | 0.536 | 0.52 | 0.628 | 0.75 | 0.721 | 1.6 | 0.951 | 2.2 | 1.059 |
| 0.20 | 0.484 | 0.36 | 0.551 | 0.54 | 0.638 | 0.80 | 0.740 | 1.7 | 0.973 | 2.4 | 1.092 |
| 0.22 | 0.516 | 0.38 | 0.565 | 0.56 | 0.647 | 0.85 | 0.758 | 1.8 | 0.994 | 2.6 | 1.123 |
| 0.24 | 0.548 | 0.40 | 0.579 | 0.58 | 0.657 | 0.90 | 0.776 | 1.9 | 1.014 | 2.8 | 1.154 |
| 0.26 | 0.580 | 0.42 | 0.593 | 0.60 | 0.666 | 0.95 | 0.791 | 2.0 | 1.034 | 3.0 | 1.184 |
| 0.28 | 0.612 | 0.44 | 0.607 | 0.65 | 0.688 | 1.0 | 0.806 | 2.25 | 1.083 | 3.5 | 1.255 |
| 0.30 | 0.644 | 0.46 | 0.621 | 0.70 | 0.710 | 1.2 | 0.865 | 2.50 | 1.130 | 4.0 | 1.324 |
| 0.35 | 0.724 | 0.48 | 0.634 | 0.75 | 0.731 | 1.4 | 0.920 | 2.75 | 1.176 | 4.5 | 1.392 |
| 0.40 | 0.804 | 0.50 | 0.648 | 0.80 | 0.752 | 1.6 | 0.973 | 3.0 | 1.221 | 5.0 | 1.460 |
| 0.45 | 0.884 | 0.60 | 0.715 | 0.85 | 0.772 | 2.0 | 1.076 | 4.0 | 1.401 | 5.5 | 1.527 |
| 0.50 | 0.964 | 0.70 | 0.782 | 0.90 | 0.792 | 3.0 | 1.328 | 5.0 | 1.579 | 6.0 | 1.594 |
| 0.55 | 1.044 | 0.80 | 0.849 | 0.95 | 0.812 | 4.0 | 1.578 | 6.0 | 1.757 | 6.5 | 1.660 |
| 0.60 | 1.124 | 0.90 | 0.915 | 1.00 | 0.832 | 5.0 | 1.828 | | | 7.0 | 1.727 |
| 0.65 | 1.204 | 1.0 | 0.982 | 2.0 | 1.215 | | | | | 8.0 | 1.861 |
| 0.70 | 1.284 | 2.0 | 1.649 | 3.0 | 1.506 | | | | | 9.0 | 1.994 |
| 0.75 | 1.364 | 3.0 | 2.316 | 4.0 | 1.977 | | | | | 10.0 | 2.127 |
| 0.80 | 1.444 | 5.0 | 3.649 | 5.0 | 2.398 | | | | | | |

| $r_{eD}=4.5$ | | $r_{eD}=5.0$ | | $r_{eD}=6.0$ | | $r_{eD}=7.0$ | | $r_{eD}=8.0$ | | $r_{eD}=9.0$ | | $r_{eD}=10.0$ | |
|---|---|---|---|---|---|---|---|---|---|---|---|---|---|
| $t_D$ | $p_D$ | $t_D$ | $p_D$ | $t_D$ | $p_D$ | $t_D$ | $p_D$ | $t_D$ | $p_D$ | $t_D$ | $p_D$ | $t_D$ | $p_D$ |
| 2.0 | 1.023 | 3.0 | 1.167 | 4.0 | 1.275 | 6.0 | 1.436 | 8.0 | 1.556 | 10.0 | 1.651 | 12.0 | 1.732 |
| 2.1 | 1.040 | 3.1 | 1.180 | 4.5 | 1.322 | 6.5 | 1.470 | 8.5 | 1.582 | 10.5 | 1.673 | 12.5 | 1.750 |
| 2.2 | 1.056 | 3.2 | 1.192 | 5.0 | 1.364 | 7.0 | 1.501 | 9.0 | 1.607 | 11.0 | 1.693 | 13.0 | 1.768 |
| 2.3 | 1.702 | 3.3 | 1.204 | 5.5 | 1.404 | 7.5 | 1.531 | 9.5 | 1.631 | 11.5 | 1.713 | 13.5 | 1.784 |
| 2.4 | 1.087 | 3.4 | 1.215 | 6.0 | 1.441 | 8.0 | 1.559 | 10.0 | 1.653 | 12.0 | 1.732 | 14.0 | 1.801 |
| 2.5 | 1.102 | 3.5 | 1.227 | 6.5 | 1.477 | 8.5 | 1.586 | 10.5 | 1.675 | 12.5 | 1.750 | 14.5 | 1.817 |
| 2.6 | 1.116 | 3.6 | 1.238 | 7.0 | 1.511 | 9.0 | 1.613 | 11.0 | 1.697 | 13.0 | 1.768 | 15.0 | 1.832 |
| 2.7 | 1.130 | 3.7 | 1.249 | 7.5 | 1.544 | 9.5 | 1.638 | 11.5 | 1.717 | 13.5 | 1.786 | 15.5 | 1.847 |
| 2.8 | 1.144 | 3.8 | 1.259 | 8.0 | 1.576 | 10.0 | 1.663 | 12.0 | 1.737 | 14.0 | 1.803 | 16.0 | 1.862 |
| 2.9 | 1.158 | 3.9 | 1.270 | 8.5 | 1.607 | 11.0 | 1.711 | 12.5 | 1.757 | 14.5 | 1.819 | 17.0 | 1.890 |
| 3.0 | 1.171 | 4.0 | 1.281 | 9.0 | 1.638 | 12.0 | 1.757 | 13.0 | 1.776 | 15.0 | 1.835 | 18.0 | 1.917 |
| 3.2 | 1.197 | 4.2 | 1.301 | 9.5 | 1.668 | 13.0 | 1.810 | 13.5 | 1.795 | 15.5 | 1.851 | 19.0 | 1.943 |
| 3.4 | 1.222 | 4.4 | 1.321 | 10.0 | 1.698 | 14.0 | 1.845 | 14.0 | 1.813 | 16.0 | 1.867 | 20.0 | 1.968 |
| 3.6 | 1.246 | 4.6 | 1.340 | 11.0 | 1.757 | 15.0 | 1.888 | 14.5 | 1.831 | 17.0 | 1.897 | 22.0 | 2.017 |
| 3.8 | 1.269 | 4.8 | 1.360 | 12.0 | 1.815 | 16.0 | 1.931 | 15.0 | 1.849 | 18.0 | 1.926 | 24.0 | 2.063 |
| 4.0 | 1.292 | 5.0 | 1.378 | 13.0 | 1.873 | 17.0 | 1.974 | 17.0 | 1.919 | 19.0 | 1.955 | 26.0 | 2.108 |
| 4.5 | 1.349 | 5.5 | 1.424 | 14.0 | 1.931 | 18.0 | 2.016 | 19.0 | 1.986 | 20.0 | 1.983 | 28.0 | 2.151 |
| 5.0 | 1.403 | 6.0 | 1.469 | 15.0 | 1.988 | 19.0 | 2.058 | 21.0 | 2.051 | 22.0 | 2.037 | 30.0 | 2.194 |
| 5.5 | 1.457 | 6.5 | 1.513 | 16.0 | 2.045 | 20.0 | 2.100 | 23.0 | 2.116 | 24.0 | 2.096 | 32.0 | 2.236 |
| 6.0 | 1.510 | 7.0 | 1.556 | 17.0 | 2.103 | 22.0 | 2.184 | 25.0 | 2.180 | 26.0 | 2.142 | 34.0 | 2.278 |
| 7.0 | 1.615 | 7.5 | 1.598 | 18.0 | 2.160 | 24.0 | 2.267 | 30.0 | 2.340 | 28.0 | 2.193 | 36.0 | 2.319 |
| 8.0 | 1.719 | 8.0 | 1.641 | 19.0 | 2.217 | 26.0 | 2.351 | 35.0 | 2.499 | 30.0 | 2.244 | 38.0 | 2.360 |
| 9.0 | 1.823 | 9.0 | 1.725 | 20.0 | 2.274 | 28.0 | 2.434 | 40.0 | 2.658 | 34.0 | 2.345 | 40.0 | 2.401 |
| 10.0 | 1.927 | 10.0 | 1.808 | 25.0 | 2.560 | 30.0 | 2.517 | 45.0 | 2.817 | 38.0 | 2.446 | 50.0 | 2.604 |
| 11.0 | 2.031 | 11.0 | 1.892 | 30.0 | 2.846 | | | | | 40.0 | 2.496 | 60.0 | 2.806 |
| 12.0 | 2.135 | 12.0 | 1.975 | | | | | | | 45.0 | 2.621 | 70.0 | 3.008 |
| 13.0 | 2.239 | 13.0 | 2.059 | | | | | | | 50.0 | 2.746 | 80.0 | 3.210 |
| 14.0 | 2.343 | 14.0 | 2.142 | | | | | | | 60.0 | 2.996 | 90.0 | 3.412 |
| 15.0 | 2.447 | 15.0 | 2.225 | | | | | | | 70.0 | 3.246 | 100.0 | 3.614 |

Notes: For $t_D$ smaller than values listed in this table for a given $r_{eD}$, reservoir is infinite acting. Find $p_D$ in Table C.1.

For $25 < t_D$ and $t_D$ larger than values in table,

$$p_D \cong \frac{(\tfrac{1}{2} + 2t_D)}{(r_{eD}^2 - 1)} - \frac{3r_{eD}^4 - 4r_{eD}^4 \ln r_{eD} - 2r_{eD}^2 - 1}{4(r_{eD}^2 - 1)^2}.$$

For Wells in bounded reservoirs with

$r_{eD}^2 \gg 1$,

$$p_D \cong \frac{2t_D}{r_{eD}^2} + \ln r_{eD} - \tfrac{3}{4}.$$

**TABLE C.3 – $p_D$ vs. $t_D$ – FINITE RADIAL SYSTEM WITH FIXED CONSTANT PRESSURE AT EXTERIOR BOUNDARY, CONSTANT RATE AT INNER BOUNDARY**

| $r_{eD}=1.5$ | | $r_{eD}=2.0$ | | $r_{eD}=2.5$ | | $r_{eD}=3.0$ | | $r_{eD}=3.5$ | | $r_{eD}=4.0$ | | $r_{eD}=6.0$ | |
|---|---|---|---|---|---|---|---|---|---|---|---|---|---|
| $t_D$ | $p_D$ | $t_D$ | $p_D$ | $t_D$ | $p_D$ | $t_D$ | $p_D$ | $t_D$ | $p_D$ | $t_D$ | $p_D$ | $t_D$ | $p_D$ |
| 0.050 | 0.230 | 0.20 | 0.424 | 0.30 | 0.502 | 0.50 | 0.617 | 0.50 | 0.620 | 1.0 | 0.802 | 4.0 | 1.275 |
| 0.055 | 0.240 | 0.22 | 0.441 | 0.35 | 0.535 | 0.55 | 0.640 | 0.60 | 0.665 | 1.2 | 0.857 | 4.5 | 1.320 |
| 0.060 | 0.249 | 0.24 | 0.457 | 0.40 | 0.564 | 0.60 | 0.662 | 0.70 | 0.705 | 1.4 | 0.905 | 5.0 | 1.361 |
| 0.070 | 0.266 | 0.26 | 0.472 | 0.45 | 0.591 | 0.70 | 0.702 | 0.80 | 0.741 | 1.6 | 0.947 | 5.5 | 1.398 |
| 0.080 | 0.282 | 0.28 | 0.485 | 0.50 | 0.616 | 0.80 | 0.738 | 0.90 | 0.774 | 1.8 | 0.986 | 6.0 | 1.432 |
| 0.090 | 0.292 | 0.30 | 0.498 | 0.55 | 0.638 | 0.90 | 0.770 | 1.0 | 0.804 | 2.0 | 1.020 | 6.5 | 1.462 |
| 0.10 | 0.307 | 0.35 | 0.527 | 0.60 | 0.659 | 1.0 | 0.799 | 1.2 | 0.858 | 2.2 | 1.052 | 7.0 | 1.490 |
| 0.12 | 0.328 | 0.40 | 0.552 | 0.70 | 0.696 | 1.2 | 0.850 | 1.4 | 0.904 | 2.4 | 1.080 | 7.5 | 1.516 |
| 0.14 | 0.344 | 0.45 | 0.573 | 0.80 | 0.728 | 1.4 | 0.892 | 1.6 | 0.945 | 2.6 | 1.106 | 8.0 | 1.539 |
| 0.16 | 0.356 | 0.50 | 0.591 | 0.90 | 0.755 | 1.6 | 0.927 | 1.8 | 0.981 | 2.8 | 1.130 | 8.5 | 1.561 |
| 0.18 | 0.367 | 0.55 | 0.606 | 1.0 | 0.778 | 1.8 | 0.955 | 2.0 | 1.013 | 3.0 | 1.152 | 9.0 | 1.580 |
| 0.20 | 0.375 | 0.60 | 0.619 | 1.2 | 0.815 | 2.0 | 0.980 | 2.2 | 1.041 | 3.4 | 1.190 | 10.0 | 1.615 |
| 0.22 | 0.381 | 0.65 | 0.630 | 1.4 | 0.842 | 2.2 | 1.000 | 2.4 | 1.065 | 3.8 | 1.232 | 12.0 | 1.667 |
| 0.24 | 0.386 | 0.70 | 0.639 | 1.6 | 0.861 | 2.4 | 1.016 | 2.6 | 1.087 | 4.5 | 1.266 | 14.0 | 1.704 |
| 0.26 | 0.390 | 0.75 | 0.647 | 1.8 | 0.876 | 2.6 | 1.030 | 2.8 | 1.106 | 5.0 | 1.290 | 16.0 | 1.730 |
| 0.28 | 0.393 | 0.80 | 0.654 | 2.0 | 0.887 | 2.8 | 1.042 | 3.0 | 1.123 | 5.5 | 1.309 | 18.0 | 1.749 |
| 0.30 | 0.396 | 0.85 | 0.660 | 2.2 | 0.895 | 3.0 | 1.051 | 3.5 | 1.153 | 6.0 | 1.325 | 20.0 | 1.762 |
| 0.35 | 0.400 | 0.90 | 0.665 | 2.4 | 0.900 | 3.5 | 1.069 | 4.0 | 1.183 | 7.0 | 1.347 | 22.0 | 1.771 |
| 0.40 | 0.402 | 0.95 | 0.669 | 2.6 | 0.905 | 4.0 | 1.080 | 5.0 | 1.225 | 8.0 | 1.361 | 24.0 | 1.777 |
| 0.45 | 0.404 | 1.0 | 0.673 | 2.8 | 0.908 | 4.5 | 1.087 | 6.0 | 1.232 | 9.0 | 1.370 | 26.0 | 1.781 |
| 0.50 | 0.405 | 1.2 | 0.682 | 3.0 | 0.910 | 5.0 | 1.091 | 7.0 | 1.242 | 10.0 | 1.376 | 28.0 | 1.784 |
| 0.60 | 0.405 | 1.4 | 0.688 | 3.5 | 0.913 | 5.5 | 1.094 | 8.0 | 1.247 | 12.0 | 1.382 | 30.0 | 1.787 |
| 0.70 | 0.405 | 1.6 | 0.690 | 4.0 | 0.915 | 6.0 | 1.096 | 9.0 | 1.250 | 14.0 | 1.385 | 35.0 | 1.789 |
| 0.80 | 0.405 | 1.8 | 0.692 | 4.5 | 0.916 | 6.5 | 1.097 | 10.0 | 1.251 | 16.0 | 1.386 | 40.0 | 1.791 |
| | | 2.0 | 0.692 | 5.0 | 0.916 | 7.0 | 1.097 | 12.0 | 1.252 | 18.0 | 1.386 | 50.0 | 1.792 |
| | | 2.5 | 0.693 | 5.5 | 0.916 | 8.0 | 1.098 | 14.0 | 1.253 | | | | |
| | | 3.0 | 0.693 | 6.0 | 0.916 | 10.0 | 1.099 | 16.0 | 1.253 | | | | |

| $r_{eD}=8.0$ | | $r_{eD}=10.0$ | | $r_{eD}=15.0$ | | $r_{eD}=20.0$ | | $r_{eD}=25.0$ | | $r_{eD}=30.0$ | | $r_{eD}=40.0$ | |
|---|---|---|---|---|---|---|---|---|---|---|---|---|---|
| $t_D$ | $p_D$ | $t_D$ | $p_D$ | $t_D$ | $p_D$ | $t_D$ | $p_D$ | $t_D$ | $p_D$ | $t_D$ | $p_D$ | $t_D$ | $p_D$ |
| 7.0 | 1.499 | 10.0 | 1.651 | 20.0 | 1.960 | 30.0 | 2.148 | 50.0 | 2.389 | 70.0 | 2.551 | 120 | 2.813 |
| 7.5 | 1.527 | 12.0 | 1.730 | 22.0 | 2.003 | 35.0 | 2.219 | 55.0 | 2.434 | 80.0 | 2.615 | 140 | 2.888 |
| 8.0 | 1.554 | 14.0 | 1.798 | 24.0 | 2.043 | 40.0 | 2.282 | 60.0 | 2.476 | 90.0 | 2.672 | 160 | 2.953 |
| 8.5 | 1.580 | 16.0 | 1.856 | 26.0 | 2.080 | 45.0 | 2.338 | 65.0 | 2.514 | 100.0 | 2.723 | 180 | 3.011 |
| 9.0 | 1.604 | 18.0 | 1.907 | 28.0 | 2.114 | 50.0 | 2.388 | 70.0 | 2.550 | 120.0 | 2.812 | 200 | 3.063 |
| 9.5 | 1.627 | 20.0 | 1.952 | 30.0 | 2.146 | 60.0 | 2.475 | 75.0 | 2.583 | 140.0 | 2.886 | 220 | 3.109 |
| 10.0 | 1.648 | 25.0 | 2.043 | 35.0 | 2.218 | 70.0 | 2.547 | 80.0 | 2.614 | 160.0 | 2.950 | 240 | 3.152 |
| 12.0 | 1.724 | 30.0 | 2.111 | 40.0 | 2.279 | 80.0 | 2.609 | 85.0 | 2.643 | 165.0 | 2.965 | 260 | 3.191 |
| 14.0 | 1.786 | 35.0 | 2.160 | 45.0 | 2.332 | 90.0 | 2.658 | 90.0 | 2.671 | 170.0 | 2.979 | 280 | 3.226 |
| 16.0 | 1.837 | 40.0 | 2.197 | 50.0 | 2.379 | 100.0 | 2.707 | 95.0 | 2.697 | 175.0 | 2.992 | 300 | 3.259 |
| 18.0 | 1.879 | 45.0 | 2.224 | 60.0 | 2.455 | 105.0 | 2.728 | 100.0 | 2.721 | 180.0 | 3.006 | 350 | 3.331 |
| 20.0 | 1.914 | 50.0 | 2.245 | 70.0 | 2.513 | 110.0 | 2.747 | 120.0 | 2.807 | 200.0 | 3.054 | 400 | 3.391 |
| 22.0 | 1.943 | 55.0 | 2.260 | 80.0 | 2.558 | 115.0 | 2.764 | 140.0 | 2.878 | 250.0 | 3.150 | 450 | 3.440 |
| 24.0 | 1.967 | 60.0 | 2.271 | 90.0 | 2.592 | 120.0 | 2.781 | 160.0 | 2.936 | 300.0 | 3.219 | 500 | 3.482 |
| 26.0 | 1.986 | 65.0 | 2.279 | 100.0 | 2.619 | 125.0 | 2.796 | 180.0 | 2.984 | 350.0 | 3.269 | 550 | 3.516 |
| 28.0 | 2.002 | 70.0 | 2.285 | 120.0 | 2.655 | 130.0 | 2.810 | 200.0 | 3.024 | 400.0 | 3.306 | 600 | 3.545 |
| 30.0 | 2.016 | 75.0 | 2.290 | 140.0 | 2.677 | 135.0 | 2.823 | 220.0 | 3.057 | 450.0 | 3.332 | 650 | 3.568 |
| 35.0 | 2.040 | 80.0 | 2.293 | 160.0 | 2.689 | 140.0 | 2.835 | 240.0 | 3.085 | 500.0 | 3.351 | 700 | 3.588 |
| 40.0 | 2.055 | 90.0 | 2.297 | 180.0 | 2.697 | 145.0 | 2.846 | 260.0 | 3.107 | 600.0 | 3.375 | 800 | 3.619 |
| 45.0 | 2.064 | 100.0 | 2.300 | 200.0 | 2.701 | 150.0 | 2.857 | 280.0 | 3.126 | 700.0 | 3.387 | 900 | 3.640 |
| 50.0 | 2.070 | 110.0 | 2.301 | 220.0 | 2.704 | 160.0 | 2.876 | 300.0 | 3.142 | 800.0 | 3.394 | 1,000 | 3.655 |
| 60.0 | 2.076 | 120.0 | 2.302 | 240.0 | 2.706 | 180.0 | 2.906 | 350.0 | 3.171 | 900.0 | 3.397 | 1,200 | 3.672 |
| 70.0 | 2.078 | 130.0 | 2.302 | 260.0 | 2.707 | 200.0 | 2.929 | 400.0 | 3.189 | 1,000 | 3.399 | 1,400 | 3.681 |
| 80.0 | 2.079 | 140.0 | 2.302 | 280.0 | 2.707 | 240.0 | 2.958 | 450.0 | 3.200 | 1,200 | 3.401 | 1,600 | 3.685 |
| | | 160.0 | 2.303 | 300.0 | 2.708 | 280.0 | 2.975 | 500.0 | 3.207 | 1,400 | 3.401 | 1,800 | 3.687 |
| | | | | | | 300.0 | 2.980 | 600.0 | 3.214 | | | 2,000 | 3.688 |
| | | | | | | 400.0 | 2.992 | 700.0 | 3.217 | | | 2,500 | 3.689 |
| | | | | | | 500.0 | 2.995 | 800.0 | 3.218 | | | | |
| | | | | | | | | 900.0 | 3.219 | | | | |

Notes: For $t_D$ smaller than values listed in this table for a given $r_{eD}$, reservoir is infinite acting. Find $p_D$ in Table C.1.

For $t_D$ larger than values listed in this table, $p_D \cong \ln r_{eD}$.

**TABLE C.3 – CONTINUED**

$r_{eD} = 50.0$

| $t_D$ | $p_D$ |
|---|---|
| 200 | 3.064 |
| 220 | 3.111 |
| 240 | 3.154 |
| 260 | 3.193 |
| 280 | 3.229 |
| 300 | 3.263 |
| 350 | 3.339 |
| 400 | 3.405 |
| 450 | 3.461 |
| 500 | 3.512 |
| 550 | 3.556 |
| 600 | 3.595 |
| 650 | 3.630 |
| 700 | 3.661 |
| 750 | 3.688 |
| 800 | 3.713 |
| 850 | 3.735 |
| 900 | 3.754 |
| 950 | 3.771 |
| 1,000 | 3.787 |
| 1,200 | 3.833 |
| 1,400 | 3.862 |
| 1,600 | 3.881 |
| 1,800 | 3.892 |
| 2,000 | 3.900 |
| 2,200 | 3.904 |
| 2,400 | 3.907 |
| 2,600 | 3.909 |
| 2,800 | 3.910 |

$r_{eD} = 60.0$

| $t_D$ | $p_D$ |
|---|---|
| 300 | 3.257 |
| 400 | 3.401 |
| 500 | 3.512 |
| 600 | 3.602 |
| 700 | 3.676 |
| 800 | 3.739 |
| 900 | 3.792 |
| 1,000 | 3.832 |
| 1,200 | 3.908 |
| 1,400 | 3.959 |
| 1,600 | 3.996 |
| 1,800 | 4.023 |
| 2,000 | 4.043 |
| 2,500 | 4.071 |
| 3,000 | 4.084 |
| 3,500 | 4.090 |
| 4,000 | 4.092 |
| 4,500 | 4.093 |
| 5,000 | 4.094 |
| 5,500 | 4.094 |

$r_{eD} = 70.0$

| $t_D$ | $p_D$ |
|---|---|
| 500 | 3.512 |
| 600 | 3.603 |
| 700 | 3.680 |
| 800 | 3.746 |
| 900 | 3.803 |
| 1,000 | 3.854 |
| 1,200 | 3.937 |
| 1,400 | 4.003 |
| 1,600 | 4.054 |
| 1,800 | 4.095 |
| 2,000 | 4.127 |
| 2,500 | 4.181 |
| 3,000 | 4.211 |
| 3,500 | 4.228 |
| 4,000 | 4.237 |
| 4,500 | 4.242 |
| 5,000 | 4.245 |
| 5,500 | 4.247 |
| 6,000 | 4.247 |
| 6,500 | 4.248 |
| 7,000 | 4.248 |
| 7,500 | 4.248 |
| 8,000 | 4.248 |

$r_{eD} = 80.0$

| $t_D$ | $p_D$ |
|---|---|
| 600 | 3.603 |
| 700 | 3.680 |
| 800 | 3.747 |
| 900 | 3.805 |
| 1,000 | 3.857 |
| 1,200 | 3.946 |
| 1,400 | 4.019 |
| 1,500 | 4.051 |
| 1,600 | 4.080 |
| 1,800 | 4.130 |
| 2,000 | 4.171 |
| 2,500 | 4.248 |
| 3,000 | 4.297 |
| 3,500 | 4.328 |
| 4,000 | 4.347 |
| 4,500 | 4.360 |
| 5,000 | 4.368 |
| 6,000 | 4.376 |
| 7,000 | 4.380 |
| 8,000 | 4.381 |
| 9,000 | 4.382 |
| 10,000 | 4.382 |
| 11,000 | 4.382 |

$r_{eD} = 90.0$

| $t_D$ | $p_D$ |
|---|---|
| 800 | 3.747 |
| 900 | 3.806 |
| 1,000 | 3.858 |
| 1,200 | 3.949 |
| 1,300 | 3.988 |
| 1,400 | 4.025 |
| 1,500 | 4.058 |
| 1,800 | 4.144 |
| 2,000 | 4.192 |
| 2,500 | 4.285 |
| 3,000 | 4.349 |
| 3,500 | 4.394 |
| 4,000 | 4.426 |
| 4,500 | 4.448 |
| 5,000 | 4.464 |
| 6,000 | 4.482 |
| 7,000 | 4.491 |
| 8,000 | 4.496 |
| 9,000 | 4.498 |
| 10,000 | 4.499 |
| 11,000 | 4.499 |
| 12,000 | 4.500 |
| 14,000 | 4.500 |

$r_{eD} = 100.0$

| $t_D$ | $p_D$ |
|---|---|
| 1,000 | 3.859 |
| 1,200 | 3.949 |
| 1,400 | 4.026 |
| 1,600 | 4.092 |
| 1,800 | 4.150 |
| 2,000 | 4.200 |
| 2,500 | 4.303 |
| 3,000 | 4.379 |
| 3,500 | 4.434 |
| 4,000 | 4.478 |
| 4,500 | 4.510 |
| 5,000 | 4.534 |
| 5,500 | 4.552 |
| 6,000 | 4.565 |
| 6,500 | 4.579 |
| 7,000 | 4.583 |
| 7,500 | 4.588 |
| 8,000 | 4.593 |
| 9,000 | 4.598 |
| 10,000 | 4.601 |
| 12,500 | 4.604 |
| 15,000 | 4.605 |

$r_{eD} = 200.0$

| $t_D$ | $p_D$ |
|---|---|
| 1,500 | 4.061 |
| 2,000 | 4.205 |
| 2,500 | 4.317 |
| 3,000 | 4.498 |
| 3,500 | 4.485 |
| 4,000 | 4.552 |
| 5,000 | 4.663 |
| 6,000 | 4.754 |
| 7,000 | 4.829 |
| 8,000 | 4.834 |
| 9,000 | 4.949 |
| 10,000 | 4.996 |
| 12,000 | 5.072 |
| 14,000 | 5.129 |
| 16,000 | 5.171 |
| 18,000 | 5.203 |
| 20,000 | 5.227 |
| 25,000 | 5.264 |
| 30,000 | 5.282 |
| 35,000 | 5.290 |
| 40,000 | 5.294 |

$r_{eD} = 300.0$

| $t_D$ | $p_D$ |
|---|---|
| 6,000 | 4.754 |
| 8,000 | 4.898 |
| 10,000 | 5.010 |
| 12,000 | 5.101 |
| 14,000 | 5.177 |
| 16,000 | 5.242 |
| 18,000 | 5.299 |
| 20,000 | 5.348 |
| 24,000 | 5.429 |
| 28,000 | 5.491 |
| 30,000 | 5.517 |
| 40,000 | 5.606 |
| 50,000 | 5.652 |
| 60,000 | 5.676 |
| 70,000 | 5.690 |
| 80,000 | 5.696 |
| 90,000 | 5.700 |
| 100,000 | 5.702 |
| 120,000 | 5.703 |
| 140,000 | 5.704 |
| 150,000 | 5.704 |

$r_{eD} = 400.0$

| $t_D$ | $p_D$ |
|---|---|
| 15,000 | 5.212 |
| 20,000 | 5.356 |
| 30,000 | 5.556 |
| 40,000 | 5.689 |
| 50,000 | 5.781 |
| 60,000 | 5.845 |
| 70,000 | 5.889 |
| 80,000 | 5.920 |
| 90,000 | 5.942 |
| 100,000 | 5.957 |
| 110,000 | 5.967 |
| 120,000 | 5.975 |
| 125,000 | 5.977 |
| 130,000 | 5.980 |
| 140,000 | 5.983 |
| 160,000 | 5.988 |
| 180,000 | 5.990 |
| 200,000 | 5.991 |
| 240,000 | 5.991 |
| 260,000 | 5.991 |

$r_{eD} = 500.0$

| $t_D$ | $p_D$ |
|---|---|
| 20,000 | 5.365 |
| 25,000 | 5.468 |
| 30,000 | 5.559 |
| 35,000 | 5.636 |
| 40,000 | 5.702 |
| 45,000 | 5.759 |
| 50,000 | 5.810 |
| 60,000 | 5.894 |
| 70,000 | 5.960 |
| 80,000 | 6.013 |
| 90,000 | 6.055 |
| 100,000 | 6.088 |
| 120,000 | 6.135 |
| 140,000 | 6.164 |
| 160,000 | 6.183 |
| 180,000 | 6.195 |
| 200,000 | 6.202 |
| 250,000 | 6.211 |
| 300,000 | 6.213 |
| 350,000 | 6.214 |
| 400,000 | 6.214 |

$r_{eD} = 600.0$

| $t_D$ | $p_D$ |
|---|---|
| 40,000 | 5.703 |
| 45,000 | 5.762 |
| 50,000 | 5.814 |
| 60,000 | 5.904 |
| 70,000 | 5.979 |
| 80,000 | 6.041 |
| 90,000 | 6.094 |
| 100,000 | 6.139 |
| 120,000 | 6.210 |
| 140,000 | 6.262 |
| 160,000 | 6.299 |
| 180,000 | 6.326 |
| 200,000 | 6.345 |
| 250,000 | 6.374 |
| 300,000 | 6.387 |
| 350,000 | 6.392 |
| 400,000 | 6.395 |
| 500,000 | 6.397 |
| 600,000 | 6.397 |

$r_{eD} = 700.0$

| $t_D$ | $p_D$ |
|---|---|
| 50,000 | 5.814 |
| 60,000 | 5.905 |
| 70,000 | 5.982 |
| 80,000 | 6.048 |
| 90,000 | 6.105 |
| 100,000 | 6.156 |
| 120,000 | 6.239 |
| 140,000 | 6.305 |
| 160,000 | 6.357 |
| 180,000 | 6.398 |
| 200,000 | 6.430 |
| 250,000 | 6.484 |
| 300,000 | 6.514 |
| 350,000 | 6.530 |
| 400,000 | 6.540 |
| 450,000 | 6.545 |
| 500,000 | 6.548 |
| 600,000 | 6.550 |
| 700,000 | 6.551 |
| 800,000 | 6.551 |

$r_{eD} = 800.0$

| $t_D$ | $p_D$ |
|---|---|
| 70,000 | 5.983 |
| 80,000 | 6.049 |
| 90,000 | 6.108 |
| 100,000 | 6.160 |
| 120,000 | 6.249 |
| 140,000 | 6.322 |
| 160,000 | 6.382 |
| 180,000 | 6.432 |
| 200,000 | 6.474 |
| 250,000 | 6.551 |
| 300,000 | 6.599 |
| 350,000 | 6.630 |
| 400,000 | 6.650 |
| 450,000 | 6.663 |
| 500,000 | 6.671 |
| 550,000 | 6.676 |
| 600,000 | 6.679 |
| 700,000 | 6.682 |
| 800,000 | 6.684 |
| 1,000,000 | 6.684 |

**TABLE C.3 – CONTINUED**

| $r_{eD}$ = 900.0 | | $r_{eD}$ = 1,000.0 | | $r_{eD}$ = 1,200.0 | | $r_{eD}$ = 1,400.0 | | $r_{eD}$ = 1,600.0 | | $r_{eD}$ = 1,800.0 | |
|---|---|---|---|---|---|---|---|---|---|---|---|
| $t_D$ | $p_D$ | $t_D$ | $p_D$ | $t_D$ | $p_D$ | $t_D$ | $p_D$ | $t_D$ | $p_D$ | $t_D$ | $p_D$ |
| $8.0(10^4)$ | 6.049 | $1.0(10^5)$ | 6.161 | $2.0(10^5)$ | 6.507 | $2.0(10^5)$ | 6.507 | $2.5(10^5)$ | 6.619 | $3.0(10^5)$ | 6.710 |
| $9.0(10^4)$ | 6.108 | $1.2(10^5)$ | 6.252 | $3.0(10^5)$ | 6.704 | $2.5(10^5)$ | 6.619 | $3.0(10^5)$ | 6.710 | $4.0(10^5)$ | 6.854 |
| $1.0(10^5)$ | 6.161 | $1.4(10^5)$ | 6.329 | $4.0(10^5)$ | 6.833 | $3.0(10^5)$ | 6.709 | $3.5(10^5)$ | 6.787 | $5.0(10^5)$ | 6.965 |
| $1.2(10^5)$ | 6.251 | $1.6(10^5)$ | 6.395 | $5.0(10^5)$ | 6.918 | $3.5(10^5)$ | 6.785 | $4.0(10^5)$ | 6.853 | $6.0(10^5)$ | 7.054 |
| $1.4(10^5)$ | 6.327 | $1.8(10^5)$ | 6.452 | $6.0(10^5)$ | 6.975 | $4.0(10^5)$ | 6.849 | $5.0(10^5)$ | 6.962 | $7.0(10^5)$ | 7.120 |
| $1.6(10^5)$ | 6.392 | $2.0(10^5)$ | 6.503 | $7.0(10^5)$ | 7.013 | $5.0(10^5)$ | 6.950 | $6.0(10^5)$ | 7.046 | $8.0(10^5)$ | 7.183 |
| $1.8(10^5)$ | 6.447 | $2.5(10^5)$ | 6.605 | $8.0(10^5)$ | 7.038 | $6.0(10^5)$ | 7.026 | $7.0(10^5)$ | 7.114 | $9.0(10^5)$ | 7.238 |
| $2.0(10^5)$ | 6.494 | $3.0(10^5)$ | 6.681 | $9.0(10^5)$ | 7.056 | $7.0(10^5)$ | 7.082 | $8.0(10^5)$ | 7.167 | $1.0(10^6)$ | 7.280 |
| $2.5(10^5)$ | 6.587 | $3.5(10^5)$ | 6.738 | $1.0(10^6)$ | 7.067 | $8.0(10^5)$ | 7.123 | $9.0(10^5)$ | 7.210 | $1.5(10^6)$ | 7.407 |
| $3.0(10^5)$ | 6.652 | $4.0(10^5)$ | 6.781 | $1.2(10^6)$ | 7.080 | $9.0(10^5)$ | 7.154 | $1.0(10^6)$ | 7.244 | $2.0(10^6)$ | 7.459 |
| $4.0(10^5)$ | 6.729 | $4.5(10^5)$ | 6.813 | $1.4(10^6)$ | 7.085 | $1.0(10^6)$ | 7.177 | $1.5(10^6)$ | 7.334 | $3.0(10^6)$ | 7.489 |
| $4.5(10^5)$ | 6.751 | $5.0(10^5)$ | 6.837 | $1.6(10^6)$ | 7.088 | $1.5(10^6)$ | 7.229 | $2.0(10^6)$ | 7.364 | $4.0(10^6)$ | 7.495 |
| $5.0(10^5)$ | 6.766 | $5.5(10^5)$ | 6.854 | $1.8(10^6)$ | 7.089 | $2.0(10^6)$ | 7.241 | $2.5(10^6)$ | 7.373 | $5.0(10^6)$ | 7.495 |
| $5.5(10^5)$ | 6.777 | $6.0(10^5)$ | 6.868 | $1.9(10^6)$ | 7.089 | $2.5(10^6)$ | 7.243 | $3.0(10^6)$ | 7.376 | $5.1(10^6)$ | 7.495 |
| $6.0(10^5)$ | 6.785 | $7.0(10^5)$ | 6.885 | $2.0(10^6)$ | 7.090 | $3.0(10^6)$ | 7.244 | $3.5(10^6)$ | 7.377 | $5.2(10^6)$ | 7.495 |
| $7.0(10^5)$ | 6.794 | $8.0(10^5)$ | 6.895 | $2.1(10^6)$ | 7.090 | $3.1(10^6)$ | 7.244 | $4.0(10^6)$ | 7.378 | $5.3(10^6)$ | 7.495 |
| $8.0(10^5)$ | 6.798 | $9.0(10^5)$ | 6.901 | $2.2(10^6)$ | 7.090 | $3.2(10^6)$ | 7.244 | $4.2(10^6)$ | 7.378 | $5.4(10^6)$ | 7.495 |
| $9.0(10^5)$ | 6.800 | $1.0(10^6)$ | 6.904 | $2.3(10^6)$ | 7.090 | $3.3(10^6)$ | 7.244 | $4.4(10^6)$ | 7.378 | $5.6(10^6)$ | 7.495 |
| $1.0(10^6)$ | 6.801 | $1.2(10^6)$ | 6.907 | $2.4(10^6)$ | 7.090 | | | | | | |

| $r_{eD}$ = 2,000.0 | | $r_{eD}$ = 2,200.0 | | $r_{eD}$ = 2,400.0 | | $r_{eD}$ = 2,600.0 | | $r_{eD}$ = 2,800.0 | | $r_{eD}$ = 3,000.0 | |
|---|---|---|---|---|---|---|---|---|---|---|---|
| $t_D$ | $p_D$ | $t_D$ | $p_D$ | $t_D$ | $p_D$ | $t_D$ | $p_D$ | $t_D$ | $p_D$ | $t_D$ | $p_D$ |
| $4.0(10^5)$ | 6.854 | $5.0(10^5)$ | 6.966 | $6.0(10^5)$ | 7.057 | $7.0(10^5)$ | 7.134 | $8.0(10^5)$ | 7.201 | $1.0(10^6)$ | 7.312 |
| $5.0(10^5)$ | 6.966 | $5.5(10^5)$ | 7.013 | $7.0(10^5)$ | 7.134 | $8.0(10^5)$ | 7.201 | $9.0(10^5)$ | 7.260 | $1.2(10^6)$ | 7.403 |
| $6.0(10^5)$ | 7.056 | $6.0(10^5)$ | 7.057 | $8.0(10^5)$ | 7.200 | $9.0(10^5)$ | 7.259 | $1.0(10^6)$ | 7.312 | $1.4(10^6)$ | 7.480 |
| $7.0(10^5)$ | 7.132 | $6.5(10^5)$ | 7.097 | $9.0(10^5)$ | 7.259 | $1.0(10^6)$ | 7.312 | $1.2(10^6)$ | 7.403 | $1.6(10^6)$ | 7.545 |
| $8.0(10^5)$ | 7.196 | $7.0(10^5)$ | 7.133 | $1.0(10^6)$ | 7.310 | $1.2(10^6)$ | 7.401 | $1.6(10^6)$ | 7.542 | $1.8(10^6)$ | 7.602 |
| $9.0(10^5)$ | 7.251 | $7.5(10^5)$ | 7.167 | $1.2(10^6)$ | 7.398 | $1.4(10^6)$ | 7.475 | $2.0(10^6)$ | 7.644 | $2.0(10^6)$ | 7.651 |
| $1.0(10^6)$ | 7.298 | $8.0(10^5)$ | 7.199 | $1.6(10^6)$ | 7.526 | $1.6(10^6)$ | 7.536 | $2.4(10^6)$ | 7.719 | $2.4(10^6)$ | 7.732 |
| $1.2(10^6)$ | 7.374 | $8.5(10^5)$ | 7.229 | $2.0(10^6)$ | 7.611 | $1.8(10^6)$ | 7.588 | $2.8(10^6)$ | 7.775 | $2.8(10^6)$ | 7.794 |
| $1.4(10^6)$ | 7.431 | $9.0(10^5)$ | 7.256 | $2.4(10^6)$ | 7.668 | $2.0(10^6)$ | 7.631 | $3.0(10^6)$ | 7.797 | $3.0(10^6)$ | 7.820 |
| $1.6(10^6)$ | 7.474 | $1.0(10^6)$ | 7.307 | $2.8(10^6)$ | 7.706 | $2.4(10^6)$ | 7.699 | $3.5(10^6)$ | 7.840 | $3.5(10^6)$ | 7.871 |
| $1.8(10^6)$ | 7.506 | $1.2(10^6)$ | 7.390 | $3.0(10^6)$ | 7.720 | $2.8(10^6)$ | 7.746 | $4.0(10^6)$ | 7.870 | $4.0(10^6)$ | 7.908 |
| $2.0(10^6)$ | 7.530 | $1.6(10^6)$ | 7.507 | $3.5(10^6)$ | 7.745 | $3.0(10^6)$ | 7.765 | $5.0(10^6)$ | 7.905 | $4.5(10^6)$ | 7.935 |
| $2.5(10^6)$ | 7.566 | $2.0(10^6)$ | 7.579 | $4.0(10^6)$ | 7.760 | $3.5(10^6)$ | 7.799 | $6.0(10^6)$ | 7.922 | $5.0(10^6)$ | 7.955 |
| $3.0(10^6)$ | 7.584 | $2.5(10^6)$ | 7.631 | $5.0(10^6)$ | 7.775 | $4.0(10^6)$ | 7.821 | $7.0(10^6)$ | 7.930 | $6.0(10^6)$ | 7.979 |
| $3.5(10^6)$ | 7.593 | $3.0(10^6)$ | 7.661 | $6.0(10^6)$ | 7.780 | $5.0(10^6)$ | 7.845 | $8.0(10^6)$ | 7.934 | $7.0(10^6)$ | 7.992 |
| $4.0(10^6)$ | 7.597 | $3.5(10^6)$ | 7.677 | $7.0(10^6)$ | 7.782 | $6.0(10^6)$ | 7.856 | $9.0(10^6)$ | 7.936 | $8.0(10^6)$ | 7.999 |
| $5.0(10^6)$ | 7.600 | $4.0(10^6)$ | 7.686 | $8.0(10^6)$ | 7.783 | $7.0(10^6)$ | 7.860 | $1.0(10^7)$ | 7.937 | $9.0(10^6)$ | 8.002 |
| $6.0(10^6)$ | 7.601 | $5.0(10^6)$ | 7.693 | $9.0(10^6)$ | 7.783 | $8.0(10^6)$ | 7.862 | $1.2(10^7)$ | 7.937 | $1.0(10^7)$ | 8.004 |
| $6.4(10^6)$ | 7.601 | $6.0(10^6)$ | 7.695 | $9.5(10^6)$ | 7.783 | $9.0(10^6)$ | 7.863 | $1.3(10^7)$ | 7.937 | $1.2(10^7)$ | 8.006 |
| | | $7.0(10^6)$ | 7.696 | | | $1.0(10^7)$ | 7.863 | | | $1.5(10^7)$ | 8.006 |
| | | $8.0(10^6)$ | 7.696 | | | | | | | | |

Some important properties of these tabulated solutions include the following:

1. For values of $t_D$ smaller than the smallest value listed for a given $r_{eD}$, the reservoir is infinite acting, and Table C.1 should be used to determine $p_D$.
2. For values of $t_D$ larger than the largest value listed for a given $r_{eD}$ (or for $t_D > r_{eD}^2$),

$$p_D \simeq \ln r_{eD}. \quad \text{(C.6)}$$

## *Example C.2 – Use of $p_D$ Solutions for Constant-Pressure Boundary*

**Problem.** An oil well is producing 300 STB/D from a water-drive reservoir that maintains pressure at the original oil/water contact at a constant 3,000 psia. Distance from the well to the oil/water contact is 900 ft. Well and reservoir properties include the following.

$\phi$ = 0.21,
$k$ = 80 md,
$c_t$ = $20 \times 10^{-6}$ psi$^{-1}$,
$r_w$ = 0.45 ft,
$\mu$ = 0.8 cp,
$h$ = 12 ft, and
$B$ = 1.25 RB/STB.

Calculate pressure in the wellbore for production after 0.1, 1.0, and 100.0 days.

**Solution.** Since $t_D = 0.000264\ kt/\phi\mu c_t r_w^2$ ($t$ in hours), then

$$t_D = \frac{(0.000264)(80)(24)\ t\ \text{days}}{(0.21)(0.8)(20 \times 10^{-6})(0.45)^2}$$

$$= 7.45 \times 10^5 t.$$

Also,

$$r_{eD} = \frac{r_e}{r_w} = \frac{900}{0.45} = 2{,}000.$$

Thus, we obtain the following.

## TABLE C.4 – $Q_{pD}$ vs. $t_D$ – INFINITE RADIAL SYSTEM, CONSTANT PRESSURE AT INNER BOUNDARY

| Dimensionless Time $t_D$ | Dimensionless Cumulative Production $Q_{pD}$ | Dimensionless Time $t_D$ | Dimensionless Cumulative Production $Q_{pD}$ | Dimensionless Time $t_D$ | Dimensionless Cumulative Production $Q_{pD}$ | Dimensionless Time $t_D$ | Dimensionless Cumulative Production $Q_{pD}$ | Dimensionless Time $t_D$ | Dimensionless Cumulative Production $Q_{pD}$ |
|---|---|---|---|---|---|---|---|---|---|
| 0.00 | 0.000 | 98 | 42.433 | 740 | 226.904 | 1,975 | 528.337 | 8,900 | 1,986.796 |
| 0.01 | 0.112 | 99 | 42.781 | 750 | 229.514 | 2,000 | 534.145 | 9,000 | 2,006.628 |
| 0.05 | 0.278 | 100 | 43.129 | 760 | 232.120 | 2,025 | 539.945 | 9,100 | 2,026.438 |
| 0.10 | 0.404 | 105 | 44.858 | 770 | 234.721 | 2,050 | 545.737 | 9,200 | 2,046.227 |
| 0.15 | 0.520 | 110 | 46.574 | 775 | 236.020 | 2,075 | 551.522 | 9,300 | 2,065.996 |
| 0.20 | 0.606 | 115 | 48.277 | 780 | 237.318 | 2,100 | 557.299 | 9,400 | 2,085.744 |
| 0.25 | 0.689 | 120 | 49.968 | 790 | 239.912 | 2,125 | 563.068 | 9,500 | 2,105.473 |
| 0.30 | 0.758 | 125 | 51.648 | 800 | 242.501 | 2,150 | 568.830 | 9,600 | 2,125.184 |
| 0.40 | 0.898 | 130 | 53.317 | 810 | 245.086 | 2,175 | 574.585 | 9,700 | 2,144.878 |
| 0.50 | 1.020 | 135 | 54.976 | 820 | 247.668 | 2,200 | 580.332 | 9,800 | 2,164.555 |
| 0.60 | 1.140 | 140 | 56.625 | 825 | 248.957 | 2,225 | 586.072 | 9,900 | 2,184.216 |
| 0.70 | 1.251 | 145 | 58.265 | 830 | 250.245 | 2,250 | 591.086 | 10,000 | 2,203.861 |
| 0.80 | 1.359 | 150 | 59.895 | 840 | 252.819 | 2,275 | 597.532 | 12,500 | 2,688.967 |
| 0.90 | 1.469 | 155 | 61.517 | 850 | 255.388 | 2,300 | 603.252 | 15,000 | 3,164.780 |
| 1 | 1.569 | 160 | 63.131 | 860 | 257.953 | 2,325 | 608.965 | 17,500 | 3,633.368 |
| 2 | 2.447 | 165 | 64.737 | 870 | 260.515 | 2,350 | 614.672 | 20,000 | 4,095.800 |
| 3 | 3.202 | 170 | 66.336 | 875 | 261.795 | 2,375 | 620.372 | 25,000 | 5,005.726 |
| 4 | 3.893 | 175 | 67.928 | 880 | 263.073 | 2,400 | 626.066 | 30,000 | 5,899.508 |
| 5 | 4.539 | 180 | 69.512 | 890 | 265.629 | 2,425 | 631.755 | 35,000 | 6,780.247 |
| 6 | 5.153 | 185 | 71.090 | 900 | 268.181 | 2,450 | 637.437 | 40,000 | 7,650.096 |
| 7 | 5.743 | 190 | 72.661 | 910 | 270.729 | 2,475 | 643.113 | 50,000 | 9,363.099 |
| 8 | 6.314 | 195 | 74.226 | 920 | 273.274 | 2,500 | 648.781 | 60,000 | 11,047.299 |
| 9 | 6.869 | 200 | 75.785 | 925 | 274.545 | 2,550 | 660.093 | 70,000 | 12,708.358 |
| 10 | 7.411 | 205 | 77.338 | 930 | 275.815 | 2,600 | 671.379 | 75,000 | 13,531.457 |
| 11 | 7.940 | 210 | 78.886 | 940 | 278.353 | 2,650 | 682.640 | 80,000 | 14,350.121 |
| 12 | 8.457 | 215 | 80.428 | 950 | 280.888 | 2,700 | 693.877 | 90,000 | 15,975.389 |
| 13 | 8.964 | 220 | 81.965 | 960 | 283.420 | 2,750 | 705.090 | 100,000 | 17,586.284 |
| 14 | 9.461 | 225 | 83.497 | 970 | 285.948 | 2,800 | 716.280 | 125,000 | 21,560.732 |
| 15 | 9.949 | 230 | 85.023 | 975 | 287.211 | 2,850 | 727.449 | $1.5(10^5)$ | $2.538(10^4)$ |
| 16 | 10.434 | 235 | 86.545 | 980 | 288.473 | 2,900 | 738.598 | $2.0(10^5)$ | $3.308(10^4)$ |
| 17 | 10.913 | 240 | 88.062 | 990 | 290.995 | 2,950 | 749.725 | $2.5(10^5)$ | $4.066(10^4)$ |
| 18 | 11.386 | 245 | 89.575 | 1,000 | 293.514 | 3,000 | 760.833 | $3.0(10^5)$ | $4.817(10^4)$ |
| 19 | 11.855 | 250 | 91.084 | 1,010 | 296.030 | 3,050 | 771.922 | $4.0(10^5)$ | $6.267(10^4)$ |
| 20 | 12.319 | 255 | 92.589 | 1,020 | 298.543 | 3,100 | 782.992 | $5.0(10^5)$ | $7.699(10^4)$ |
| 21 | 12.778 | 260 | 94.090 | 1,025 | 299.799 | 3,150 | 794.042 | $6.0(10^5)$ | $9.113(10^4)$ |
| 22 | 13.233 | 265 | 95.588 | 1,030 | 301.053 | 3,200 | 805.075 | $7.0(10^5)$ | $1.051(10^5)$ |
| 23 | 13.684 | 270 | 97.081 | 1,040 | 303.560 | 3,250 | 816.090 | $8.0(10^5)$ | $1.189(10^5)$ |
| 24 | 14.131 | 275 | 98.571 | 1,050 | 306.065 | 3,300 | 827.088 | $9.0(10^5)$ | $1.326(10^5)$ |
| 25 | 14.573 | 280 | 100.057 | 1,060 | 308.567 | 3,350 | 838.067 | $1.0(10^6)$ | $1.462(10^5)$ |
| 26 | 15.013 | 285 | 101.540 | 1,070 | 311.066 | 3,400 | 849.028 | $1.5(10^6)$ | $2.126(10^5)$ |
| 27 | 15.450 | 290 | 103.019 | 1,075 | 312.314 | 3,450 | 859.974 | $2.0(10^6)$ | $2.781(10^5)$ |
| 28 | 15.883 | 295 | 104.495 | 1,080 | 313.562 | 3,500 | 870.903 | $2.5(10^6)$ | $3.427(10^5)$ |
| 29 | 16.313 | 300 | 105.968 | 1,090 | 316.055 | 3,550 | 881.816 | $3.0(10^6)$ | $4.064(10^5)$ |
| 30 | 16.742 | 305 | 107.437 | 1,100 | 318.545 | 3,600 | 892.712 | $4.0(10^6)$ | $5.313(10^5)$ |
| 31 | 17.167 | 310 | 108.904 | 1,110 | 321.032 | 3,650 | 903.594 | $5.0(10^6)$ | $6.544(10^5)$ |
| 32 | 17.590 | 315 | 110.367 | 1,120 | 323.517 | 3,700 | 914.459 | $6.0(10^6)$ | $7.761(10^5)$ |
| 33 | 18.011 | 320 | 111.827 | 1,125 | 324.760 | 3,750 | 926.309 | $7.0(10^6)$ | $8.965(10^5)$ |
| 34 | 18.429 | 325 | 113.284 | 1,130 | 326.000 | 3,800 | 936.144 | $8.0(10^6)$ | $1.016(10^6)$ |
| 35 | 18.845 | 330 | 114.738 | 1,140 | 328.480 | 3,850 | 946.966 | $9.0(10^6)$ | $1.134(10^6)$ |
| 36 | 19.259 | 335 | 116.189 | 1,150 | 330.958 | 3,900 | 957.773 | $1.0(10^7)$ | $1.252(10^6)$ |
| 37 | 19.671 | 340 | 117.638 | 1,160 | 333.433 | 3,950 | 968.566 | $1.5(10^7)$ | $1.828(10^6)$ |
| 38 | 20.080 | 345 | 119.083 | 1,170 | 335.906 | 4,000 | 979.344 | $2.0(10^7)$ | $2.398(10^6)$ |
| 39 | 20.488 | 350 | 120.526 | 1,175 | 337.142 | 4,050 | 990.108 | $2.5(10^7)$ | $2.961(10^6)$ |
| 40 | 20.894 | 355 | 121.966 | 1,180 | 338.376 | 4,100 | 1,000.858 | $3.0(10^7)$ | $3.517(10^6)$ |
| 41 | 21.298 | 360 | 123.403 | 1,190 | 340.843 | 4,150 | 1,011.595 | $4.0(10^7)$ | $4.610(10^6)$ |
| 42 | 21.701 | 365 | 124.838 | 1,200 | 343.308 | 4,200 | 1,022.318 | $5.0(10^7)$ | $5.689(10^6)$ |
| 43 | 22.101 | 370 | 126.270 | 1,210 | 345.770 | 4,250 | 1,033.028 | $6.0(10^7)$ | $6.758(10^6)$ |
| 44 | 22.500 | 375 | 127.699 | 1,220 | 348.230 | 4,300 | 1,043.724 | $7.0(10^7)$ | $7.816(10^6)$ |
| 45 | 22.897 | 380 | 129.126 | 1,225 | 349.460 | 4,350 | 1,054.409 | $8.0(10^7)$ | $8.866(10^6)$ |
| 46 | 23.291 | 385 | 130.550 | 1,230 | 350.688 | 4,400 | 1,065.082 | $9.0(10^7)$ | $9.911(10^6)$ |
| 47 | 23.684 | 390 | 131.972 | 1,240 | 353.144 | 4,450 | 1,075.743 | $1.0(10^8)$ | $1.095(10^7)$ |
| 48 | 24.076 | 395 | 133.391 | 1,250 | 355.597 | 4,500 | 1,086.390 | $1.5(10^8)$ | $1.604(10^7)$ |
| 49 | 24.466 | 400 | 134.808 | 1,260 | 358.048 | 4,550 | 1,097.024 | $2.0(10^8)$ | $2.108(10^7)$ |
| 50 | 24.855 | 405 | 136.223 | 1,270 | 360.496 | 4,600 | 1,107.646 | $2.5(10^8)$ | $2.607(10^7)$ |
| 51 | 25.244 | 410 | 137.635 | 1,275 | 361.720 | 4,650 | 1,118.257 | $3.0(10^8)$ | $3.100(10^7)$ |
| 52 | 25.633 | 415 | 139.045 | 1,280 | 362.942 | 4,700 | 1,128.854 | $4.0(10^8)$ | $4.071(10^7)$ |
| 53 | 26.020 | 420 | 140.453 | 1,290 | 365.386 | 4,750 | 1,139.439 | $5.0(10^8)$ | $5.03\ (10^7)$ |
| 54 | 26.406 | 425 | 141.859 | 1,300 | 367.828 | 4,800 | 1,150.012 | $6.0(10^8)$ | $5.98\ (10^7)$ |
| 55 | 26.791 | 430 | 143.262 | 1,310 | 370.267 | 4,850 | 1,160.574 | $7.0(10^8)$ | $6.928(10^7)$ |
| 56 | 27.174 | 435 | 144.664 | 1,320 | 372.704 | 4,900 | 1,171.125 | $8.0(10^8)$ | $7.865(10^7)$ |
| 57 | 27.555 | 440 | 146.064 | 1,325 | 373.922 | 4,950 | 1,181.666 | $9.0(10^8)$ | $8.797(10^7)$ |
| 58 | 27.935 | 445 | 147.461 | 1,330 | 375.139 | 5,000 | 1,192.198 | $1.0(10^9)$ | $9.725(10^7)$ |
| 59 | 28.314 | 450 | 148.856 | 1,340 | 377.572 | 5,100 | 1,213.222 | $1.5(10^9)$ | $1.429(10^8)$ |
| 60 | 28.691 | 455 | 150.249 | 1,350 | 380.003 | 5,200 | 1,234.203 | $2.0(10^9)$ | $1.880(10^8)$ |
| 61 | 29.068 | 460 | 151.460 | 1,360 | 382.432 | 5,300 | 1,255.141 | $2.5(10^9)$ | $2.328(10^8)$ |
| 62 | 29.443 | 465 | 153.029 | 1,370 | 384.859 | 5,400 | 1,276.037 | $3.0(10^9)$ | $2.771(10^8)$ |

Notes: For $t_D < 0.01$, $Q_{pD} = 2\sqrt{t_D/\pi}$.

For $t_D \geq 200$, $Q_{pD} \cong \dfrac{-4.29881 + 2.02566\, t_D}{\ln t_D}$.

**TABLE C.4 – CONTINUED**

| Dimensionless Time $t_D$ | Dimensionless Cumulative Production $Q_{pD}$ | Dimensionless Time $t_D$ | Dimensionless Cumulative Production $Q_{pD}$ | Dimensionless Time $t_D$ | Dimensionless Cumulative Production $Q_{pD}$ | Dimensionless Time $t_D$ | Dimensionless Cumulative Production $Q_{pD}$ | Dimensionless Time $t_D$ | Dimensionless Cumulative Production $Q_{pD}$ |
|---|---|---|---|---|---|---|---|---|---|
| 63 | 29.818 | 470 | 154.416 | 1,375 | 386.070 | 5,500 | 1,296.893 | $4.0(10^{9})$ | $3.645(10^{8})$ |
| 64 | 30.192 | 475 | 155.801 | 1,380 | 387.283 | 5,600 | 1,317.709 | $5.0(10^{9})$ | $4.510(10^{8})$ |
| 65 | 30.565 | 480 | 157.184 | 1,390 | 389.705 | 5,700 | 1,338.486 | $6.0(10^{9})$ | $5.368(10^{8})$ |
| 66 | 30.937 | 485 | 158.565 | 1,400 | 392.125 | 5,800 | 1,359.225 | $7.0(10^{9})$ | $6.220(10^{8})$ |
| 67 | 31.308 | 490 | 159.945 | 1,410 | 394.543 | 5,900 | 1,379.927 | $8.0(10^{9})$ | $7.066(10^{8})$ |
| 68 | 31.679 | 495 | 161.322 | 1,420 | 396.959 | 6,000 | 1,400.593 | $9.0(10^{9})$ | $7.909(10^{8})$ |
| 69 | 32.048 | 500 | 162.698 | 1,425 | 398.167 | 6,100 | 1,421.224 | $1.0(10^{10})$ | $8.747(10^{8})$ |
| 70 | 32.417 | 510 | 165.444 | 1,430 | 399.373 | 6,200 | 1,441.820 | $1.5(10^{10})$ | $1.288(10^{9})$ |
| 71 | 32.785 | 520 | 168.183 | 1,440 | 401.786 | 6,300 | 1,462.383 | $2.0(10^{10})$ | $1.697(10^{9})$ |
| 72 | 33.151 | 525 | 169.549 | 1,450 | 404.197 | 6,400 | 1,482.912 | $2.5(10^{10})$ | $2.103(10^{9})$ |
| 73 | 33.517 | 530 | 170.914 | 1,460 | 406.606 | 6,500 | 1,503.408 | $3.0(10^{10})$ | $2.505(10^{9})$ |
| 74 | 33.883 | 540 | 173.639 | 1,470 | 409.013 | 6,600 | 1,523.872 | $4.0(10^{10})$ | $3.299(10^{9})$ |
| 75 | 34.247 | 550 | 176.357 | 1,475 | 410.214 | 6,700 | 1,544.305 | $5.0(10^{10})$ | $4.087(10^{9})$ |
| 76 | 34.611 | 560 | 179.069 | 1,480 | 411.418 | 6,800 | 1,564.706 | $6.0(10^{10})$ | $4.868(10^{9})$ |
| 77 | 34.974 | 570 | 181.774 | 1,490 | 413.820 | 6,900 | 1,585.077 | $7.0(10^{10})$ | $5.643(10^{9})$ |
| 78 | 35.336 | 580 | 184.473 | 1,500 | 416.220 | 7,000 | 1,605.418 | $8.0(10^{10})$ | $6.414(10^{9})$ |
| 79 | 35.697 | 590 | 187.166 | 1,525 | 422.214 | 7,100 | 1,625.729 | $9.0(10^{10})$ | $7.183(10^{9})$ |
| 80 | 36.058 | 600 | 189.852 | 1,550 | 428.196 | 7,200 | 1,646.011 | $1.0(10^{11})$ | $7.948(10^{9})$ |
| 81 | 36.418 | 610 | 192.533 | 1,575 | 434.168 | 7,300 | 1,666.265 | $1.5(10^{11})$ | $1.17(10^{10})$ |
| 82 | 36.777 | 620 | 195.208 | 1,600 | 440.128 | 7,400 | 1,686.490 | $2.0(10^{11})$ | $1.55(10^{10})$ |
| 83 | 37.136 | 625 | 196.544 | 1,625 | 446.077 | 7,500 | 1,706.688 | $2.5(10^{11})$ | $1.92(10^{10})$ |
| 84 | 37.494 | 630 | 197.878 | 1,650 | 452.016 | 7,600 | 1,726.859 | $3.0(10^{11})$ | $2.29(10^{10})$ |
| 85 | 37.851 | 640 | 200.542 | 1,675 | 457.945 | 7,700 | 1,747.002 | $4.0(10^{11})$ | $3.02(10^{10})$ |
| 86 | 38.207 | 650 | 203.201 | 1,700 | 463.863 | 7,800 | 1,767.120 | $5.0(10^{11})$ | $3.75(10^{10})$ |
| 87 | 38.563 | 660 | 205.854 | 1,725 | 469.771 | 7,900 | 1,787.212 | $6.0(10^{11})$ | $4.47(10^{10})$ |
| 88 | 38.919 | 670 | 208.502 | 1,750 | 475.669 | 8,000 | 1,807.278 | $7.0(10^{11})$ | $5.19(10^{10})$ |
| 89 | 39.272 | 675 | 209.825 | 1,775 | 481.558 | 8,100 | 1,827.319 | $8.0(10^{11})$ | $5.89(10^{10})$ |
| 90 | 39.626 | 680 | 211.145 | 1,800 | 487.437 | 8,200 | 1,847.336 | $9.0(10^{11})$ | $6.58(10^{10})$ |
| 91 | 39.979 | 690 | 213.784 | 1,825 | 493.307 | 8,300 | 1,867.329 | $1.0(10^{12})$ | $7.28(10^{10})$ |
| 92 | 40.331 | 700 | 216.417 | 1,850 | 499.167 | 8,400 | 1,887.298 | $1.5(10^{12})$ | $1.08(10^{11})$ |
| 93 | 40.684 | 710 | 219.046 | 1,875 | 505.019 | 8,500 | 1,907.243 | $2.0(10^{12})$ | $1.42(10^{11})$ |
| 94 | 41.034 | 720 | 221.670 | 1,900 | 510.861 | 8,600 | 1,927.166 | | |
| 95 | 41.385 | 725 | 222.980 | 1,925 | 516.695 | 8,700 | 1,947.065 | | |
| 96 | 41.735 | 730 | 224.289 | 1,950 | 522.520 | 8,800 | 1,966.942 | | |
| 97 | 42.084 | | | | | | | | |

| $t$ (days) | $t_D$ | $p_D$ | Source | $p_{wf}$ (psia) |
|---|---|---|---|---|
| 0.1 | $7.45\times10^4$ | 6.015 | $0.5(\ln t_D + 0.80907)$ | 2,735 |
| 1.0 | $7.45\times10^5$ | 7.161 | Table C.3 | 2,684 |
| 100.0 | $7.45\times10^7$ | 7.601 | $\ln r_{eD}$ | 2,665 |

Note that for $r_{eD}=2{,}000$, the reservoir is infinite acting at $t_D=7.45\times10^4$. This means that Table C.1 rather than Table C.3 is to be used to determine $p_D$; however, for this $t_D$, Eq. C.3, which "extends" Table C.1, is appropriate. For $t_D=7.45\times10^5$, $p_D$ is found in Table C.3; for $t_D=7.45\times10^7$, $p_D$ is calculated from Eq. C.6. Note also that

$$p_{wf}=p_i-\frac{qB\mu}{0.00708\,kh}p_D$$

$$=3{,}000-\frac{(300)(1.25)(0.8)}{(0.00708)(80)(12)}p_D$$

$$=3{,}000-44.14\,p_D.$$

## Constant Pressure at Inner Boundary, No Flow Across Outer Boundary

This solution of the diffusivity equation models radial flow of a slightly compressible liquid in a homogeneous reservoir of uniform thickness; reservoir at uniform pressure $p_i$ before production; no flow across the outer boundary (at $r=r_e$); and constant BHP $p_{wf}$ at the single producing well (centered in the reservoir) with wellbore radius $r_w$. The solution – pressure as a function of time and radius for fixed values of $r_e$, $r_w$, and rock and fluid properties – is expressed most conveniently in terms of dimensionless variables and parameters:

$$p_D=f(t_D,r_D,r_{eD}),$$

where

$$p_D=\frac{p_i-p}{p_i-p_{wf}},$$

$$t_D=\frac{0.000264\,kt}{\phi\mu c_t r_w^2},$$

$$r_D=r/r_w,$$

$$r_{eD}=r_e/r_w.$$

For this problem, instantaneous rate $q$ and cumulative production $Q_p$ are of more practical importance than $p_D$, and these quantities can be derived from the fundamental solutions, $p_D$. In Appendix B, we showed that a dimensionless production rate, $q_D$, and dimensionless cumulative production, $Q_{pD}$, can be defined as

$$q_D=\frac{qB\mu}{0.00708\,kh\,(p_i-p_{wf})},$$

and

$$Q_{pD}=\int_{o}^{t_D} q_D \, dt_D$$

$$=\frac{B}{1.119\,\phi c_t h r_w^2\,(p_i-p_{wf})}\,Q_p.$$

Dimensionless cumulative production, $Q_{pD}$, is presented in Tables C.4 and C.5. Table C.4 is for infinite-acting reservoirs, and Table C.5 is for finite reservoirs with $1.5 \le r_{eD} \le 1\times10^6$. For $r_{eD}\ge 20$, values for both $q_D$ and $Q_{pD}$ are given. In Table C.5, for values of $t_D$ smaller than the smallest value for a given $r_{eD}$, the reservoir is infinite-acting and Table C.4 should be used. For values of $t_D$ larger than those in the table for a given value of $r_{eD}$, the reservoir has reached steady state, and

$$Q_{pD}=(r_{eD}^2-1)/2.$$

For $r_{eD}=\infty$ and $t_D\ge 200$, $q_D$ can be approximated as[3]

$$Q_{pD}=\frac{(-4.29881+2.02566\,t_D)}{\ln t_D}.$$

Since both $q_D$ and $Q_{pD}$ are based on solutions ($p_D$) that, for $r_D=1$ ($r=r_w$), depend only on $t_D$ and $r_{eD}$, $q_D$ and $Q_{pD}$ also should depend only on $t_D$ and $r_{eD}$. Table C.5 confirms this expectation. For given $r_{eD}$ and $t_D$, $Q_{pD}$ is determined uniquely.

Although the $Q_{pD}$ solutions can be used to model individual well problems, they more often are used to model water influx from an aquifer of radius $r_e$ into a petroleum reservoir of radius $r_w$ for a fixed reservoir pressure, $p_{wf}$. Superposition is used to model a variable pressure history, as illustrated in Example C.4.

## *Example C.3 – Use of $Q_{pD}$ Solutions*

**Problem.** An oil well is produced with a constant

TABLE C.5 – $Q_{pD}$ vs. $t_D$ – FINITE RADIAL SYSTEM WITH CLOSED EXTERIOR BOUNDARY, CONSTANT PRESSURE AT INNER BOUNDARY

| $r_{eD}=1.5$ | | $r_{eD}=2.0$ | | $r_{eD}=2.5$ | | $r_{eD}=3.0$ | | $r_{eD}=3.5$ | | $r_{eD}=4.0$ | |
|---|---|---|---|---|---|---|---|---|---|---|---|
| $t_D$ | $Q_{pD}$ | $t_D$ | $Q_{pD}$ | $t_D$ | $Q_{pD}$ | $t_D$ | $Q_{pD}$ | $t_D$ | $Q_{pD}$ | $t_D$ | $Q_{pD}$ |
| 0.05 | 0.276 | 0.05 | 0.278 | 0.10 | 0.408 | 0.30 | 0.755 | 1.00 | 1.571 | 2.00 | 2.442 |
| 0.06 | 0.304 | 0.075 | 0.345 | 0.15 | 0.509 | 0.40 | 0.895 | 1.20 | 1.761 | 2.20 | 2.598 |
| 0.07 | 0.330 | 0.10 | 0.404 | 0.20 | 0.599 | 0.50 | 1.023 | 1.40 | 1.940 | 2.40 | 2.748 |
| 0.08 | 0.354 | 0.125 | 0.458 | 0.25 | 0.681 | 0.60 | 1.143 | 1.60 | 2.111 | 2.60 | 2.893 |
| 0.09 | 0.375 | 0.150 | 0.507 | 0.30 | 0.758 | 0.70 | 1.256 | 1.80 | 2.273 | 2.80 | 3.034 |
| 0.10 | 0.395 | 0.175 | 0.553 | 0.35 | 0.829 | 0.80 | 1.363 | 2.00 | 2.427 | 3.00 | 3.170 |
| 0.11 | 0.414 | 0.200 | 0.597 | 0.40 | 0.897 | 0.90 | 1.465 | 2.20 | 2.574 | 3.25 | 3.334 |
| 0.12 | 0.431 | 0.225 | 0.638 | 0.45 | 0.962 | 1.00 | 1.563 | 2.40 | 2.715 | 3.50 | 3.493 |
| 0.13 | 0.446 | 0.250 | 0.678 | 0.50 | 1.024 | 1.25 | 1.791 | 2.60 | 2.849 | 3.75 | 3.645 |
| 0.14 | 0.461 | 0.275 | 0.715 | 0.55 | 1.083 | 1.50 | 1.997 | 2.80 | 2.976 | 4.00 | 3.792 |
| 0.15 | 0.474 | 0.300 | 0.751 | 0.60 | 1.140 | 1.75 | 2.184 | 3.00 | 3.098 | 4.25 | 3.932 |
| 0.16 | 0.486 | 0.325 | 0.785 | 0.65 | 1.195 | 2.00 | 2.353 | 3.25 | 3.242 | 4.50 | 4.068 |
| 0.17 | 0.497 | 0.350 | 0.817 | 0.70 | 1.248 | 2.25 | 2.507 | 3.50 | 3.379 | 4.75 | 4.198 |
| 0.18 | 0.507 | 0.375 | 0.848 | 0.75 | 1.299 | 2.50 | 2.646 | 3.75 | 3.507 | 5.00 | 4.323 |
| 0.19 | 0.517 | 0.400 | 0.877 | 0.80 | 1.348 | 2.75 | 2.772 | 4.00 | 3.628 | 5.50 | 4.560 |
| 0.20 | 0.525 | 0.425 | 0.905 | 0.85 | 1.395 | 3.00 | 2.886 | 4.25 | 3.742 | 6.00 | 4.779 |
| 0.21 | 0.533 | 0.450 | 0.932 | 0.90 | 1.440 | 3.25 | 2.990 | 4.50 | 3.850 | 6.50 | 4.982 |
| 0.22 | 0.541 | 0.475 | 0.958 | 0.95 | 1.484 | 3.50 | 3.084 | 4.75 | 3.951 | 7.00 | 5.169 |
| 0.23 | 0.548 | 0.500 | 0.983 | 1.0 | 1.526 | 3.75 | 3.170 | 5.00 | 4.047 | 7.50 | 5.343 |
| 0.24 | 0.554 | 0.550 | 1.028 | 1.1 | 1.605 | 4.00 | 3.247 | 5.50 | 4.222 | 8.00 | 5.504 |
| 0.25 | 0.559 | 0.600 | 1.070 | 1.2 | 1.679 | 4.25 | 3.317 | 6.00 | 4.378 | 8.50 | 5.653 |
| 0.26 | 0.565 | 0.650 | 1.108 | 1.3 | 1.747 | 4.50 | 3.381 | 6.50 | 4.516 | 9.00 | 5.790 |
| 0.28 | 0.574 | 0.700 | 1.143 | 1.4 | 1.811 | 4.75 | 3.439 | 7.00 | 4.639 | 9.50 | 5.917 |
| 0.30 | 0.582 | 0.750 | 1.174 | 1.5 | 1.870 | 5.00 | 3.491 | 7.50 | 4.749 | 10 | 6.035 |
| 0.32 | 0.588 | 0.800 | 1.203 | 1.6 | 1.924 | 5.50 | 3.581 | 8.00 | 4.846 | 11 | 6.246 |
| 0.34 | 0.594 | 0.900 | 1.253 | 1.7 | 1.975 | 6.00 | 3.656 | 8.50 | 4.932 | 12 | 6.425 |
| 0.36 | 0.599 | 1.000 | 1.295 | 1.8 | 2.022 | 6.50 | 3.717 | 9.00 | 5.009 | 13 | 6.580 |
| 0.38 | 0.603 | 1.1 | 1.330 | 2.0 | 2.106 | 7.00 | 3.767 | 9.50 | 5.078 | 14 | 6.712 |
| 0.40 | 0.606 | 1.2 | 1.358 | 2.2 | 2.178 | 7.50 | 3.809 | 10.00 | 5.138 | 15 | 6.825 |
| 0.45 | 0.613 | 1.3 | 1.382 | 2.4 | 2.241 | 8.00 | 3.843 | 11 | 5.241 | 16 | 6.922 |
| 0.50 | 0.617 | 1.4 | 1.402 | 2.6 | 2.294 | 9.00 | 3.894 | 12 | 5.321 | 17 | 7.004 |
| 0.60 | 0.621 | 1.6 | 1.432 | 2.8 | 2.340 | 10.00 | 3.928 | 13 | 5.385 | 18 | 7.076 |
| 0.70 | 0.623 | 1.7 | 1.444 | 3.0 | 2.380 | 11.00 | 3.951 | 14 | 5.435 | 20 | 7.189 |
| 0.80 | 0.624 | 1.8 | 1.453 | 3.4 | 2.444 | 12.00 | 3.967 | 15 | 5.476 | 22 | 7.272 |
| | | 2.0 | 1.468 | 3.8 | 2.491 | 14.00 | 3.985 | 16 | 5.506 | 24 | 7.332 |
| | | 2.5 | 1.487 | 4.2 | 2.525 | 16.00 | 3.993 | 17 | 5.531 | 26 | 7.377 |
| | | 3.0 | 1.495 | 4.6 | 2.551 | 18.00 | 3.997 | 18 | 5.551 | 30 | 7.434 |
| | | 4.0 | 1.499 | 5.0 | 2.570 | 20.00 | 3.999 | 20 | 5.579 | 34 | 7.464 |
| | | 5.0 | 1.500 | 6.0 | 2.599 | 22.00 | 3.999 | 25 | 5.611 | 38 | 7.481 |
| | | | | 7.0 | 2.613 | 24.00 | 4.000 | 30 | 5.621 | 42 | 7.490 |
| | | | | 8.0 | 2.619 | | | 35 | 5.624 | 46 | 7.494 |
| | | | | 9.0 | 2.622 | | | 40 | 5.625 | 50 | 7.497 |
| | | | | 10.0 | 2.624 | | | | | | |

BHP of 2,000 psia for 1.0 hour from a reservoir initially at 2,500 psia. The reservoir is finite; there is no flow across the outer boundary. Other rock, fluid, and well properties include the following.

$B$ = 1.2 RB/STB,
$\mu$ = 1 cp,
$k$ = 0.294 md,
$r_w$ = 0.5 ft,
$h$ = 15 ft,
$\phi$ = 0.15,
$c_t$ = $20\times10^{-6}$ psi$^{-1}$, and
$r_e$ = 1,000 ft.

Calculate the cumulative production, $Q_p$, in barrels.

**Solution.** We will calculate $t_D$ and $r_{eD}$ and read $Q_{pD}$ from either Table C.4 or Table C.5.

$$t_D=\frac{0.000264\,kt}{\phi\mu c_t r_w^2}$$

$$=\frac{(0.000264)(0.294)(1)}{(0.15)(1)(20\times10^{-6})(0.5)^2}=103,$$

$$r_{eD}=\frac{1,000}{0.5}=2,000.$$

There is no entry in Table C.5 at this $t_D$. Thus, the reservoir is infinite acting, and from Table C.4, $Q_{pD}=44.3$. Then,

$$Q_p=1.119\,\phi c_t h r_w^2\,(p_i-p_{wf})\,Q_{pD}/B$$

$$=(1.119)(0.15)(20\times10^{-6})(15)(0.5)^2$$

$$\cdot(2,500-2,000)(44.3)/1.2$$

$$=0.233\text{ STB}.$$

---

### *Example C.4 – Analysis of Variable Pressure History With $Q_{pD}$ Solution*

**Problem.** A well is completed in a reservoir with an initial pressure of 6,000 psi. The well can be considered centered in the cylindrical reservoir; there is no flow across the outer boundary. Reservoir, fluid, and well properties include the following.

**TABLE C.5 – CONTINUED**

| $r_{eD}=4.5$ | | $r_{eD}=5.0$ | | $r_{eD}=6.0$ | | $r_{eD}=7.0$ | | $r_{eD}=8.0$ | | $r_{eD}=9.0$ | | $r_{eD}=10.0$ | |
|---|---|---|---|---|---|---|---|---|---|---|---|---|---|
| $t_D$ | $Q_{pD}$ | $t_D$ | $Q_{pD}$ | $t_D$ | $Q_{pD}$ | $t_D$ | $Q_{pD}$ | $t_D$ | $Q_{pD}$ | $t_D$ | $Q_{pD}$ | $t_D$ | $Q_{pD}$ |
| 2.5 | 2.835 | 3.0 | 3.195 | 6.0 | 5.148 | 9.00 | 6.861 | 9 | 6.861 | 10 | 7.417 | 15 | 9.965 |
| 3.0 | 3.196 | 3.5 | 3.542 | 6.5 | 5.440 | 9.50 | 7.127 | 10 | 7.398 | 15 | 9.945 | 20 | 12.32 |
| 3.5 | 3.537 | 4.0 | 3.875 | 7.0 | 5.724 | 10 | 7.389 | 11 | 7.920 | 20 | 12.26 | 22 | 13.22 |
| 4.0 | 3.859 | 4.5 | 4.193 | 7.5 | 6.002 | 11 | 7.902 | 12 | 8.431 | 22 | 13.13 | 24 | 14.09 |
| 4.5 | 4.165 | 5.0 | 4.499 | 8.0 | 6.273 | 12 | 8.397 | 13 | 8.930 | 24 | 13.98 | 26 | 14.95 |
| 5.0 | 4.454 | 5.5 | 4.792 | 8.5 | 6.537 | 13 | 8.876 | 14 | 9.418 | 26 | 14.79 | 28 | 15.78 |
| 5.5 | 4.727 | 6.0 | 5.074 | 9.0 | 6.795 | 14 | 9.341 | 15 | 9.895 | 28 | 15.59 | 30 | 16.59 |
| 6.0 | 4.986 | 6.5 | 5.345 | 9.5 | 7.047 | 15 | 9.791 | 16 | 10.361 | 30 | 16.35 | 32 | 17.38 |
| 6.5 | 5.231 | 7.0 | 5.605 | 10.0 | 7.293 | 16 | 10.23 | 17 | 10.82 | 32 | 17.10 | 34 | 18.16 |
| 7.0 | 5.464 | 7.5 | 5.854 | 10.5 | 7.533 | 17 | 10.65 | 18 | 11.26 | 34 | 17.82 | 36 | 18.91 |
| 7.5 | 5.684 | 8.0 | 6.094 | 11 | 7.767 | 18 | 11.06 | 19 | 11.70 | 36 | 18.52 | 38 | 19.65 |
| 8.0 | 5.892 | 8.5 | 6.325 | 12 | 8.220 | 19 | 11.46 | 20 | 12.13 | 38 | 19.19 | 40 | 20.37 |
| 8.5 | 6.089 | 9.0 | 6.547 | 13 | 8.651 | 20 | 11.85 | 22 | 12.95 | 40 | 19.85 | 42 | 21.07 |
| 9.0 | 6.276 | 9.5 | 6.760 | 14 | 9.063 | 22 | 12.58 | 24 | 13.74 | 42 | 20.48 | 44 | 21.76 |
| 9.5 | 6.453 | 10 | 6.965 | 15 | 9.456 | 24 | 13.27 | 26 | 14.50 | 44 | 21.09 | 46 | 22.42 |
| 10 | 6.621 | 11 | 7.350 | 16 | 9.829 | 26 | 13.92 | 28 | 15.23 | 46 | 21.69 | 48 | 23.07 |
| 11 | 6.930 | 12 | 7.706 | 17 | 10.19 | 28 | 14.53 | 30 | 15.92 | 48 | 22.26 | 50 | 23.71 |
| 12 | 7.208 | 13 | 8.035 | 18 | 10.53 | 30 | 15.11 | 34 | 17.22 | 50 | 22.82 | 52 | 24.33 |
| 13 | 7.457 | 14 | 8.339 | 19 | 10.85 | 35 | 16.39 | 38 | 18.41 | 52 | 23.36 | 54 | 24.94 |
| 14 | 7.680 | 15 | 8.620 | 20 | 11.16 | 40 | 17.49 | 40 | 18.97 | 54 | 23.89 | 56 | 25.53 |
| 15 | 7.880 | 16 | 8.879 | 22 | 11.74 | 45 | 18.43 | 45 | 20.26 | 56 | 24.39 | 58 | 26.11 |
| 16 | 8.060 | 18 | 9.338 | 24 | 12.26 | 50 | 19.24 | 50 | 21.42 | 58 | 24.88 | 60 | 26.67 |
| 18 | 8.365 | 20 | 9.731 | 25 | 12.50 | 60 | 20.51 | 55 | 22.46 | 60 | 25.36 | 65 | 28.02 |
| 20 | 8.611 | 22 | 10.07 | 31 | 13.74 | 70 | 21.45 | 60 | 23.40 | 65 | 26.48 | 70 | 29.29 |
| 22 | 8.809 | 24 | 10.35 | 35 | 14.40 | 80 | 22.13 | 70 | 24.98 | 70 | 27.52 | 75 | 30.49 |
| 24 | 8.968 | 26 | 10.59 | 39 | 14.93 | 90 | 22.63 | 80 | 26.26 | 75 | 28.48 | 80 | 31.61 |
| 26 | 9.097 | 28 | 10.80 | 51 | 16.05 | 100 | 23.00 | 90 | 27.28 | 80 | 29.36 | 85 | 32.67 |
| 28 | 9.200 | 30 | 10.98 | 60 | 16.56 | 120 | 23.47 | 100 | 28.11 | 85 | 30.18 | 90 | 33.66 |
| 30 | 9.283 | 34 | 11.26 | 70 | 16.91 | 140 | 23.71 | 120 | 29.31 | 90 | 30.93 | 95 | 34.60 |
| 34 | 9.404 | 38 | 11.46 | 80 | 17.41 | 160 | 23.85 | 140 | 30.08 | 95 | 31.63 | 100 | 35.48 |
| 38 | 9.481 | 42 | 11.61 | 90 | 17.27 | 180 | 23.92 | 160 | 30.58 | 100 | 32.27 | 120 | 38.51 |
| 42 | 9.532 | 46 | 11.71 | 100 | 17.36 | 200 | 23.96 | 180 | 30.91 | 120 | 34.39 | 140 | 40.89 |
| 46 | 9.565 | 50 | 11.79 | 110 | 17.41 | 500 | 24.00 | 200 | 31.12 | 140 | 35.92 | 160 | 42.75 |
| 50 | 9.586 | 60 | 11.91 | 120 | 17.45 | | | 240 | 31.34 | 160 | 37.04 | 180 | 44.21 |
| 60 | 9.612 | 70 | 11.96 | 130 | 17.46 | | | 280 | 31.43 | 180 | 37.85 | 200 | 45.36 |
| 70 | 9.621 | 80 | 11.98 | 140 | 17.48 | | | 320 | 31.47 | 200 | 38.44 | 240 | 46.95 |
| 80 | 9.623 | 90 | 11.99 | 150 | 17.49 | | | 360 | 31.49 | 240 | 39.17 | 280 | 47.94 |
| 90 | 9.624 | 100 | 12.00 | 160 | 17.49 | | | 400 | 31.50 | 280 | 39.56 | 320 | 48.54 |
| 100 | 9.625 | 120 | 12.00 | 180 | 17.50 | | | 500 | 31.50 | 320 | 39.77 | 360 | 48.91 |
| | | | | 200 | 17.50 | | | | | 360 | 39.88 | 400 | 49.14 |
| | | | | 220 | 17.50 | | | | | 400 | 39.94 | 440 | 49.28 |
| | | | | | | | | | | 440 | 39.97 | 480 | 49.36 |
| | | | | | | | | | | 480 | 39.98 | | |

$k = 31.6$ md,
$\phi = 0.21$,
$r_w = 0.33$ ft,
$r_e = 3{,}300$ ft,
$B_o = 1.25$ RB/STB,
$h = 20$ ft,
$\mu_o = 0.8$ cp, and
$c_t = 20 \times 10^{-6}$ psi$^{-1}$.

The well produced for 6 months with flowing BHP $p_{wf}$ of 5,500 psi, for 6 more months with $p_{wf} = 4{,}500$ psi, and for 6 more months with $p_{wf} = 5{,}000$ psi. Calculate cumulative production after 18 months of production from this well.

**Solution.** Superposition is required to solve this problem. We can calculate the cumulative production by adding the cumulative production from three wells, each beginning to produce when $p_{wf}$ is changed and each producing with pressure drawdown equal to the difference in pressures before and after the change:

Well 1 produces for $(18 - 0) = 18$ months with $(p_i - p_{wf\,1}) = 6{,}000 - 5{,}500 = 500$ psi.

Well 2 produces for $(18\text{-}6) = 12$ months with $(p_{wf\,1} - p_{wf\,2}) = 5{,}500 - 4{,}500 = 1{,}000$ psi.

Well 3 produces for $(18 - 12) = 6$ months with $(p_{wf\,2} - p_{wf\,3}) = 4{,}500 - 5{,}000 = -500$ psi.

For this well, $r_{eD} = 3{,}300/0.33 = 10{,}000$.

$$t_D = \frac{0.000264\,kt}{\phi \mu c_t r_w^2}$$

**TABLE C-5 – (CONTINUED)**
**$q_D$ and $Q_{pD}$ vs. $t_D$ – FINITE RADIAL SYSTEM WITH CLOSED EXTERIOR BOUNDARY – CONSTANT PRESSURE AT INNER BOUNDARY**

| $r_{eD} = 20$ | | | $r_{eD} = 50$ | | | $r_{eD} = 100$ | | |
|---|---|---|---|---|---|---|---|---|
| $t_D$ | $q_D$ | $Q_{pD}$ | $t_D$ | $q_D$ | $Q_{pD}$ | $t_D$ | $q_D$ | $Q_{pD}$ |
| 100 | 0.3394 | 42.91 | 600 | 0.2652 | 189.0 | 2,000 | 0.2304 | 532 |
| 130 | 0.3174 | 52.76 | 800 | 0.2915 | 241 | 3,000 | 0.2179 | 757 |
| 160 | 0.2975 | 61.98 | 1,000 | 0.2393 | 290 | 4,000 | 0.2070 | 969 |
| 200 | 0.2728 | 73.38 | 1,300 | 0.2220 | 359 | 5,000 | 0.1967 | 1,171 |
| 240 | 0.2502 | 83.83 | 1,600 | 0.2060 | 473 | 6,000 | 0.1869 | 1,363 |
| 300 | 0.2197 | 97.91 | 2,000 | 0.1865 | 502 | 8,000 | 0.1686 | 1,718 |
| 400 | 0.1770 | 117.7 | 2,400 | 0.1682 | 573 | $1 \times 10^4$ | 0.1536 | 2,000 |
| 500 | 0.1426 | 133.6 | 3,000 | 0.1543 | 667 | $1.3 \times 10^4$ | 0.1304 | $2.46 \times 10^3$ |
| 600 | 0.1148 | 146.4 | 4,000 | 0.1133 | 795 | $1.6 \times 10^4$ | 0.1118 | $2.82 \times 10^3$ |
| 700 | 0.0925 | 156.7 | 5,000 | 0.0833 | 895 | $2 \times 10^4$ | 0.0910 | $3.23 \times 10^3$ |
| 800 | 0.0745 | 165.1 | 6,000 | 0.0682 | 974 | $2.4 \times 10^4$ | 0.0741 | $3.56 \times 10^3$ |
| 1,000 | 0.0483 | 177.1 | 8,000 | 0.0418 | 1,082 | $3 \times 10^4$ | 0.0645 | $3.94 \times 10^3$ |
| 1,300 | 0.0483 | 187.8 | $1 \times 10^4$ | 0.0254 | 1,148 | $4 \times 10^4$ | 0.0326 | $4.37 \times 10^3$ |
| 1,600 | 0.0132 | 193.4 | $1.3 \times 10^4$ | 0.0120 | 1,201 | $5 \times 10^4$ | 0.0195 | $4.62 \times 10^3$ |
| 2,000 | 0.0056 | 196.9 | $1.6 \times 10^4$ | 0.0056 | 1,227 | $6 \times 10^4$ | 0.0117 | $4.77 \times 10^3$ |
| 3,000 | 0.0006 | 199.2 | $2 \times 10^4$ | 0.0021 | 1,241 | $8 \times 10^4$ | 0.0042 | $4.92 \times 10^3$ |
| | | | $2.4 \times 10^4$ | 0.0006 | 1,246 | $1 \times 10^5$ | 0.0015 | $4.97 \times 10^3$ |
| | | | $3 \times 10^4$ | 0.0002 | 1,249 | $1.1 \times 10^5$ | 0.0009 | $4.98 \times 10^3$ |

| $r_{eD} = 200$ | | | $r_{eD} = 500$ | | | $r_{eD} = 1{,}000$ | | |
|---|---|---|---|---|---|---|---|---|
| $t_D$ | $q_D$ | $Q_{pD}$ | $t_D$ | $q_D$ | $Q_{pD}$ | $t_D$ | $q_D$ | $Q_{pD}$ |
| $1 \times 10^4$ | 0.1943 | $2.19 \times 10^3$ | $1 \times 10^5$ | 0.1566 | $1.75 \times 10^4$ | $3 \times 10^4$ | 0.1773 | $5.89 \times 10^3$ |
| $1.3 \times 10^4$ | 0.1860 | $2.77 \times 10^3$ | $1.3 \times 10^5$ | 0.1498 | $2.21 \times 10^4$ | $4 \times 10^4$ | 0.1729 | $7.64 \times 10^3$ |
| $1.6 \times 10^4$ | 0.1820 | $3.33 \times 10^3$ | $1.6 \times 10^5$ | 0.1435 | $2.65 \times 10^4$ | $5 \times 10^4$ | 0.1697 | $9.35 \times 10^3$ |
| $2 \times 10^4$ | 0.1742 | $4.04 \times 10^3$ | $2 \times 10^5$ | 0.1354 | $3.21 \times 10^4$ | $1 \times 10^5$ | 0.1604 | $1.76 \times 10^4$ |
| $2.4 \times 10^4$ | 0.1668 | $4.72 \times 10^3$ | $2.4 \times 10^5$ | 0.1277 | $3.73 \times 10^4$ | $2 \times 10^5$ | 0.1518 | $3.32 \times 10^4$ |
| $3 \times 10^4$ | 0.1562 | $5.69 \times 10^3$ | $3 \times 10^5$ | 0.1170 | $4.47 \times 10^4$ | $3 \times 10^5$ | 0.1464 | $4.80 \times 10^4$ |
| $4 \times 10^4$ | 0.1401 | $7.17 \times 10^3$ | $4 \times 10^5$ | 0.1012 | $5.56 \times 10^4$ | $4 \times 10^5$ | 0.1416 | $6.24 \times 10^4$ |
| $5 \times 10^4$ | 0.1236 | $8.50 \times 10^3$ | $5 \times 10^5$ | 0.0875 | $6.50 \times 10^4$ | $5 \times 10^5$ | 0.1371 | $7.64 \times 10^4$ |
| $6 \times 10^4$ | 0.1126 | $9.68 \times 10^3$ | $6 \times 10^5$ | 0.0756 | $7.31 \times 10^4$ | $6 \times 10^5$ | 0.1327 | $8.98 \times 10^4$ |
| $8 \times 10^4$ | 0.0905 | $1.17 \times 10^4$ | $8 \times 10^5$ | 0.0565 | $8.62 \times 10^4$ | $7 \times 10^5$ | 0.1285 | $1.03 \times 10^5$ |
| $1 \times 10^5$ | 0.0728 | $1.33 \times 10^4$ | $1 \times 10^6$ | 0.0422 | $9.60 \times 10^4$ | $8 \times 10^5$ | 0.1244 | $1.16 \times 10^5$ |
| $1.3 \times 10^5$ | 0.0524 | $1.52 \times 10^4$ | $1.3 \times 10^6$ | 0.0273 | $1.06 \times 10^5$ | $9 \times 10^5$ | 0.1204 | $1.23 \times 10^5$ |
| $1.6 \times 10^5$ | 0.0378 | $1.65 \times 10^4$ | $1.6 \times 10^6$ | 0.0176 | $1.13 \times 10^5$ | $1 \times 10^6$ | 0.1166 | $1.40 \times 10^5$ |
| $2 \times 10^5$ | 0.0244 | $1.78 \times 10^4$ | $2 \times 10^6$ | 0.0098 | $1.18 \times 10^5$ | $1.4 \times 10^6$ | 0.1024 | $1.83 \times 10^5$ |
| $2.4 \times 10^5$ | 0.0138 | $1.86 \times 10^4$ | $2.4 \times 10^6$ | 0.0055 | $1.21 \times 10^5$ | $2 \times 10^6$ | 0.0844 | $2.39 \times 10^5$ |
| $3 \times 10^5$ | 0.0082 | $1.92 \times 10^4$ | $3 \times 10^6$ | 0.0023 | $1.23 \times 10^5$ | $2.4 \times 10^6$ | 0.0741 | $2.71 \times 10^5$ |
| $4 \times 10^5$ | 0.0028 | $1.97 \times 10^4$ | $4 \times 10^6$ | 0.0005 | $1.25 \times 10^5$ | $3 \times 10^6$ | 0.0610 | $3.11 \times 10^5$ |
| $5 \times 10^5$ | 0.0009 | $1.99 \times 10^4$ | $5 \times 10^6$ | 0.0001 | $1.25 \times 10^5$ | $4 \times 10^6$ | 0.0442 | $3.63 \times 10^5$ |
| | | | | | | $5 \times 10^6$ | 0.0320 | $4.01 \times 10^5$ |
| | | | | | | $7 \times 10^6$ | 0.0167 | $4.48 \times 10^5$ |
| | | | | | | $8.4 \times 10^6$ | 0.0106 | $4.67 \times 10^5$ |
| | | | | | | $1 \times 10^7$ | 0.0063 | $4.80 \times 10^5$ |
| | | | | | | $1.4 \times 10^7$ | 0.0017 | $4.95 \times 10^5$ |
| | | | | | | $2 \times 10^7$ | 0.0004 | $4.99 \times 10^5$ |
| | | | | | | $3 \times 10^7$ | 0.0000 | $5.00 \times 10^5$ |

Notes: 1. For $t_D$ smaller than values listed in this table for a given $r_{eD}$, reservoir is infinite acting. Find $Q_{pD}$ in Table C.4.
2. For $t_D$ larger than values listed in this table, $Q_{pD} \cong ((r_{eD})^2 - 1)/2$.
3. For $t_D$ larger than values listed in this table, $q_D \cong 0.0$.

$$= \frac{(0.000264)(31.6)(730\ \text{hr/month})(t\ \text{months})}{(0.21)(0.8)(2\times10^{-5})(0.33)^2}$$

$$= 1.664\times10^{7}\ (t\ \text{months}).$$

For Well 1, $t_D = (1.664\times10^7)(18) = 3.0\times10^8$.

From Table C.5,

$$Q_{pD} = 2.54\times10^7,$$

$$Q_p = 1.119\ \phi c_t h r_w^2\ (p_i - p_{wf})\,Q_{pD}/B$$

$$= (1.119)(0.21)(2\times10^{-5})(20)(0.33)^2$$

$$\cdot (p_i - p_{wf})\,Q_{pD}/1.25$$

$$= 8.19\times10^{-6}\ (p_i - p_{wf})\,Q_{pD}$$

$$= (8.19\times10^{-6})(500)(2.54\times10^7)$$

$$= 1.04\times10^5\ \text{STB}.$$

For Well 2, $t_D = (1.664\times10^7)(12) = 2.0\times10^8$.

$$Q_{pD} = 1.89\times10^7,$$

$$Q_p = (8.19\times10^{-6})(1{,}000)(1.89\times10^7)$$

$$= 1.55\times10^5\ \text{STB}.$$

For Well 3, $t_D = (1.664\times10^7)(6) = 1.0\times10^8$.

$$Q_{pD} = 1.06\times10^7,$$

$$Q_p = (8.19\times10^{-6})(-500)(1.06\times10^7)$$

$$= -0.434\times10^5.$$

Then

$$Q_p = Q_{p1} + Q_{p2} + Q_{p3}$$

$$= 1.04\times10^5 + 1.55\times10^5 - 0.434\times10^5$$

$$= 2.16\times10^5\ \text{STB}.$$

## Exercises

C.1 An oil well is producing from a reservoir ($r_e = 900$ ft) with no fluid crossing the outer boundary of the reservoir. Initial reservoir pressure is 3,000 psia. Pressure in the wellbore is maintained at 2,000 psia. Well and reservoir properties include the following:

$\phi$ = 0.21,
$k$ = 80 md,
$c_t$ = $20\times10^{-6}$ psi$^{-1}$,
$r_w$ = 0.45 ft,
$\mu$ = 0.8 cp,
$h$ = 12 ft, and
$B$ = 1.2 RB/STB.

Calculate instantaneous rate and cumulative production after 1.0 day of production.

**TABLE C.5 – (CONTINUED)**

| $r_{eD}$ = 2,000 | | | $r_{eD}$ = 4,000 | | | $r_{eD}$ = 1.0 ($10^4$) | | |
|---|---|---|---|---|---|---|---|---|
| $t_D$ | $q_D$ | $Q_{pD}$ | $t_D$ | $q_D$ | $Q_{pD}$ | $t_D$ | $q_D$ | $Q_{pD}$ |
| $1\times10^5$ | 0.1604 | | $9\times10^5$ | 0.1366 | | $3\times10^6$ | 0.1263 | $4.06\times10^5$ |
| $2\times10^5$ | 0.1520 | | $1\times10^6$ | 0.1356 | | $4\times10^6$ | 0.1240 | $5.31\times10^5$ |
| $3\times10^5$ | 0.1475 | | $1.3\times10^6$ | 0.1333 | | $5\times10^6$ | 0.1222 | $6.76\times10^5$ |
| $4\times10^5$ | 0.1445 | $6.27\times10^4$ | $1.6\times10^6$ | 0.1315 | $2.26\times10^5$ | $6\times10^6$ | 0.1210 | $7.76\times10^5$ |
| $5\times10^5$ | 0.1422 | $7.70\times10^4$ | $2\times10^6$ | 0.1296 | $2.78\times10^5$ | $8\times10^6$ | 0.1188 | $1.02\times10^6$ |
| $6\times10^5$ | 0.1404 | $9.11\times10^4$ | $2.4\times10^6$ | 0.1280 | $3.30\times10^5$ | $1\times10^7$ | 0.1174 | $1.25\times10^6$ |
| $7\times10^5$ | 0.1389 | $1.05\times10^5$ | $3\times10^6$ | 0.1262 | $4.06\times10^5$ | $1.2\times10^7$ | 0.1162 | $1.49\times10^6$ |
| $8\times10^5$ | 0.1375 | $1.19\times10^5$ | $4\times10^6$ | 0.1237 | $5.31\times10^5$ | $1.4\times10^7$ | 0.1152 | $1.72\times10^6$ |
| $9\times10^5$ | 0.1363 | $1.33\times10^5$ | $5\times10^6$ | 0.1215 | $6.54\times10^5$ | $1.6\times10^7$ | 0.1143 | $1.95\times10^6$ |
| $1\times10^6$ | 0.1352 | $1.46\times10^5$ | $6\times10^6$ | 0.1194 | $7.74\times10^5$ | $1.8\times10^7$ | 0.1135 | $2.17\times10^6$ |
| $1.3\times10^6$ | 0.1320 | $1.86\times10^5$ | $8\times10^6$ | 0.1155 | $1.01\times10^6$ | $2\times10^7$ | 0.1128 | $2.40\times10^6$ |
| $1.6\times10^6$ | 0.1291 | $2.25\times10^5$ | $1\times10^7$ | 0.1118 | $1.24\times10^6$ | $2.4\times10^7$ | 0.1115 | $2.85\times10^6$ |
| $2\times10^6$ | 0.1254 | $2.76\times10^5$ | $1.2\times10^7$ | 0.1081 | $1.46\times10^6$ | $3\times10^7$ | 0.1098 | $3.51\times10^6$ |
| $2.4\times10^6$ | 0.1216 | $3.26\times10^5$ | $1.4\times10^7$ | 0.1046 | $1.67\times10^6$ | $4\times10^7$ | 0.1071 | $4.60\times10^6$ |
| $3\times10^6$ | 0.1166 | $3.97\times10^5$ | $1.6\times10^7$ | 0.1012 | $1.87\times10^6$ | $5\times10^7$ | 0.1050 | $5.66\times10^6$ |
| $4\times10^6$ | 0.1084 | $5.10\times10^5$ | $1.8\times10^7$ | 0.0979 | $2.07\times10^6$ | $7\times10^7$ | 0.0998 | $7.70\times10^6$ |
| $5\times10^6$ | 0.1008 | $6.14\times10^5$ | $2\times10^7$ | 0.0948 | $2.27\times10^6$ | $8\times10^7$ | 0.0975 | $8.69\times10^6$ |
| $7\times10^6$ | 0.0872 | $8.02\times10^5$ | $2.3\times10^7$ | 0.0902 | $2.54\times10^6$ | $9\times10^7$ | 0.0952 | $9.65\times10^6$ |
| $1\times10^7$ | 0.0701 | $1.04\times10^6$ | $3\times10^7$ | 0.0803 | $3.14\times10^6$ | $1\times10^8$ | 0.0930 | $1.06\times10^7$ |
| $1.3\times10^7$ | 0.0563 | $1.23\times10^6$ | $4\times10^7$ | 0.0681 | $3.88\times10^6$ | $1.2\times10^8$ | 0.0887 | $1.24\times10^7$ |
| $1.7\times10^7$ | 0.0421 | $1.42\times10^6$ | $5\times10^7$ | 0.0577 | $4.51\times10^6$ | $1.4\times10^8$ | 0.0846 | $1.41\times10^7$ |
| $2\times10^7$ | 0.0339 | $1.53\times10^6$ | $7\times10^7$ | 0.0415 | $5.49\times10^6$ | $1.7\times10^8$ | 0.0788 | $1.66\times10^7$ |
| $2.4\times10^7$ | 0.0253 | $1.65\times10^6$ | $8\times10^7$ | 0.0352 | $5.87\times10^6$ | $2\times10^8$ | 0.0734 | $1.89\times10^7$ |
| $3\times10^7$ | 0.0164 | $1.78\times10^6$ | $9\times10^7$ | 0.0298 | $6.20\times10^6$ | $2.4\times10^8$ | 0.0668 | $2.17\times10^7$ |
| $4\times10^7$ | 0.0079 | $1.89\times10^6$ | $1\times10^8$ | 0.0252 | $6.47\times10^6$ | $3\times10^8$ | 0.0580 | $2.54\times10^7$ |
| $5\times10^7$ | 0.0038 | $1.95\times10^6$ | $1.2\times10^8$ | 0.0181 | $6.90\times10^6$ | $4\times10^8$ | 0.0458 | $3.06\times10^7$ |
| $7\times10^7$ | 0.0009 | $1.99\times10^6$ | $1.4\times10^8$ | 0.0130 | $7.21\times10^6$ | $5\times10^8$ | 0.0362 | $3.47\times10^7$ |
| $1\times10^8$ | 0.0001 | $2.00\times10^6$ | $1.7\times10^8$ | 0.0079 | $7.52\times10^6$ | $6\times10^8$ | 0.0286 | $3.79\times10^7$ |
| $1.3\times10^8$ | 0.0000 | $2.00\times10^6$ | $2\times10^8$ | 0.0048 | $7.71\times10^6$ | $7\times10^8$ | 0.0226 | $4.04\times10^7$ |
| | | | $2.3\times10^8$ | 0.0029 | $7.82\times10^6$ | $8\times10^8$ | 0.0178 | $4.24\times10^7$ |
| | | | $2.6\times10^8$ | 0.0018 | $7.89\times10^6$ | $1\times10^9$ | 0.0111 | $4.53\times10^7$ |
| | | | $3\times10^8$ | 0.0009 | $7.94\times10^6$ | $1.4\times10^9$ | 0.0043 | $4.82\times10^7$ |
| | | | $4\times10^8$ | 0.0002 | $7.99\times10^6$ | $2\times10^9$ | 0.0011 | $4.96\times10^7$ |
| | | | $5\times10^8$ | 0.0000 | $8.00\times10^6$ | $3\times10^9$ | 0.0001 | $5.00\times10^7$ |

Notes: 1. For $t_D$ smaller than values listed in this table for a given $r_{eD}$, reservoir is infinite acting. Find $Q_{pD}$ in Table C.4.
2. For $t_D$ larger than values listed in this table, $Q_{pD} \cong [(r_{eD})^2 - 1]/2$.
3. For $t_D$ larger than values listed in this table, $q_D \cong 0.0$.

C.2 A single oil well is producing in the center of a circular, full water-drive reservoir. The pressure at the oil/water contact is constant at 3,340 psia and the radial distance to the oil/water contact is 1,500 ft. The well produced for the first 15 days at a rate of 500 STB/D, the next 14 days at 300 STB/D, and the last day at 200 STB/D. Calculate the cumulative production and wellbore pressure at the end of 30 days.

$\phi$ = 0.20,
$k$ = 75 md,
$c_t$ = $17.5\times10^{-6}$ psi$^{-1}$,
$r_w$ = 0.5 ft,
$\mu$ = 0.75 cp,
$h$ = 15 ft, and
$B$ = 1.2 RB/STB.

C.3 An oil reservoir is initially at a pressure, $p_i$, of 2,734 psia. If the boundary pressure is suddenly lowered to 2,724 psia and held there, calculate the cumulative water influx into the reservoir after 100, 200, 400, and 800 days. Reservoir and aquifer properties include the following.

$\phi$ = 0.2,
$k$ = 83 md,
$c_t$ = $8\times10^{-6}$ psi$^{-1}$,
$r_w$ = 3000 ft (reservoir radius),
$r_e$ = 30,000 ft (aquifer radius),
$\mu$ = 0.62 cp, and
$h$ = 40 ft.

C.4 If, in Exercise C.3, the reservoir boundary pressure suddenly dropped to 2,704 psia at the end of 100 days, calculate the total water influx at 400 days total elapsed time.

## References


1. van Everdingen, A.F. and Hurst, W.: "The Application of the Laplace Transformation to Flow Problems in Reservoirs," *Trans.*, AIME (1949) **186**, 305-324.
2. Chatas, A.T.: "A Practical Treatment of Nonsteady-State Flow Problems in Reservoir Systems," *Pet. Eng.* (Aug. 1953) B-44 through B-56.
3. Edwardson, M.J. *et al.*: "Calculation of Formation Temperature Disturbances Caused by Mud Circulation," *J. Pet. Tech.* (April 1962) 416-426; *Trans.*, AIME, **225**.


**TABLE C.5 – (CONTINUED)**

| $r_{eD}=2.5(10^4)$ | | | $r_{eD}=1.0(10^5)$ | | | $r_{eD}=2.5(10^5)$ | | | $r_{eD}=1.0(10^6)$ | | |
|---|---|---|---|---|---|---|---|---|---|---|---|
| $t_D$ | $q_D$ | $Q_{pD}$ | $t_D$ | $q_D$ | $Q_{pD}$ | $t_D$ | $q_D$ | $Q_{pD}$ | $t_D$ | $q_D$ | $Q_{pD}$ |
| $3\times10^{7}$ | 0.1103 | | $1.4\times10^{8}$ | 0.1017 | | $2\times10^{9}$ | 0.0897 | | $2\times10^{10}$ | 0.0813 | |
| $4\times10^{7}$ | 0.1086 | | $2\times10^{8}$ | 0.1000 | | $3\times10^{9}$ | 0.0881 | | $3\times10^{10}$ | 0.0800 | |
| $6\times10^{7}$ | 0.1064 | $6.77\times10^{6}$ | $2.4\times10^{8}$ | 0.0990 | | $4\times10^{9}$ | 0.0870 | | $4\times10^{10}$ | 0.0791 | |
| $7\times10^{7}$ | 0.1054 | $7.83\times10^{6}$ | $3\times10^{8}$ | 0.0980 | $3.10\times10^{7}$ | $5\times10^{9}$ | 0.0861 | $4.51\times10^{8}$ | $6\times10^{10}$ | 0.0778 | |
| $8\times10^{7}$ | 0.1047 | $8.86\times10^{6}$ | $3.5\times10^{8}$ | 0.0971 | $3.59\times10^{7}$ | $6\times10^{9}$ | 0.0854 | $5.37\times10^{8}$ | $8\times10^{10}$ | 0.0770 | |
| $9\times10^{7}$ | 0.1041 | $9.93\times10^{7}$ | $4\times10^{8}$ | 0.0966 | $4.07\times10^{7}$ | $7\times10^{9}$ | 0.0849 | $6.22\times10^{8}$ | $1\times10^{11}$ | 0.0763 | $7.95\times10^{10}$ |
| $1\times10^{8}$ | 0.1035 | $1.10\times10^{7}$ | $5\times10^{8}$ | 0.0956 | $5.03\times10^{7}$ | $8\times10^{9}$ | 0.0844 | $7.07\times10^{8}$ | $1.3\times10^{11}$ | 0.0756 | $1.02\times10^{10}$ |
| $1.4\times10^{8}$ | 0.1016 | $1.51\times10^{7}$ | $6\times10^{8}$ | 0.0948 | $5.98\times10^{7}$ | $1\times10^{10}$ | 0.0836 | $8.75\times10^{8}$ | $1.6\times10^{11}$ | 0.0750 | $1.25\times10^{10}$ |
| $2\times10^{8}$ | 0.0993 | $2.11\times10^{7}$ | $7\times10^{8}$ | 0.0941 | $6.93\times10^{7}$ | $1.4\times10^{10}$ | 0.0824 | $1.21\times10^{9}$ | $2\times10^{11}$ | 0.0743 | $1.55\times10^{10}$ |
| $2.6\times10^{8}$ | 0.0973 | $2.70\times10^{7}$ | $8\times10^{8}$ | 0.0935 | $7.87\times10^{7}$ | $2\times10^{10}$ | 0.0809 | $1.70\times10^{9}$ | $2.4\times10^{11}$ | 0.0737 | $1.84\times10^{10}$ |
| $3\times10^{8}$ | 0.0960 | $3.09\times10^{7}$ | $8.4\times10^{8}$ | 0.0933 | $8.24\times10^{7}$ | $3\times10^{10}$ | 0.0787 | $2.49\times10^{9}$ | $3\times10^{11}$ | 0.0730 | $2.28\times10^{10}$ |
| $3.3\times10^{8}$ | 0.0950 | $3.37\times10^{7}$ | $9\times10^{8}$ | 0.0930 | $8.80\times10^{7}$ | $4\times10^{10}$ | 0.0766 | $3.27\times10^{9}$ | $4\times10^{11}$ | 0.0719 | $3.21\times10^{10}$ |
| $3.6\times10^{8}$ | 0.0940 | $3.66\times10^{7}$ | $1\times10^{9}$ | 0.0925 | $9.73\times10^{7}$ | $4.4\times10^{10}$ | 0.0757 | $3.58\times10^{9}$ | $5\times10^{11}$ | 0.0709 | $3.72\times10^{10}$ |
| $4\times10^{8}$ | 0.0927 | $4.03\times10^{7}$ | $1.4\times10^{9}$ | 0.0911 | $1.34\times10^{8}$ | $4.7\times10^{10}$ | 0.0751 | $3.80\times10^{9}$ | $6\times10^{11}$ | 0.0697 | $4.42\times10^{10}$ |
| $4.4\times10^{8}$ | 0.0915 | $4.40\times10^{7}$ | $2\times10^{9}$ | 0.0896 | $1.80\times10^{8}$ | $5\times10^{10}$ | 0.0745 | $4.03\times10^{9}$ | $7\times10^{11}$ | 0.0686 | $5.11\times10^{10}$ |
| $5\times10^{8}$ | 0.0896 | $4.94\times10^{7}$ | $3\times10^{9}$ | 0.0877 | $2.77\times10^{8}$ | $5.4\times10^{10}$ | 0.0737 | $4.32\times10^{9}$ | $8\times10^{11}$ | 0.0676 | $6.32\times10^{10}$ |
| $5.4\times10^{8}$ | 0.0804 | $5.30\times10^{7}$ | $4\times10^{9}$ | 0.0861 | $3.64\times10^{8}$ | $6\times10^{10}$ | 0.0725 | $4.76\times10^{9}$ | $1\times10^{12}$ | 0.0656 | $7.13\times10^{10}$ |
| $6\times10^{8}$ | 0.0866 | $5.82\times10^{7}$ | $5\times10^{9}$ | 0.0845 | $4.49\times10^{8}$ | $7\times10^{10}$ | 0.0705 | $5.48\times10^{9}$ | $1.2\times10^{12}$ | 0.0636 | $8.42\times10^{10}$ |
| $6.4\times10^{8}$ | 0.0855 | $6.17\times10^{7}$ | $6\times10^{9}$ | 0.0829 | $5.33\times10^{8}$ | $7.4\times10^{10}$ | 0.0698 | $5.76\times10^{9}$ | $1.4\times10^{12}$ | 0.0617 | $9.67\times10^{10}$ |
| $7\times10^{8}$ | 0.0837 | $6.67\times10^{7}$ | $7\times10^{9}$ | 0.0814 | $6.15\times10^{8}$ | $8\times10^{10}$ | 0.0686 | $6.17\times10^{9}$ | $1.5\times10^{12}$ | 0.0607 | $1.03\times10^{12}$ |
| $7.4\times10^{8}$ | 0.0826 | $7.01\times10^{7}$ | $8\times10^{9}$ | 0.0799 | $6.95\times10^{8}$ | $8.4\times10^{10}$ | 0.0679 | $6.42\times10^{9}$ | | | |
| $8\times10^{8}$ | 0.0809 | $7.50\times10^{7}$ | $9\times10^{9}$ | 0.0784 | $7.75\times10^{8}$ | $9\times10^{10}$ | 0.0668 | $6.85\times10^{9}$ | $>1.5\times10^{12}$ not determined. | | |
| $8.4\times10^{8}$ | 0.0798 | $7.82\times10^{7}$ | $1\times10^{10}$ | 0.0770 | $8.52\times10^{8}$ | $1\times10^{11}$ | 0.0658 | $7.51\times10^{9}$ | | | |
| $9\times10^{8}$ | 0.0782 | $8.29\times10^{7}$ | $1.3\times10^{10}$ | 0.0728 | $1.08\times10^{9}$ | $1.1\times10^{11}$ | 0.0632 | $8.15\times10^{9}$ | | | |
| $1\times10^{9}$ | 0.0756 | $9.06\times10^{7}$ | $1.6\times10^{10}$ | 0.0689 | $1.29\times10^{9}$ | $1.3\times10^{11}$ | 0.0593 | $9.38\times10^{9}$ | | | |
| $1.3\times10^{9}$ | 0.0683 | $1.12\times10^{8}$ | $2\times10^{10}$ | 0.0639 | $1.56\times10^{9}$ | $1.6\times10^{11}$ | 0.0551 | $1.11\times10^{10}$ | | | |
| $1.6\times10^{9}$ | 0.0616 | $1.32\times10^{8}$ | $2.4\times10^{10}$ | 0.0594 | $1.80\times10^{9}$ | $2\times10^{11}$ | 0.0494 | $1.32\times10^{10}$ | | | |
| $2\times10^{9}$ | 0.0538 | $1.55\times10^{8}$ | $3\times10^{10}$ | 0.0531 | $2.14\times10^{9}$ | $2.4\times10^{11}$ | 0.0443 | $1.51\times10^{10}$ | | | |
| $2.4\times10^{9}$ | 0.0469 | $1.75\times10^{8}$ | $4\times10^{10}$ | 0.0441 | $2.62\times10^{9}$ | $3\times10^{11}$ | 0.0376 | $1.75\times10^{10}$ | | | |
| $3\times10^{9}$ | 0.0382 | $2.00\times10^{8}$ | $5\times10^{10}$ | 0.0366 | $3.03\times10^{9}$ | $4\times10^{11}$ | 0.0286 | $2.08\times10^{10}$ | | | |
| $4\times10^{9}$ | 0.0272 | $2.33\times10^{8}$ | $6\times10^{10}$ | 0.0304 | $3.36\times10^{9}$ | $5\times10^{11}$ | 0.0217 | $2.33\times10^{10}$ | | | |
| $5\times10^{9}$ | 0.0193 | $2.56\times10^{8}$ | $7\times10^{10}$ | 0.0253 | $3.64\times10^{9}$ | $7\times10^{11}$ | 0.0126 | $2.67\times10^{10}$ | | | |
| $6\times10^{9}$ | 0.0138 | $2.72\times10^{8}$ | $8\times10^{10}$ | 0.0210 | $3.87\times10^{9}$ | $1\times10^{12}$ | 0.0055 | $2.92\times10^{10}$ | | | |
| $8\times10^{9}$ | 0.0070 | $2.92\times10^{8}$ | $9\times10^{10}$ | 0.0174 | $4.06\times10^{9}$ | $1.3\times10^{12}$ | 0.0024 | $3.04\times10^{10}$ | | | |
| $1\times10^{10}$ | 0.0035 | $3.02\times10^{8}$ | $1\times10^{11}$ | 0.0145 | $4.22\times10^{9}$ | $1.6\times10^{12}$ | 0.0011 | $3.09\times10^{10}$ | | | |
| $1.4\times10^{10}$ | 0.0009 | $3.10\times10^{8}$ | $1.3\times10^{11}$ | 0.0083 | $4.55\times10^{9}$ | $2\times10^{12}$ | 0.0004 | $3.11\times10^{10}$ | | | |
| $2\times10^{10}$ | 0.0001 | $3.12\times10^{8}$ | $1.6\times10^{11}$ | 0.0048 | $4.74\times10^{9}$ | | | | | | |
| $3\times10^{10}$ | 0.0000 | $3.12\times10^{8}$ | $2\times10^{11}$ | 0.0023 | $4.88\times10^{9}$ | | | | | | |
| | | | $2.4\times10^{11}$ | 0.0011 | $4.94\times10^{9}$ | | | | | | |
| | | | $3\times10^{11}$ | 0.0004 | $4.98\times10^{9}$ | | | | | | |
| | | | $3.4\times10^{11}$ | 0.0002 | $4.99\times10^{9}$ | | | | | | |
| | | | $4\times10^{11}$ | 0.0001 | $5.00\times10^{9}$ | | | | | | |

Notes: 1. For $t_D$ smaller than values listed in this table for a given $r_{eD}$, reservoir is infinite acting. Find $Q_{pD}$ in Table C.4.
2. For $t_D$ larger than values listed in this table, $Q_{pD}\cong[(r_{eD})^2-1]/2$.
3. For $t_D$ larger than values listed in this table, $q_D\cong0.0$.

# Appendix D
# Rock and Fluid Property Correlations

## Introduction

Pressure transient test analysis requires knowledge of reservoir fluid properties such as viscosities, compressibilities, and formation volume factors. In addition, formation compressibility is a rock property frequently required for test analysis. For most of these properties, laboratory analysis provides the most accurate answer; however, in many cases, laboratory results are not available, and the test analyst must use empirical correlations of experimental data.

This appendix provides a summary of correlations that have proved useful for test analysis. These correlations are selected from those presented by Earlougher;[1] his collection of correlations is probably the best and the most complete in print at the time of this writing. We assume that the reader of this text has completed a study of the fundamentals of reservoir fluid properties. Such studies provide detail on the meaning, origin, and applicability of these correlations. Readers not familiar with this basis are referred to texts by Amyx *et al.*[2] and McCain.[3] In this appendix, we simply reproduce figures from Earlougher's collection and illustrate the use of each with an example.

## Pseudocritical Temperature and Pressure of Liquid Hydrocarbons

Pseudocritical temperature ($T_{pc}$) and pressure ($p_{pc}$) of undersaturated crude oils can be estimated from an approximate correlation developed by Trube[4] and presented in Fig. D-1. These pseudocritical properties are required to estimate compressibility of an undersaturated crude oil, as illustrated by Example D.5.

To use Trube's correlation, one must know the specific gravity of the undersaturated reservoir liquid corrected to 60°F. Example D.1 illustrates use of the correlation.

---

### *Example D.1 – Estimation of Pseudocritical Temperature and Pressure for Undersaturated Crude Oil*

**Problem:** Estimate the pseudocritical temperature, $T_{pc}$, and pressure, $p_{pc}$, of an undersaturated crude oil with gravity of 30°API (specific gravity = 0.876 at 60°F).

**Solution:** From Fig. D-1, $T_{pc} = 1160°R$ and $p_{pc} = 285$ psia.

---

## Bubble-Point Pressure of Crude Oil

The test analyst may need to estimate the saturation or bubble-point pressure of a crude oil in some circumstances – e.g., to determine whether a reservoir is saturated or undersaturated at a certain pressure. Standing's correlation[5] is useful for this estimate; Fig. D-2 shows this correlation.

To use Standing's correlation, one must know solution GOR, gas gravity, stock-tank oil gravity, and reservoir temperature. Example D.2 illustrates use of this correlation.

---

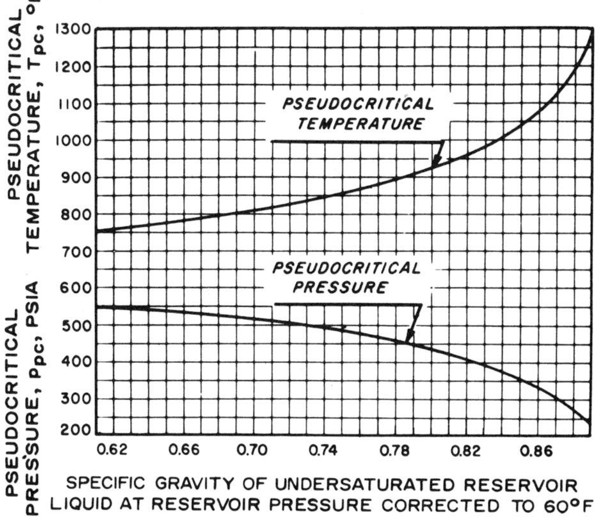


**Fig. D-1** – Approximate correlation of liquid pseudocritical pressure and temperature with specific gravity.[1]

### *Example D.2 – Estimation of Bubble-Point of Crude Oil*

**Problem.** Estimate the bubble-point pressure of a crude oil from a reservoir producing at a GOR of 350 scf/STB (believed to be solution gas only) with gas gravity 0.75 and oil gravity 30° API from a reservoir with temperature 200°F.

**Solution.** On Fig. D-2, we start on the left by extending a horizontal line from the assumed GOR of 350 scf/STB to the line for a gas gravity of 0.75; from this point, we draw a vertical line that terminates at the line for a stock-tank oil gravity of 30° API; we then draw a horizontal line from that point that terminates at the intersection with the 200°F reservoir temperature line. Finally, we draw a vertical line from this point that intersects the bubble-point pressure scale at 1,930 psia. Thus, the estimated bubble-point or saturation pressure of the crude oil is 1,930 psia.

## Solution GOR

Frequently, the saturation pressure of a reservoir oil is known, but solution GOR at saturation pressure (required for some test analyses) is unknown. Fig. D-2 also can be used for this estimate.

To determine solution GOR, one must know bubble-point pressure, reservoir temperature, stock-tank oil gravity, and gas gravity. Example D.3 illustrates use of Standing's[5] correlation for estimating solution GOR.

### *Example D.3 – Estimation of Solution GOR*

**Problem.** Estimate the solution GOR of a crude oil from a reservoir with bubble-point pressure 1,930 psia, temperature 200°F, oil gravity 30°API, and gas gravity 0.75.

**Solution.** In Fig. D-2, we start at the lower right by extending a vertical line from the bubble-point pressure to the 200°F reservoir-temperature line. We next draw a horizontal line to the left that terminates at the intersection with the line for tank-oil gravity of 30°API. We then draw a vertical line from this point to intersect the 0.75 gas-gravity line. Finally, we draw a horizontal line from that point to the GOR scale on the left and read the GOR to be 350 scf/STB.

## Oil Formation Volume Factor

Fig. D-3 can be used to estimate the formation

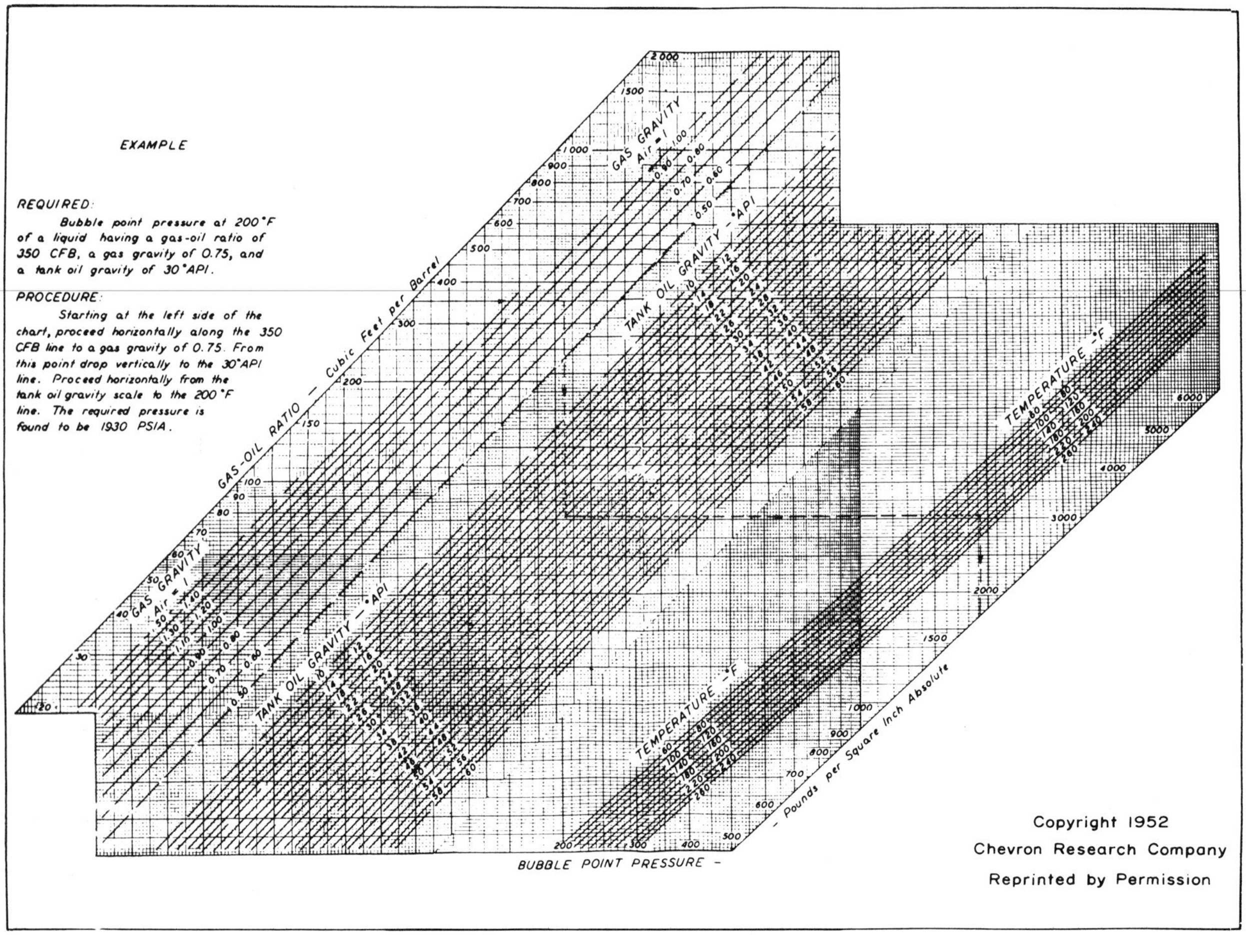


**Fig. D-2** – Bubble-point pressure or dissolved GOR of oil.[1]

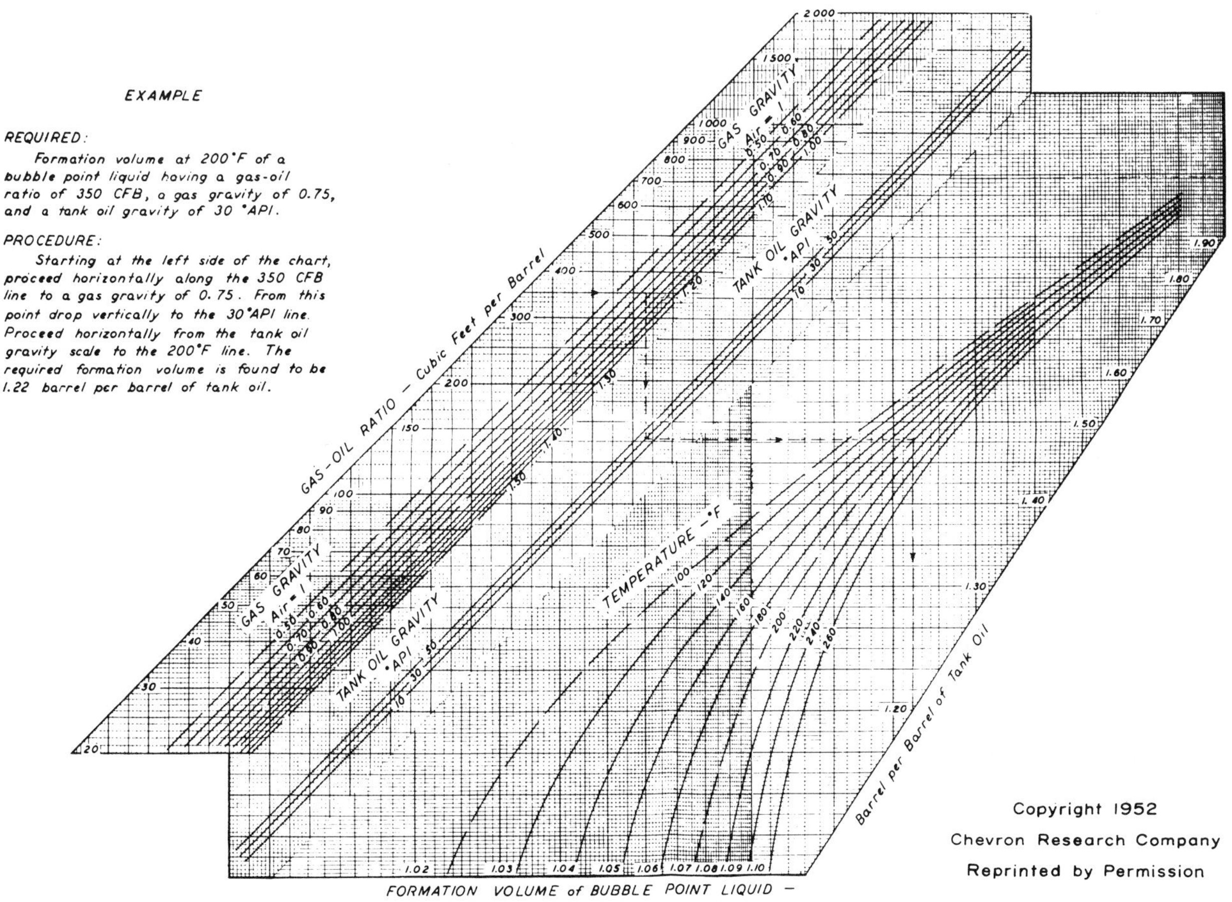


Fig. D-3 – Oil formation volume factor.[1]

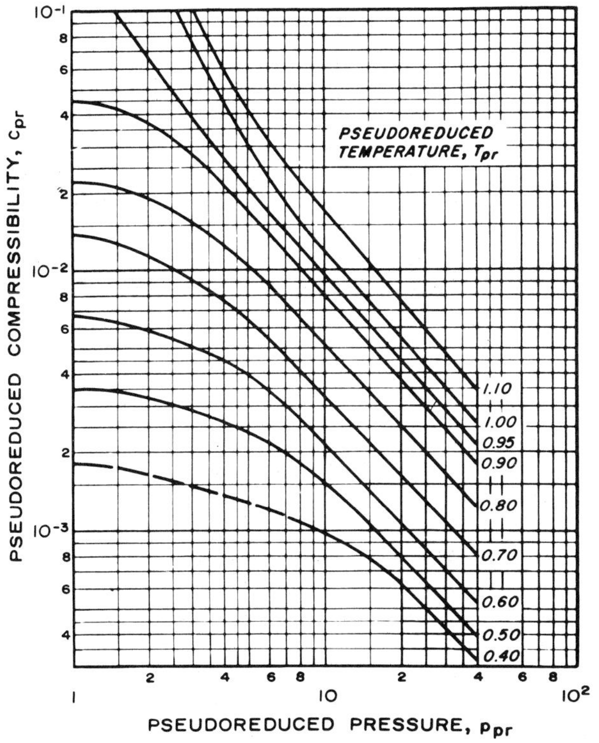


Fig. D-4 – Correlation of pseudoreduced compressibility for an undersaturated oil.[1]

volume factor $B_o$ of saturated crude oil. For this estimate, one must know solution GOR, gas gravity, stock-tank oil gravity, and reservoir temperature. Example D.4 illustrates use of this correlation, which also was developed by Standing.

---

## *Example D.4 – Estimation of Oil Formation Volume Factor*

**Problem.** Estimate the formation volume factor of a saturated crude oil from a reservoir with solution GOR of 350 scf/STB, gas gravity of 0.75, tank oil gravity of 30°API, and reservoir temperature of 200°F.

**Solution.** In Fig. D-3, we start at the upper left and extend a horizontal line from a solution GOR of 350 scf/STB to intersect the line for gas gravity of 0.75. We then extend a vertical line from this point to intersect the 30°API tank-oil-gravity line. From this point of intersection, we draw a horizontal line to intersect the line for a reservoir temperature of 200°F. Finally, we draw a vertical line from that point to the formation-volume-factor scale at the lower right and read $B_o = 1.22$ RB/STB.

---

## Compressibility of Undersaturated Oil

The correlation in Fig. D-4, developed by Trube,[4] can be used to estimate the compressibility, $c_o$, of an

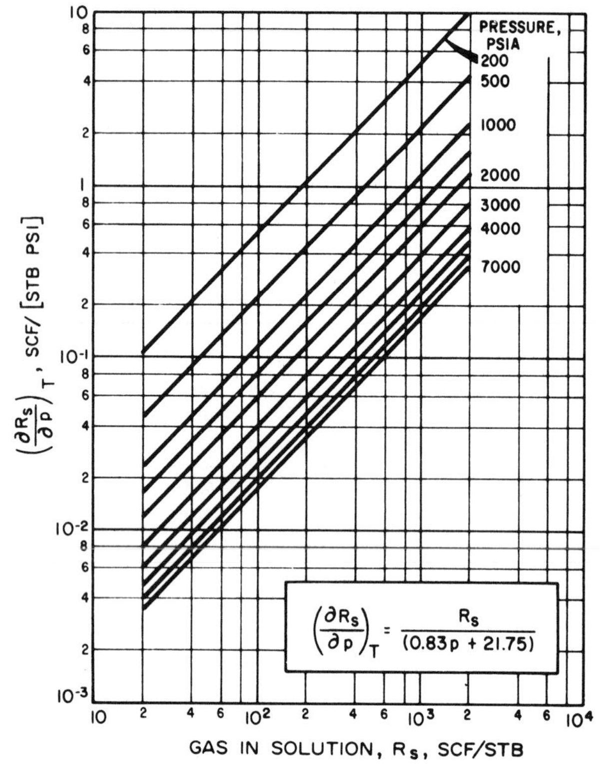


Fig. D-5 – Change of gas in solution in oil with pressure vs. gas in solution.[1]

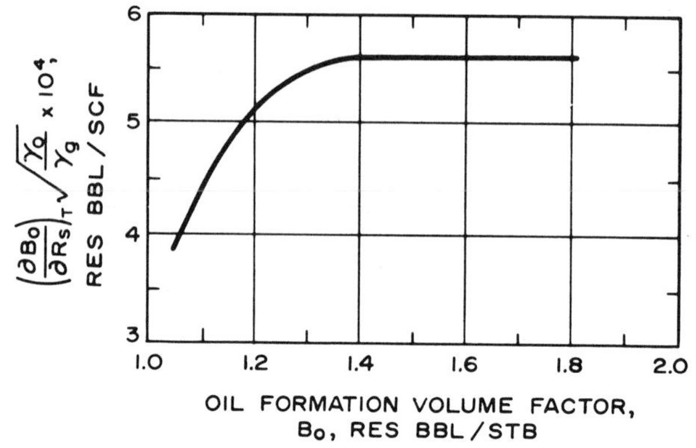


Fig. D-6 – Change of oil formation volume factor with gas in solution vs. oil formation volume factor.[1]

undersaturated crude oil.

The basic information required to use this figure is specific gravity of the undersaturated crude oil, reservoir temperature, and reservoir pressure. Given these values, we can estimate pseudocritical temperature ($T_{pc}$) and pressure ($p_{pc}$), pseudoreduced temperature ($T_{pr} = T/T_{pc}$) and pressure ($p_{pr} = p/p_{pc}$) and, thus, pseudoreduced compressibility ($c_{pr}$), which leads to oil compressibility ($c_o$) from the definition

$$c_o = c_{pr}/p_{pc}.$$

Example D.5 illustrates this sequence of calculations.

---

### *Example D.5 – Estimation of Undersaturated Oil Compressibility*

**Problem.** Estimate the compressibility, $c_o$, of an undersaturated crude oil with 30°API gravity (0.876 specific gravity) at a reservoir temperature of 200°F and pressure of 5,000 psia.

**Solution.** In Example D.1, we found that the pseudocritical temperature of a 30°API oil was $T_{pc} = 1180°R$ and that the pseudocritical pressure was $p_{pc} = 275$ psia. Thus,

$$T_{pr} = T/T_{pc} = (200 + 460)/1{,}160 = 0.569,$$

and

$$p_{pr} = p/p_{pc} = 5{,}000/285 = 17.5.$$

From Fig. D-4, then,

$$c_{pr} = 0.001.$$

By definition, $c_o = c_{pr}/p_{pc}$, so

$$c_o = 0.001/285 = 3.51 \times 10^{-6} \text{ psi}^{-1}.$$

---

## Compressibility of Saturated Crude Oil

The apparent compressibility of saturated crude oil is significantly higher than that of undersaturated oil. The reason is that a pressure drop results in evolution of gas from the oil; the total volume of oil remaining actually *decreases* with pressure decline (although the density of the remaining liquid oil actually decreases slightly). The net result is that the *total* volume of oil and evolved gas becomes greater as pressure drops, leading to an apparent compressibility of the system that is appreciably higher than that of liquid oil alone. In equation form,

$$c_o = -\frac{1}{B_o}\frac{dB_o}{dp} + \frac{B_g}{B_o}\frac{dR_s}{dp} \quad \text{(D.1)}$$

The first term accounts for the volume change in the liquid caused by (1) vaporization of some of the liquid and (2) increase in density of the remainder of the liquid. The derivative $dB_o/dp$ is a positive number, so vaporization dominates. The second term accounts for the volume occupied by gas evolved as pressure decreases (or dissolved as pressure increases). The derivative $dR_s/dp$ is positive, so this term is positive. Further, its numerical value is greater than that of the term $-(1/B_o)(dB_o/dp)$. Ramey[6] proposed correlations that lead to an estimate of $c_o$ for a saturated oil; these correlations are given in Figs. D-5 and D-6. To use these correlations, we must note that Eq. D.1 can be written as

$$c_o = -\frac{1}{B_o}\frac{dB_o}{dp} + \frac{B_g}{B_o}\frac{dR_s}{dp}$$

$$= \frac{1}{B_o}\frac{dR_s}{dp}\left(B_g - \frac{dB_o}{dR_s}\right). \quad \text{(D.2)}$$

Fig. D-5 provides an estimate of $dR_s/dp$; Fig. D-6 estimates $dB_o/dR_s$.

To estimate $c_o$, one needs values of reservoir pressure, $p$, solution GOR, $R_s$ (which, in turn, can be estimated from Fig. D-2), tank-oil specific gravity ($\gamma_o$) and gas gravity ($\gamma_g$), oil formation volume factor, $B_o$ (which can be estimated from Fig. D-3), and gas formation volume factor, $B_g$. ($B_g$ can be calculated from reservoir temperature, pressure, and gas gravity, which leads to the real-gas-law deviation factor $z$: $B_g = 0.00504\ Tz/p$ RB/scf.) Example D.6 illustrates use of these correlations.

---

## *Example D.6 – Estimation of Saturated Oil Compressibility*

**Problem.** Pressure in an oil reservoir has dropped below the initial bubble point to 2,500 psia. Reservoir fluid characteristics include the following:

$\gamma_o$ = 0.825 (40°API),
$\gamma_g$ = 0.7,
$T$ = 200°F = 660°R, and
$z$ = 0.851.

Estimate the apparent compressibility of this oil at its current saturation pressure of 2,500 psia.

**Solution.** To evaluate $c_o$, we must determine each term in Eq. D.2.

$$c_o = \frac{1}{B_o}\frac{dR_s}{dp}\left(B_g - \frac{dB_o}{dR_s}\right).$$

From knowledge of $p$, $\gamma_g$, $\gamma_o$, and $T$, we can estimate $R_s$ from Fig. D-2, as in Example D.3; the result is $R_s = 640$ scf/STB. From knowledge of $R_s$, $\gamma_g$, $\gamma_o$, and $T$, we can estimate $B_o$ from Fig. D-3, as in Example D.4; the result is $B_o = 1.36$ RB/STB.

We can calculate $B_g$ since $T$, $p$, and $z$ are known:

$$B_g = 0.00504\ Tz/p$$

$$= (0.00504)(660)(0.851)/2{,}500$$

$$= 0.001132 \text{ RB/scf}.$$

From the inset in Fig. D-5,

$$\frac{dR_s}{dp} = \frac{R_s}{(0.83p + 21.75)}$$

$$= \frac{640}{[(0.83)(2{,}500) + 21.75]}$$

$$= 0.3052 \text{ scf/STB-psi}.$$

(This result also could be read from the curves plotted in Fig. D-5.)

From Fig. D-6,

$$\frac{dB_o}{dR_s}\sqrt{\frac{\gamma_o}{\gamma_g}} \times 10^4 = 5.6.$$

Thus,

$$\frac{dB_o}{dR_s} = 5.6 \times 10^{-4}\sqrt{\frac{\gamma_g}{\gamma_o}}$$

$$= (5.6 \times 10^{-4})\sqrt{\frac{0.7}{0.825}}$$

$$= 0.516 \times 10^{-3} \text{ RB/scf}.$$

We now can calculate $c_o$:

$$c_o = \frac{1}{B_o}\frac{dR_s}{dp}\left[B_g - \frac{dB_o}{dR_s}\right]$$

$$= \left(\frac{1}{1.36}\right)(0.3052)(0.001132 - 0.000516)$$

$$= 0.138 \times 10^{-3} \text{ psi}^{-1}$$

---

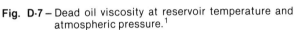

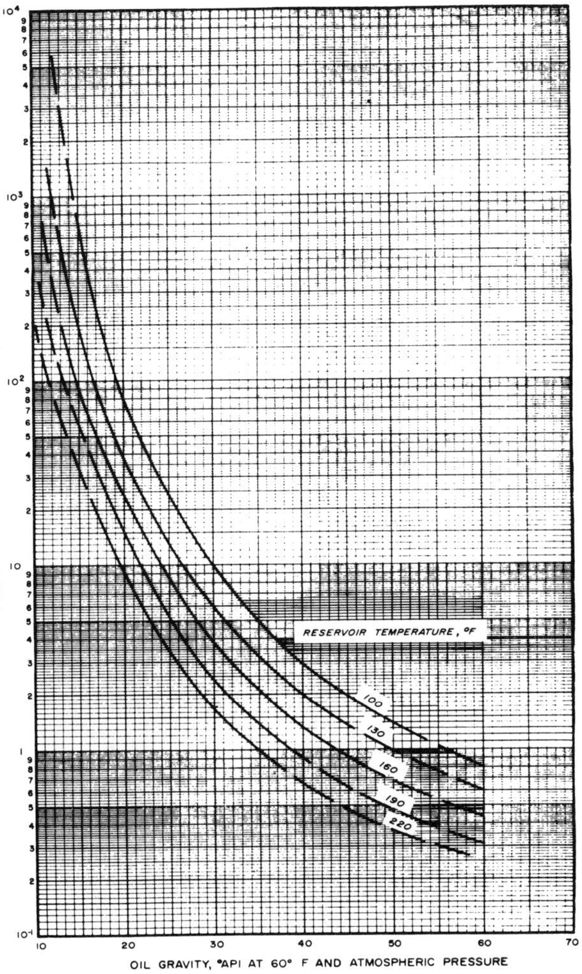


**Fig. D-7** – Dead oil viscosity at reservoir temperature and atmospheric pressure.[1]

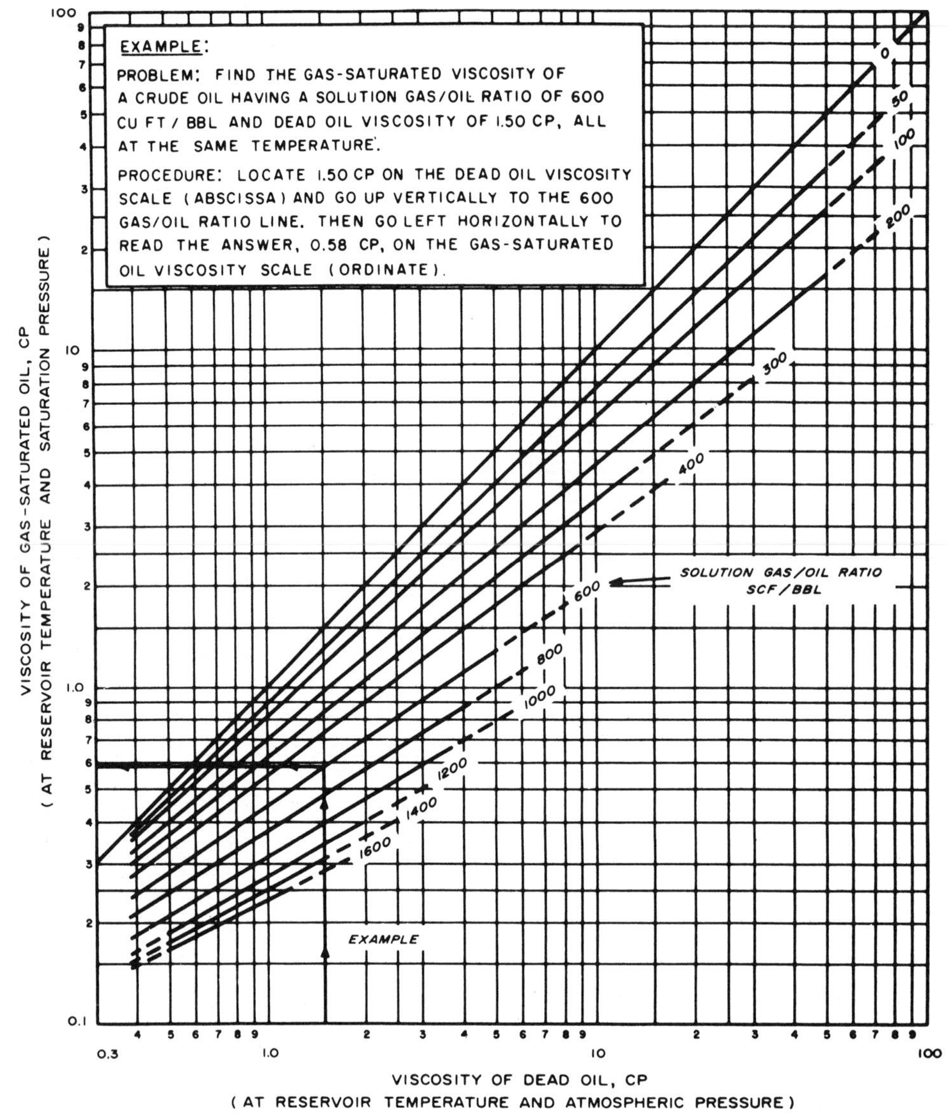
EXAMPLE:
PROBLEM: FIND THE GAS-SATURATED VISCOSITY OF A CRUDE OIL HAVING A SOLUTION GAS/OIL RATIO OF 600 CU FT / BBL AND DEAD OIL VISCOSITY OF 1.50 CP, ALL AT THE SAME TEMPERATURE.
PROCEDURE: LOCATE 1.50 CP ON THE DEAD OIL VISCOSITY SCALE ( ABSCISSA ) AND GO UP VERTICALLY TO THE 600 GAS/OIL RATIO LINE. THEN GO LEFT HORIZONTALLY TO READ THE ANSWER, 0.58 CP, ON THE GAS-SATURATED OIL VISCOSITY SCALE ( ORDINATE ).
SOLUTION GAS/OIL RATIO SCF/BBL
0
50
100
200
300
400
600
800
1000
1200
1400
1600
EXAMPLE
VISCOSITY OF GAS-SATURATED OIL, CP
( AT RESERVOIR TEMPERATURE AND SATURATION PRESSURE )
VISCOSITY OF DEAD OIL, CP
( AT RESERVOIR TEMPERATURE AND ATMOSPHERIC PRESSURE )
0.1
1.0
10
100
0.3

**Fig. D-8** – Viscosity of gas-saturated crude oil at reservoir temperature and pressure. Dead oil viscosity from laboratory data or from Fig. D-7.[1]

## Oil Viscosity

Viscosity of saturated crude oil can be estimated from Beal's correlation[7] (Fig. D-7) combined with Chew and Connally's data[8] (Fig. D-8). Viscosity estimates for undersaturated oil require the same figures plus an additional correlation of Beal's, Fig. D-9. Fig. D-7 provides an estimate of dead (gas-free) oil viscosity, $\mu_{od}$; Fig. D-8 provides an estimate of gas-saturated oil viscosity, $\mu_{ob}$, at saturation pressure; and Fig. D-9 provides an estimate of viscosity increase above bubble-point pressure.

These viscosity estimates require knowledge of reservoir temperature, oil gravity, solution GOR, and, in the case of an undersaturated oil, bubble-point and reservoir pressures. Example D.7 illustrates this estimation procedure.

---

### *Example D.7 – Estimation of Oil Viscosity*

**Problem.** Estimate the viscosity of an undersaturated oil at a reservoir pressure of 5,000 psia and at the oil's saturation pressure of 1,930 psia. Solution GOR is 350 scf/STB, oil gravity is 30°API, and reservoir temperature is 200°F.

**Solution.** From Fig. D-7, the dead oil viscosity, $\mu_{od}$, is 2.15 cp. From Fig. D-8, the viscosity of gas-saturated oil ($\mu_{ob}$) at the bubble point is 1.0 cp. Finally, using data from Fig. D-9, the viscosity of the undersaturated oil ($\mu_o$) at a pressure of 5,000 psi is

$$\mu_o = 1 + (0.067)\left(\frac{5{,}000 - 1{,}930}{1{,}000}\right) = 1.21 \text{ cp.}$$

---

## Solubility of Gas in Water

Solubility of natural gas in water can be estimated from correlations of Dodson and Standing,[9] presented in Figs. D-10 and D-11. Fig. D-10 gives the solubility of natural gas in pure (nonsaline) water; Fig. D-11 provides a means of correcting solubility in pure water for brine salinity.

To estimate solubility of gas in water, one must know reservoir temperature and pressure and total

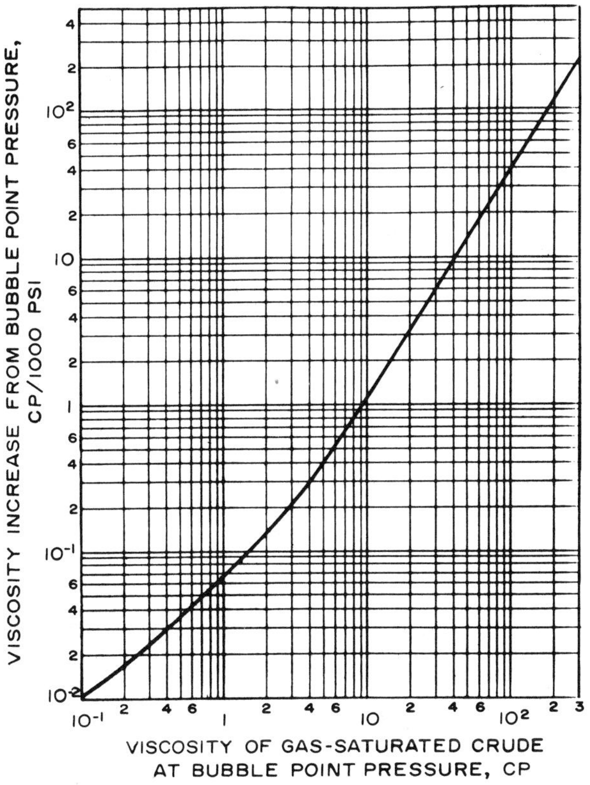


Fig. D-9 – Rate of increase in oil viscosity above bubble-point pressure.[1]

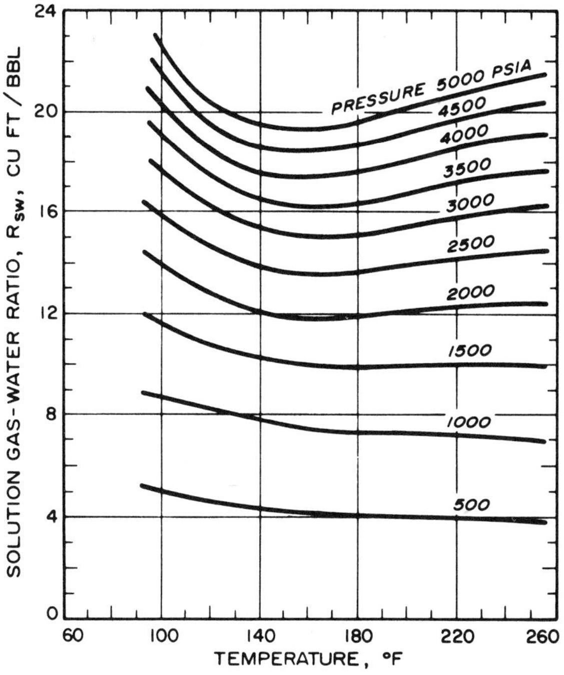


Fig. D-10 – Solubility of natural gas in pure water.[1]

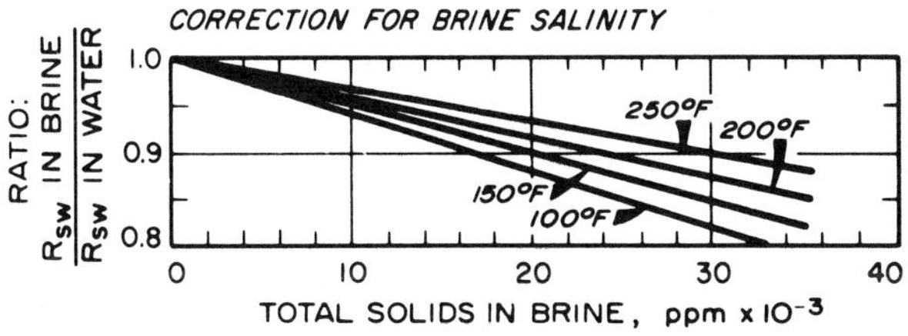


Fig. D-11 – Correction of natural gas solubility for dissolved solids.[1]

solids content of the water. Example D.8 illustrates this estimation procedure.

---

## *Example D.8 – Estimation of Gas Solubility in Water*

**Problem.** Estimate the solubility of natural gas in a formation water with 20,000 ppm dissolved solids, reservoir temperature of 200°F, and reservoir pressure of 5,000 psia.

**Solution.** From Fig. D-10, the solubility ($R_{swp}$) of natural gas in *pure* water at 200°F and 5,000 psia is 20.2 scf/STB. From Fig. D-11, the correction factor ($R_{sw}/R_{swp}$) for 20,000-ppm salinity at 200°F is 0.92. Thus,

$$R_{sw} = (R_{swp})(R_{sw}/R_{swp})$$

$$= (20.2)(0.92)$$

$$= 18.6 \text{ scf/STB.}$$

---

## Water Formation Volume Factor

The formation volume factor $B_w$ can be estimated by using Fig. D-12, an additional correlation developed by Dodson and Standing,[9] along with Figs. D-10 and D-11. This estimate requires the same information used to estimate the solubility of gas in water (reservoir temperature and pressure and total solids content of the water). Example D.9 illustrates the estimation procedure.

---

## *Example D.9 – Estimation of Water Formation Volume Factor*

**Problem.** Estimate the formation volume factor of brine with 20,000 ppm dissolved solids at a temperature of 200°F and pressure of 5,000 psia.

**Solution.** From Fig. D-12, the formation volume factor of pure water is 1.021 RB/STB; for gas-saturated pure water, it is 1.030 RB/STB. Brine with 20,000 ppm dissolved solids contains less dissolved gas than pure water; from Fig. D-11, the ratio of solubility in brine to that in pure water is 0.92. Using this value to interpolate between volume factors for pure and gas-saturated water, then,

$$B_w = 1.021 + (1.030 - 1.021)(0.92)$$

$$= 1.029 \text{ RB/STB.}$$

---

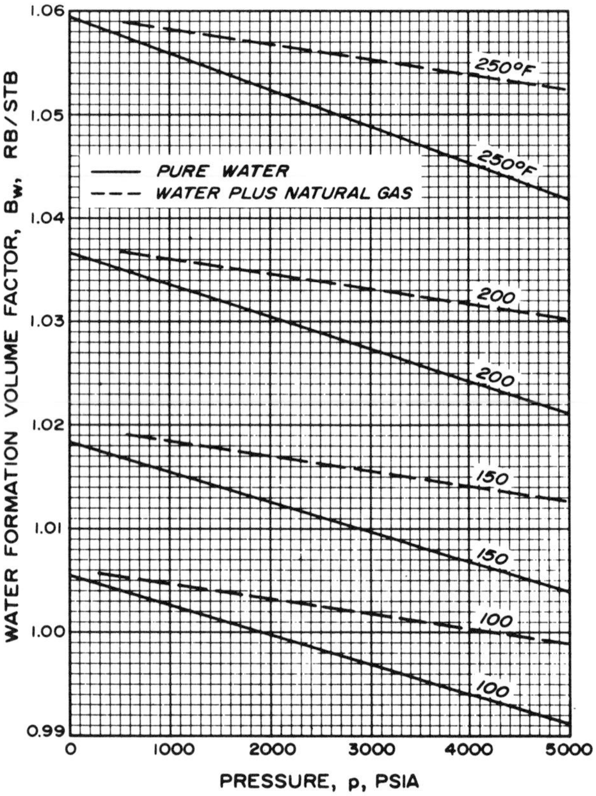


Fig. D-12 – Formation volume factor for pure water, gas-free and gas-saturated.[1]

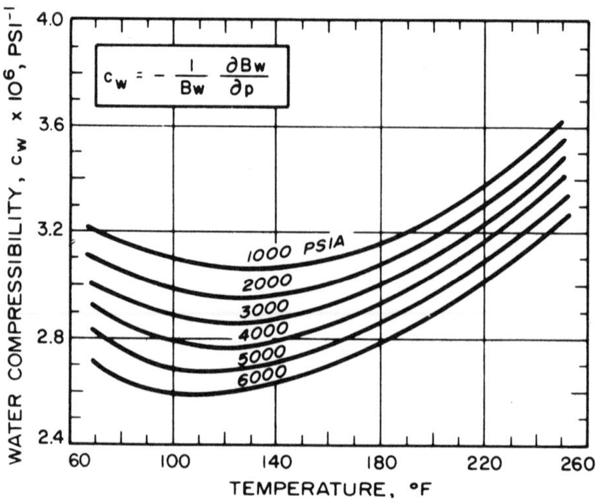


Fig. D-13 – Compressibility of gas-free water.[1]

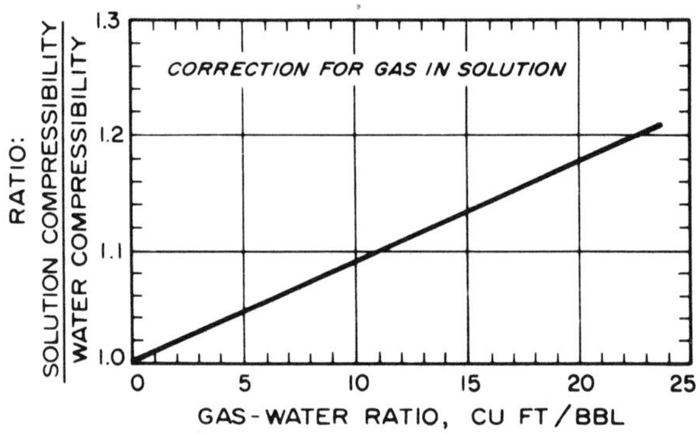


Fig. D-14 – Effect of dissolved gas on water compressibility.[1]

## Compressibility of Water in Undersaturated Reservoirs

The formation water that occurs in an undersaturated oil reservoir will not release gas as pressure is decreased; this would lead to formation of or increase in a gas saturation in the reservoir. In such a case, we will assume that the formation water is saturated with gas at reservoir pressure. One implication of this is that, as gas is released from solution in the water, it is assumed to be redissolved in the undersaturated oil.

For this assumed system behavior, Dodson and Standing's[9] correlations (given in Figs. D-13 and D-14) can be used to estimate the compressibility of water in an undersaturated oil reservoir.

These water-compressibility estimates require knowledge of reservoir temperature and pressure and formation-water salinity. Example D.10 illustrates the calculation procedure.

---

### *Example D.10 – Estimation of Water Compressibility in an Undersaturated Reservoir*

**Problem.** Estimate the compressibility of a formation water containing 20,000 ppm dissolved solids in a reservoir with temperature 200°F and pressure 5,000 psia. Oil in the reservoir has a saturation pressure of 1,930 psia.

**Solution.** We found in Example D.8 that solubility of gas in water at the stated conditions is 18.6 scf/STB. From Fig. D-13, the compressibility ($c_{wp}$) of pure (gas-free) water is $2.96 \times 10^{-6}$ psi$^{-1}$. At a gas/water ratio of 18.6 scf/STB, the correction factor for gas in solution is 1.16 (Fig. D-14). Thus,

$$c_w = c_{wp}(c_w/c_{wp})$$
$$= (2.96 \times 10^{-6})(1.16)$$
$$= 3.43 \times 10^{-6} \text{ psi}^{-1}.$$

---

## Compressibility of Water in a Saturated Reservoir

In a saturated reservoir, gas released from solution in the formation water either will begin to form or will increase a gas saturation as reservoir pressure is lowered; as Ramey[6] pointed out, this dramatically increases the apparent water compressibility. In this case, the water compressibility is calculated from

$$c_w = -\frac{1}{B_w}\frac{dB_w}{dp} + \frac{B_g}{B_w}\frac{dR_{sw}}{dp}. \quad \text{..........} \quad \text{(D.3)}$$

The term $-(1/B_w)(dB_w/dp)$ is still determined

using Fig. D-13 (for gas-free water) and Fig. D-14 (to correct for the effect of gas in solution). Ramey's correlation,[6] presented in Fig. D-15, is used to estimate $dR_{sw}/dp$ for fresh water, and Fig. D-11 is used to correct for the effect of salinity on $dR_{sw}/dp$.

This compressibility estimate requires knowledge of formation-water salinity, reservoir temperature and pressure, and formation volume factor of the gas dissolved in the water. Example D.11 illustrates this estimation procedure.

---

## *Example D.11 – Estimation of Water Compressibility in a Saturated Reservoir*

**Problem.** Estimate the apparent compressibility of formation water containing 30,000 ppm dissolved solids in an undersaturated oil reservoir at 200°F and

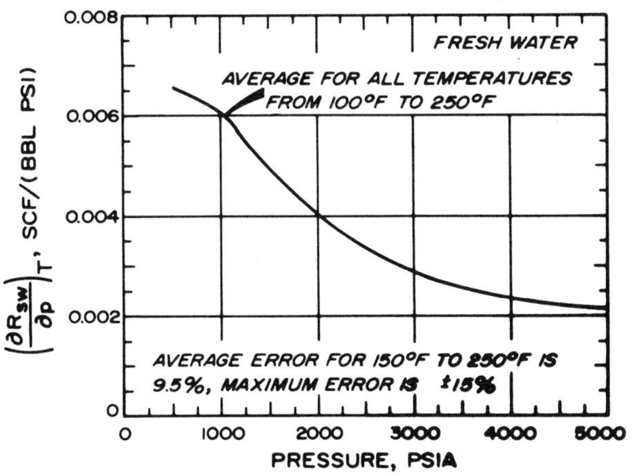


Fig. D-15 – Change of natural gas in solution in formation water with pressure vs. pressure.[1]

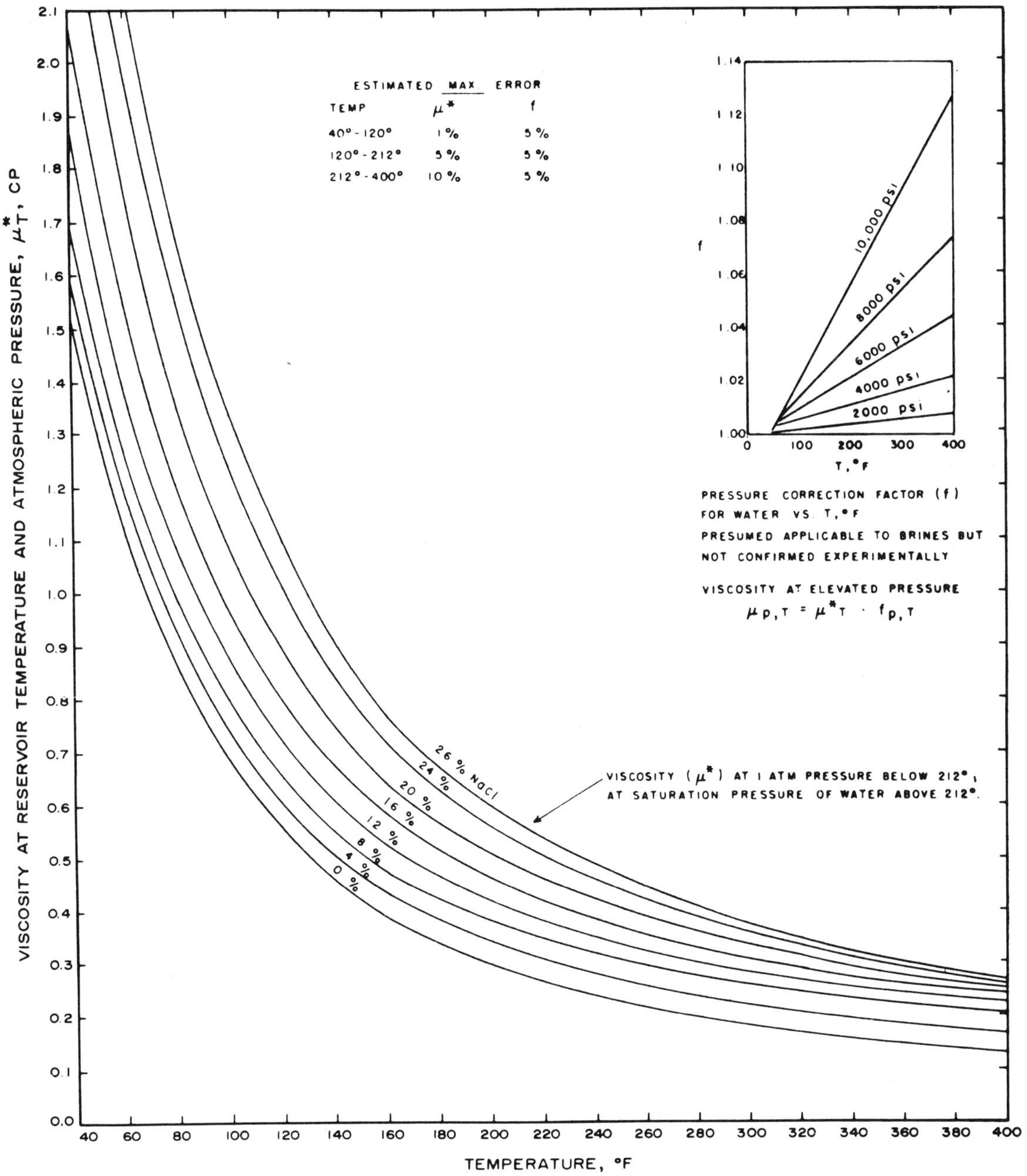


Fig. D-16 – Water viscosity at various salinities and temperatures.[1]

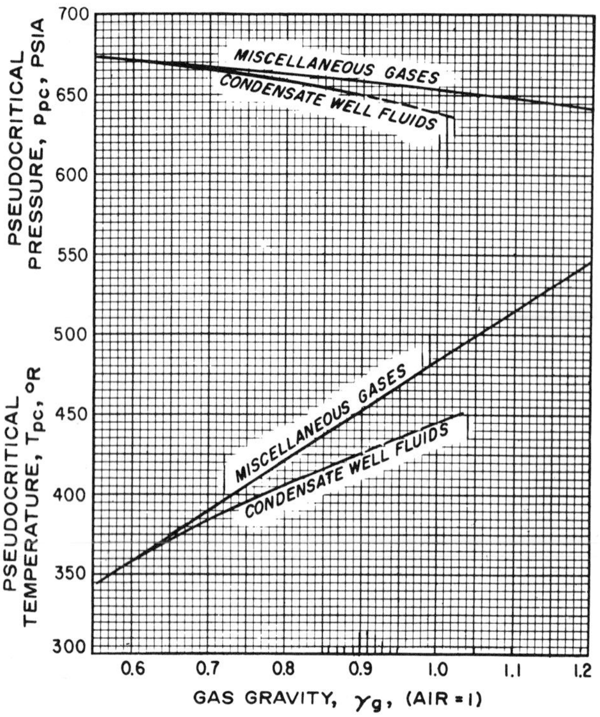


**Fig. D-17** – Correlation of pseudocritical properties of condensate well fluids and miscellaneous natural gases with fluid gravity.[1]

2,500 psia. Formation volume factor of the 0.7-gravity gas is 0.001132 RB/scf.

**Solution.** From Fig. D-15, for fresh water, $dR_{sw}/dp = 0.0033$. The correction factor for the effect of salinity (Fig. D-11) is 0.875; thus, for brine, $dR_{sw}/dp = (0.0033)(0.875) = 0.0029$ scf/STB-psi. From Fig. D-12, using procedures outlined in Example D.9, $B_w = 1.033$ RB/STB. From Figs. D-13 and D-14, using procedures of Example D.10,

$$-\frac{1}{B_w}\frac{dB_w}{dp} = (3.13 \times 10^{-6})(1.11)$$

$$= 3.47 \times 10^{-6} \text{ psi}^{-1}.$$

Thus,

$$c_w = -\frac{1}{B_w}\frac{dB_w}{dp} + \frac{B_g}{B_w}\frac{dR_{sw}}{dp}$$

$$= 3.47 \times 10^{-6} + \frac{(0.001132)(0.0029)}{(1.033)}$$

$$= 6.65 \times 10^{-6} \text{ psi}^{-1}.$$

## Water Viscosity

Fig. D-16, first presented in the literature by Matthews and Russell,[10] allows estimates of water viscosity as a function of reservoir temperature, salinity, and reservoir pressure. Example D.12 illustrates use of this correlation.

### *Example D.12 – Estimation of Water Viscosity*

**Problem.** Estimate the viscosity of water containing 20,000 ppm (2%) dissolved solids at 200°F and 5,000 psia.

**Solution.** From Fig. D-16, the viscosity $\mu_T{}^*$ of water with 2% NaCl at 200°F and atmospheric pressure is 0.32 cp. The correction factor $f$ for 5,000-psia pressure is 1.016 (inset at upper right of figure). Thus,

$$\mu_w = \mu_T{}^* f = (0.32)(1.016) = 0.33 \text{ cp}.$$

## Pseudocritical Properties of Gas

Pseudocritical temperature, $T_{pc}$, and pressure, $p_{pc}$, are useful quantities that allow application of generalized correlations of gas properties needed in applications. The most accurate values of these quantities are calculated from compositions of natural gas mixtures.[2,3] Approximate values can be determined from a correlation developed by Brown *et al.*[11] This correlation, presented in Fig. D-17, is based on gas gravity, $\gamma_g$; values of critical properties depend on whether the gas is from a gas condensate reservoir (the "condensate well fluids" curve) or from a relatively dry gas reservoir (the "miscellaneous gases" curve). Example D.13 illustrates use of this correlation.

### *Example D.13 – Estimation of Pseudocritical Gas Properties*

**Problem.** Estimate $T_{pc}$ and $p_{pc}$ for a dry gas of gravity ($\gamma_g$) 0.7.

**Solution.** From Fig. D-17, $T_{pc} = 390$°R and $p_{pc} = 665$ psia.

## Gas-Law Deviation Factor ($z$-Factor) and Gas Formation Volume Factor

Application of the real gas law,

$$pV = znRT, \quad \text{(D.4)}$$

to relate pressure, volume, and temperature for gases requires values of the deviation factor $z$. One application of particular importance is in calculating gas formation volume factor, $B_g$, from

$$B_g = 0.00504\, Tz/p \text{ (RB/scf)}. \quad \text{(D.5)}$$

Fig. D-18, developed by Standing and Katz,[12] can be used to estimate $z$. One must know pseudoreduced pressure, $p_{pr}$, and pseudoreduced temperature, $T_{pr}$, to determine $z$. By definition,

$$p_{pr} = p/p_{pc} \quad \text{(D.6)}$$

and

$$T_{pr} = T/T_{pc}. \quad \text{(D.7)}$$

In these definitions, pressures must be expressed in psia (psig + 14.7) and temperatures in degrees Rankine (°F + 460).

To use this correlation, one must know reservoir

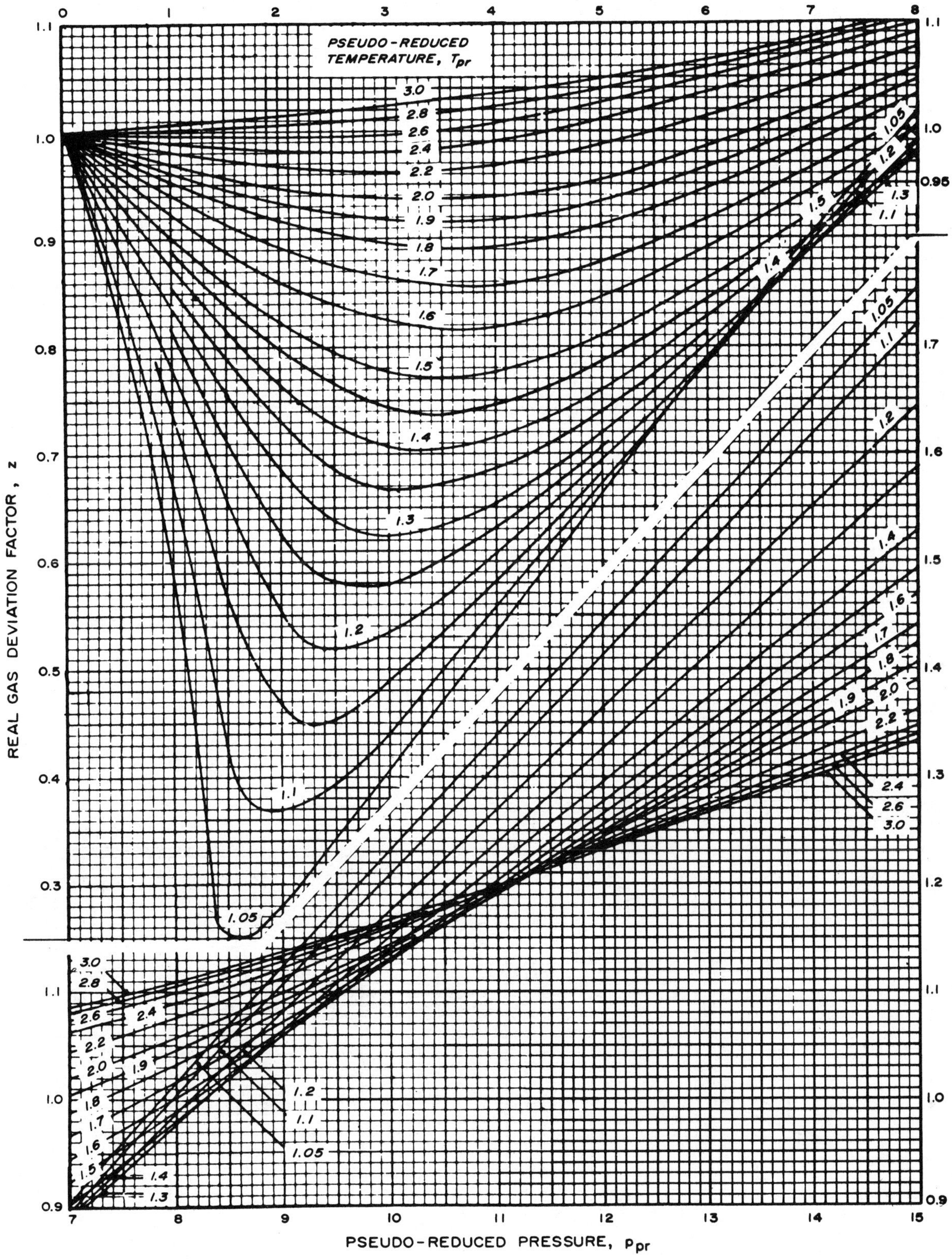


**Fig. D-18** – Gas-law deviation factor for natural gases as a function of pseudoreduced pressure and temperature.[1]

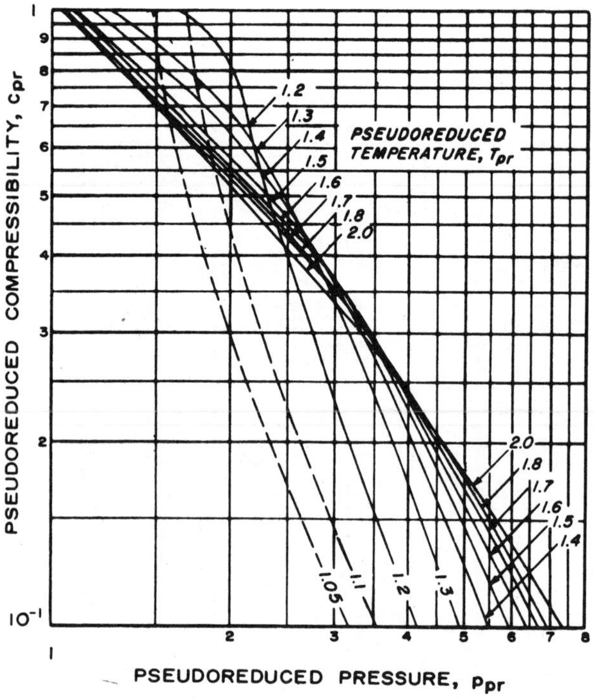


Fig. D-19A – Correlation of pseudoreduced compressibility for natural gases.[1]

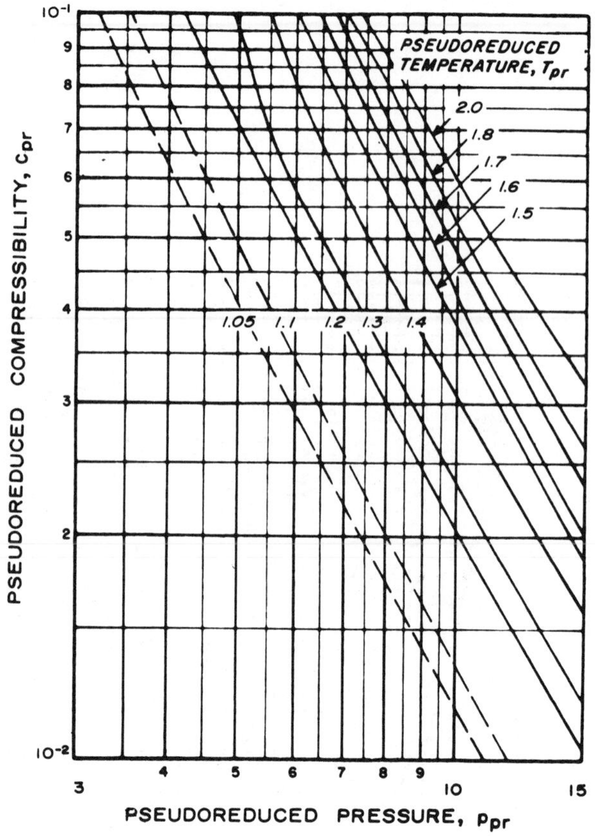


Fig. D-19B – Correlation of pseudoreduced compressibility for natural gases.[1]

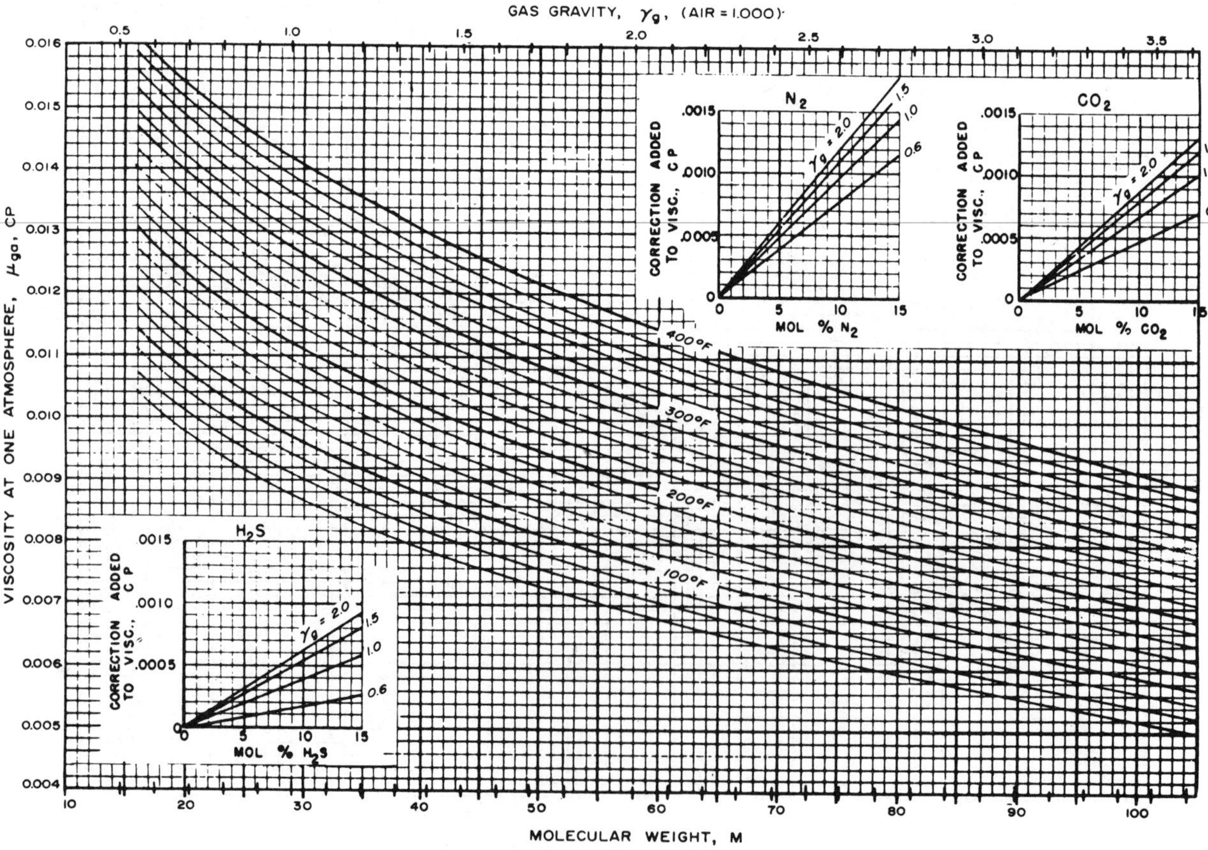


Fig. D-20 – Viscosity of natural gases at 1 atm.[1]

temperature and pressure and pseudocritical temperature and pressure (from either composition or gas gravity). Example D.14 illustrates use of this correlation.

---

### *Example D.14 – Estimation of Gas-Law Deviation Factor and Gas Formation Volume Factor*

**Problem.** Estimate $z$ for 0.7-gravity gas in a reservoir with temperature 200°F and pressure 2,500 psia. Use this value of $z$ to determine the gas formation volume factor $B_g$.

**Solution.** In Example D.13, we found that $T_{pc} = 390°\text{R}$ and $p_{pc} = 665$ psia. Then,

$$T_{pr} = T/T_{pc} = (200 + 460)/390 = 1.69,$$

and

$$p_{pr} = p/p_{pc} = 2{,}500/665 = 3.76.$$

From Fig. D-18, $z = 0.851$. Then,

$$B_g = 0.00504\ Tz/p$$
$$= (0.00504)(660)(0.851)/2{,}500$$
$$= 0.001132 \text{ RB/scf}.$$

---

## Gas Compressibility

Figs. D-19A and D-19B (developed by Trube[13]) lead to estimates of gas compressibility, $c_g$. In these figures (which cover different ranges of the independent variables), pseudoreduced compressibility, $c_{pr}$, is plotted as a function of pseudoreduced pressure, $p_{pr}$, with the parameter pseudoreduced temperature, $T_{pr}$. Pseudoreduced compressibility is defined as $c_{pr} = c_g p_{pc}$; thus, gas compressibility is found from the relation

$$c_g = c_{pr}/p_{pc}. \qquad \text{(D.8)}$$

Use of Figs. D-19A and D-19B requires knowledge of reservoir temperature and pressure and pseudocritical temperature and pressure of the gas (from either composition or gravity). Example D.15 illustrates use of this correlation.

---

### *Example D.15 – Estimation of Gas Compressibility*

**Problem.** Estimate the compressibility of a 0.7-gravity gas at a reservoir temperature of 200°F and pressure of 2,500 psia.

**Solution.** In Example D.14, we found that $T_{pr} = 1.69$, $p_{pr} = 3.76$, and $p_{pc} = 665$ psia for these conditions. From Fig. D-19A, $c_{pr} = 0.26$. Thus,

$$c_g = c_{pr}/p_{pc} = 0.26/665$$
$$= 0.00039 \text{ psi}^{-1}.$$

---

## Gas Viscosity

Figs. D-20, D-21A, and D-21B (from the work of Carr *et al.*[14]) can be used to estimate gas viscosity at

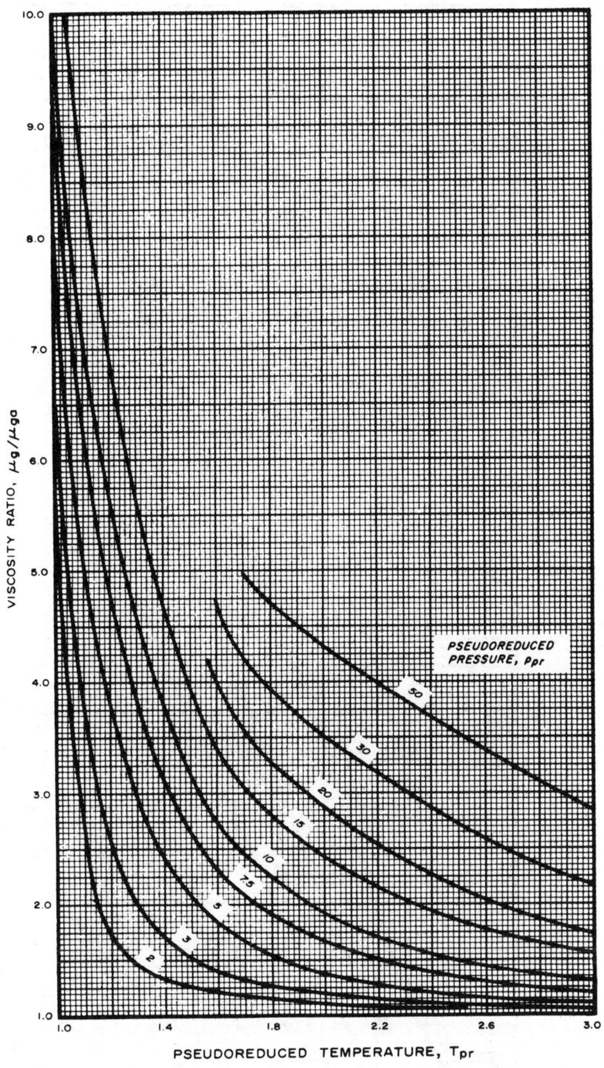


**Fig. D-21A** – Effect of temperature and pressure on gas viscosity.[1]

reservoir conditions. From knowledge of reservoir temperature and gas gravity (or its equivalent, molecular weight), we can estimate the viscosity of a hydrocarbon gas, $\mu_{g1}$, at atmospheric pressure. Insets in Fig. D-20 allow corrections to this viscosity for nonhydrocarbon components of the gas. Figs. D-21A and D-21B (two different ways of plotting the same data) permit calculation of gas viscosity at reservoir temperature and pressure, given viscosity at atmospheric pressure, pseudoreduced temperature, $T_{pr}$, and pseudoreduced pressure, $p_{pr}$. Example D.16 illustrates application of these figures.

---

### *Example D.16 – Estimation of Gas Viscosity*

**Problem.** Estimate the viscosity of a 0.7-gravity hydrocarbon gas (no $H_2S$, $N_2$, or $CO_2$) at 200°F and 2,500 psia.

**Solution.** From Fig. D-20, the viscosity $\mu_{g1}$ of 0.7-gravity gas [molecular weight = 0.7 × 28.97 = 20.3 lbm/(lbm-mole)] at 200°F and atmospheric pressure

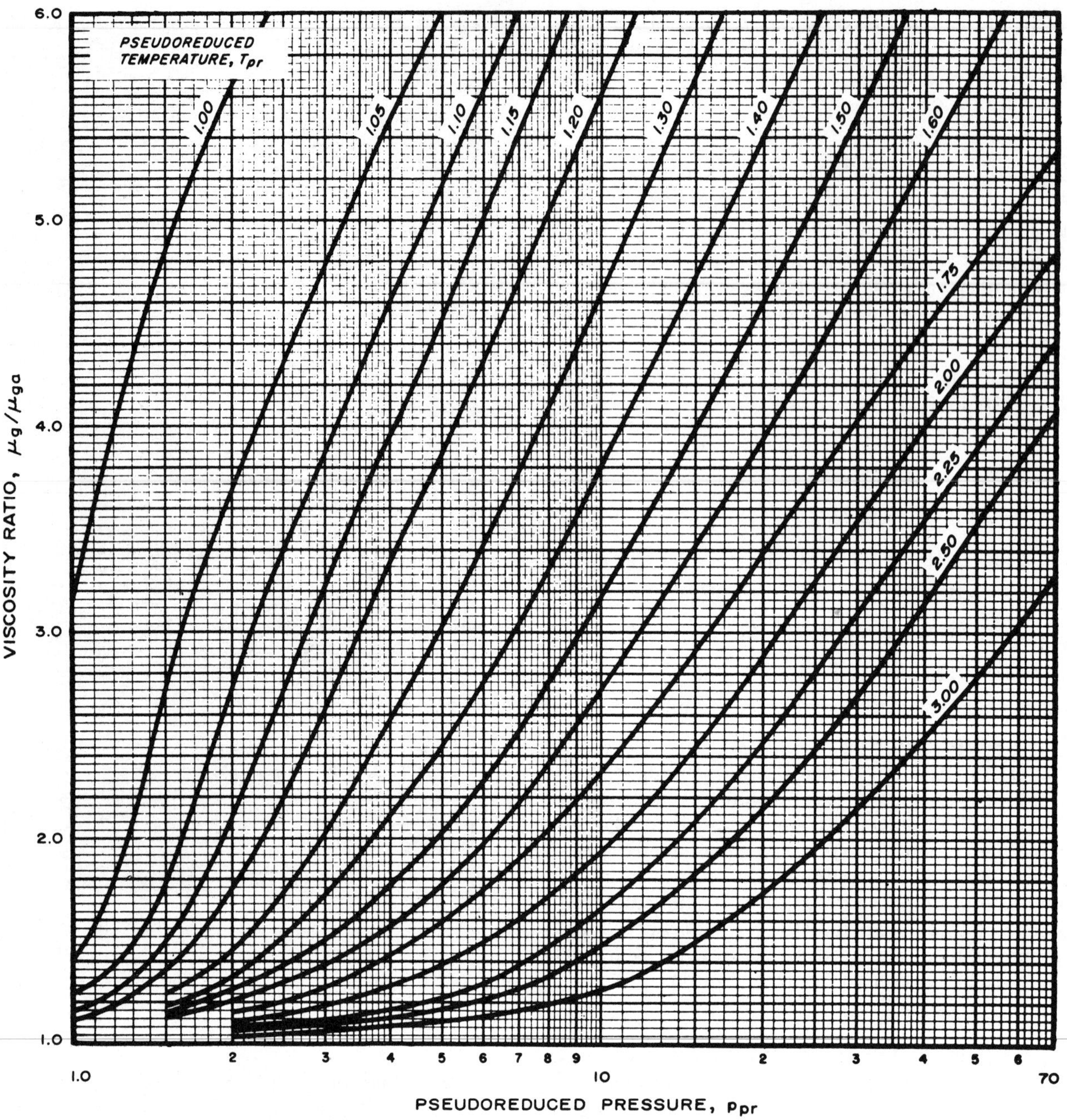


**Fig. D-21B** – Effect of pressure and temperature on gas viscosity.[1]

is 0.01225 cp. In Example D.14, we found that, at these conditions, $T_{pr}=1.69$ and $p_{pr}=3.76$. Thus, from Fig. D-21A or D-21B, $\mu_g/\mu_{g1}=1.45$. At 200°F and 2,500 psia, gas viscosity is

$$\mu_g=(\mu_g/\mu_{g1})\mu_{g1}$$

$$=(0.01225)(1.45)$$

$$=0.0178 \text{ cp.}$$

---

## Formation Compressibility

Formation compressibility, $c_f$, is defined as

$$c_f=-\frac{1}{V_p}\left(\frac{\partial V_p}{\partial p}\right)_T, \quad \ldots\ldots\ldots\ldots\ldots\ldots \text{(D.9)}$$

where $V_p$ = pore volume of porous medium.

Formation compressibility is a complex function of rock type, porosity, pore pressure, overburden pressure, and, in general, the stresses in different directions in the formation. No reliable correlation of this quantity with the controlling variables has been presented in the literature; indeed, laboratory determinations of $c_f$ are difficult, and many reported values of this quantity are doubtless erroneous because conditions in the field were not duplicated in the laboratory. A much-used correlation, developed by Hall,[15] is presented in Fig. D-22. This correlation relates $c_f$ to a single variable – porosity. As reported by Earlougher,[1] this correlation is known to be incorrect by an order of magnitude or more in specific situations. Thus, while the correlation is easy to use, the result may be seriously in error for any given application.

Use of this correlation is illustrated in Example D.17. The result may be of no greater accuracy than simply assuming $c_f \simeq 4 \times 10^{-6}$ psi$^{-1}$, since only one of the many variables affecting $c_f$ has been taken into account.

---

## *Example D.17 – Estimation of Formation Compressibility*

**Problem.** Estimate the formation compressibility $c_f$ for a reservoir with 20% porosity.

**Solution.** From Fig. D-22, $c_f = 3.6 \times 10^{-6}$ psi$^{-1}$.

---

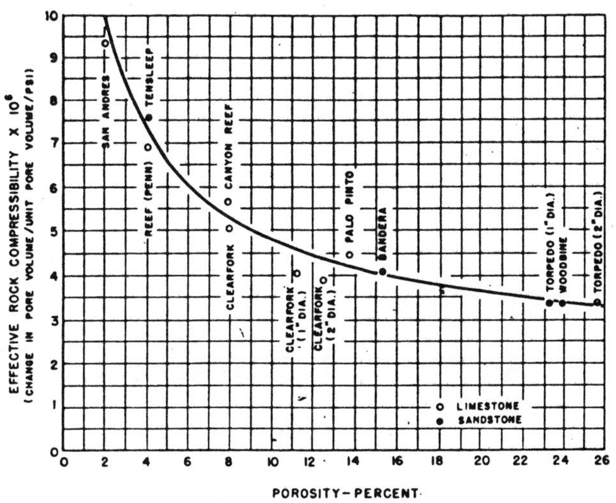


**Fig. D-22** – Formation compressibility.[15]

## Exercises

Results of pressure transient test analysis sometimes are combined with rock and fluid properties to calculate the following quantities:

Total reservoir flow rate,
$q_{Rt} = q_o B_o + q_w B_w + (q_g - R_s q_o / 1{,}000) B_g$.
Total mobility,
$\lambda_t = k_o/\mu_o + k_w/\mu_w + k_g/\mu_g$.
Total compressibility,
$c_t = c_o S_o + c_w S_w + c_g S_g + c_f$.

The following exercises require calculation of $q_{Rt}$, $\lambda_t$, and $c_t$ for two cases.

D.1. Calculate $q_{Rt}$, $\lambda_t$, and $c_t$ for an undersaturated oil reservoir with the following properties.

$q_o$ = 100 STB/D,
$q_w$ = 20 STB/D,
$q_g$ = $q_o R_s$ (reservoir produces dissolved gas only),
Reservoir pressure = 4,000 psia,
Reservoir temperature = 220°F,
$R_s$ = 400 scf/STB,
$\gamma_g$ = 0.7,
$\gamma_o$ = 0.85,
Water salinity = 25,000 ppm (2.5% NaCl),
$k_o$ = 20 md,
$k_w$ = 0.93 md,
$k_g$ = 0 (no free-gas saturation),
$\phi$ = 0.18,
$S_o$ = 0.65,
$S_w$ = 0.35, and
$S_g$ = 0.

D.2. Calculate $q_{Rt}$, $\lambda_t$, and $c_t$ for a saturated oil reservoir with the following properties:

$q_o$ = 100 STB/D,
$q_w$ = 5 STB/D,
$q_g$ = 250 Mscf/D,
Oil gravity = 38°API,
Gas gravity = 0.8,
Reservoir pressure = 2,000 psia,
Reservoir temperature = 200°F,
$k_o$ = 100 md,
$k_w$ = 3.3 md,
$k_g$ = 7.25 md,
Water salinity = 27,500 ppm,
$S_w$ = 0.25,
$S_g$ = 0.05,
$S_o$ = 0.70, and
$\phi$ = 0.18.

## References

1. Earlougher, R.C. Jr.: *Advances in Well Test Analysis*, Monograph Series, SPE, Dallas (1977) **5**.
2. Amyx, J.W., Bass, D.M. Jr., and Whiting, R.L.: *Petroleum Reservoir Engineering: Physical Properties*, McGraw-Hill Book Co., Inc., New York City (1960).
3. McCain, W.D. Jr.: *The Properties of Petroleum Fluids*, Petroleum Publishing Co., Tulsa (1973).
4. Trube, A.S.: "Compressibility of Undersaturated Hydrocarbon Reservoir Fluids," *Trans.*, AIME (1957) **210**, 341-344.
5. Standing, M.B.: *Volumetric and Phase Behavior of Oil Field Hydrocarbon Systems*, Reinhold Publishing Corp., New York City (1952).
6. Ramey, H.J. Jr.: "Rapid Methods for Estimating Reservoir Compressibilities," *J. Pet. Tech.* (April 1964) 447-454; *Trans.*, AIME, **231**.
7. Beal, C.: "The Viscosity of Air, Water, Natural Gas, Crude Oil and Its Associated Gases at Oil-Field Temperatures and Pressures," *Trans.*, AIME (1946) **165**, 94-115.
8. Chew, J. and Connally, C.A. Jr.: "A Viscosity Correlation for Gas-Saturated Crude Oils," *Trans.*, AIME (1959) **216**, 23-25.
9. Dodson, C.R. and Standing, M.B.: "Pressure-Volume-Temperature and Solubility Relations for Natural-Gas-Water Mixtures," *Drill. and Prod. Prac.*, API (1944) 173-179.
10. Matthews, C.S. and Russell, D.G.: *Pressure Buildup and Flow Tests in Wells*, Monograph Series, SPE, Dallas (1967) **1**, Appendix G.
11. Brown, G.G., Katz, D.L., Oberfell, G.G., and Alden, R.C.: *Natural Gasoline and the Volatile Hydrocarbons*, Natural Gasoline Assn. of America, Tulsa (1948).
12. Standing, M.B. and Katz, D.L.: "Density of Natural Gases," *Trans.*, AIME (1942) **146**, 140-149.
13. Trube, A.S.: "Compressibility of Natural Gases," *Trans.*, AIME (1957) **210**, 355-357.
14. Carr, N.L., Kobayashi, R., and Burrows, D.B.: "Viscosity of Hydrocarbon Gases Under Pressure," *Trans.*, AIME (1954) **201**, 264-272.
15. Hall, H.N.: "Compressibility of Reservoir Rocks," *Trans.*, AIME (1953) **198**, 309-311.

# Appendix E
# A General Theory of Well Testing

The purpose of this appendix is to summarize an approach to well test analysis that is not limited either to wells in infinite-acting reservoirs or to wells centered in cylindrical reservoirs. This theory is stated more clearly in terms of dimensionless variables, which we have largely avoided in the text because they are abstract and thus are less easily understood than other approaches. For some applications, though, use of dimensionless variables is essential – and general theory is such an area.

We have seen that for $t \leq 948\ \phi\mu c_t r_e^2/k$ (infinite-acting reservoir) and for constant-rate flow for a well centered in a cylindrical reservoir,

$$p_i - p_{wf} = -70.6\frac{qB\mu}{kh}\left[\ln\left(\frac{1{,}688\ \phi\mu c_t r_w^2}{kt}\right) - 2s\right], \quad \text{(1.11)}$$

and for $t > 948\ \phi\mu c_t r_e^2/k$ (pseudosteady state),

$$p_i - p_{wf} = 141.2\frac{qB\mu}{kh}\left[\frac{0.000527\ kt}{\phi\mu c_t r_e^2} + \ln\left(\frac{r_e}{r_w}\right) - 0.75 + s\right]. \quad \text{(1.12 modified)}$$

We can generalize by writing these equations in terms of dimensionless variables:

$$\frac{0.00708\ kh\ (p_i - p_{wf})}{qB\mu} = \frac{1}{2}\left(\ln\frac{0.000264\ kt}{\phi\mu c_t r_w^2} + 0.809\right) + s,$$

or

$$\frac{0.00708\ kh\ (p_i - p_{wf})}{qB\mu} = \frac{1}{2}\left(\ln t_D + 0.809\right) + s$$

$$= p_D + s,\ t_D \leq 0.25\ r_{De}^2, \quad \text{(E.1)}$$

and

$$\frac{0.00708\ kh\ (p_i - p_{wf})}{qB\mu} = \left[\frac{2}{(r_e/r_w)^2} \cdot \left(\frac{0.000264\ kt}{\phi\mu c_t r_w^2}\right) + \ln\left(\frac{r_e}{r_w}\right) - \frac{3}{4} + s\right],$$

or

$$\frac{0.00708\ kh\ (p_i - p_{wf})}{qB\mu} = \left(\frac{2t_D}{r_{De}^2} + \ln r_{De} - 0.75\right) + s$$

$$= p_D + s,\ t_D > 0.25\ r_{De}^2. \quad \text{(E.2)}$$

Even more generally, we can write

$$\frac{0.00708\ kh\ (p_i - p_{wf})}{qB\mu} = p_D(t_D) + s. \quad \text{(E.3)}$$

Thus, given a value $t_D$, there exists a rule for determining $p_D$ for a well centered in a cylindrical reservoir – either Eq. E.1 or Eq. E.2. Thus, this general expression includes transient flow and pseudosteady-state flow as special cases.

The value of this generalization becomes clearer if now we reconsider a subject we introduced in Chap. 3 – $n$ changes in rate in the producing history of a well (Fig. E-1).

The total pressure drawdown at the well, expressed in terms of dimensionless variables, is found by superposition, as we showed in Chap. 3 for the special case of transient flow. Here, note that rate $q_1$ acts for time, $t$; $(q_2 - q_1)$ for time $(t - t_1)$, ... ; and $(q_n - q_{n-1})$ for time $(t - t_{n-1})$. Thus, we write, in general,

$$\frac{0.00708\ kh\ [p_i - p_{wf}(t_D)]}{B\mu} = (q_1 - 0)[p_D(t_D - 0) + s] + (q_2 - q_1)[p_D(t_D - t_{D1}) + s] + \ldots + (q_n - q_{n-1})[p_D(t_D - t_{D,n-1}) + s],$$

or, more compactly,

$$\frac{0.00708\ kh\ [p_i - p_{wf}(t_D)]}{B\mu} = \left[\sum_{j=1}^{n} \Delta q_j\ p_D\left(t_D - t_{Dj}\right)\right] + q_n s, \quad \text{(E.4)}$$

where

$\Delta q_j = q_j - q_{j-1}$ (and $q_0 \equiv 0$),

$t_{D0} \equiv 0$.

Eq. E.4 is general – i.e., it applies to a reservoir in which, for some values of $t_D - t_{Dj}$, $p_D$ can be the pseudosteady-state solution, and, for other values of $t_D - t_{Dj}$, $p_D$ can be the transient solution.

As an example, consider a pressure buildup test in a cylindrical reservoir. Let $q_1 = q$; $q_2 = 0$; $t_{D1} = t_{pD} + \Delta t_D$; and $t_{D2} - t_{D1} = \Delta t_D$. Then,

$$\frac{0.00708\, kh\, (p_i - p_{ws})}{B\mu} = q p_D (t_{pD} + \Delta t_D) - q p_D (\Delta t_D).$$

If and only if flow is transient for total time $t_{pD} + \Delta t_D$, then

$$\frac{0.00708\, kh\, (p_i - p_{ws})}{B\mu} = \frac{1}{2} q \left[\ln (t_{pD} + \Delta t_D) + 0.809\right] - \frac{1}{2} q (\ln \Delta t_D + 0.809)$$

or

$$p_i - p_{ws} = 70.6 \frac{qB\mu}{kh} \ln \left(\frac{t_{pD} + \Delta t_D}{\Delta t_D}\right)$$

$$= 162.6 \frac{qB\mu}{kh} \log \left(\frac{t_p + \Delta t}{\Delta t}\right).$$

In fact, the arguments leading to Eq. E.4 understate its generality. For constant-rate flow in cylindrical reservoirs, $p_D$ can be calculated for all $t_D$ from simple equations – but the method is not restricted to cylindrical reservoirs. It applies to *any* drainage configuration for which $p_D$ can be calculated as a function of $t_D$ (using finite-difference simulation or any other convenient means).

We now examine a useful method[1-3] for determining $p_D$ as a function of $t_D$ for more general reservoir shapes; this method uses the Matthews-Brons-Hazebroek functions[4] developed for use in determining average reservoir pressure. We start by noting that for a pressure buildup test in a reservoir of general shape,

$$\frac{0.00708\, kh\, (p_i - p_{ws})}{qB\mu} = p_D (t_D + \Delta t_D) - p_D (\Delta t_D). \quad \text{(E.5)}$$

For $\Delta t_D$ sufficiently small (e.g., $\Delta t_D \le 0.25\, r_{De}^2$ in a cylindrical reservoir), flow will be transient regardless of drainage-area configuration, and $p_D\, (\Delta t_D) = 1/2(\ln \Delta t_D + 0.809)$. If we now add and subtract $1/2 \ln (t_D + \Delta t_D)$ on the right side of Eq. E.5 written at these sufficiently small values of $\Delta t_D$, the result is

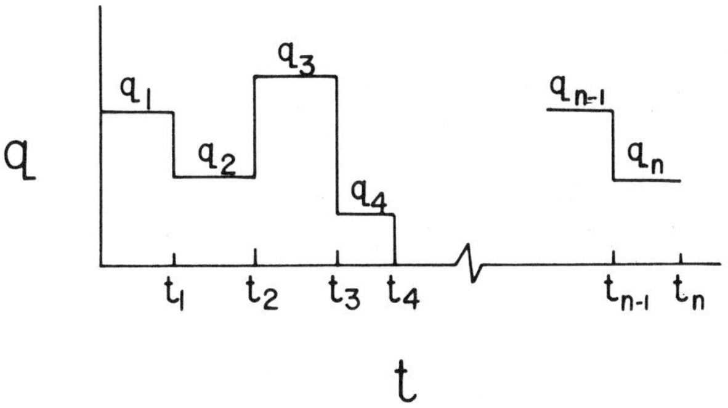


**Fig. E-1** – *n* rate changes in well's producing history.

$$\frac{0.00708\, kh\, (p_i - p_{ws})}{qB\mu} = \frac{1}{2} \ln \left(\frac{t + \Delta t}{\Delta t}\right) - \frac{1}{2}\left[\ln (t_D + \Delta t_D) + 0.809\right] + p_D (t_D + \Delta t_D). \quad \text{(E.6)}$$

Now, for some small values of $\Delta t_D \le \Delta t_{Ds}$, $\ln (t_D + \Delta t_D) \simeq \ln (t_D)$ and $p_D (t_D + \Delta t_D) \simeq p_D (t_D)$.

For $\Delta t_D \le \Delta t_{Ds}$, then,

$$\frac{0.00708\, kh\, (p_i - p_{ws})}{qB\mu} = \frac{1}{2} \ln \left(\frac{t + \Delta t}{\Delta t}\right) + p_D (t_D) - \frac{1}{2}\left[\ln (2.246\, t_D)\right]$$

$$= \frac{1}{2} \ln \left(\frac{t + \Delta t}{\Delta t}\right) + \text{constant}. \quad \text{(E.7)}$$

The implication of Eq. E.7 is simply that for sufficiently small $\Delta t$, a plot of $p_{ws}$ vs. $\ln [(t + \Delta t)/\Delta t]$ or $\log [(t + \Delta t)/\Delta t]$ will be linear and will have a slope, $m$, related to permeability. This linear relationship exists, of course, only for sufficiently small values of $\Delta t$. Once we have established the linear relationship, though, we can *extrapolate* it to larger times. In particular, we can extrapolate $p_{ws}$ to $(t + \Delta t)/\Delta t = 1$. Matthews *et al.*[4] did this and chose to call the extrapolated pressure $p^*$. Eq. E.7 shows that

$$\frac{0.00708\, kh}{qB\mu} (p_i - p^*) = p_D (t_D) - \frac{1}{2} \ln (2.246\, t_D). \quad \text{(E.8)}$$

A material balance shows that

$$c_t V_p (p_i - \bar{p}) = 5.615\, qBt/24$$

$$= c_t A h \phi (p_i - \bar{p}), \quad \text{(E.9)}$$

where $A$ is the area drained by a well (square feet). (Eq. E.9 is valid regardless of drainage-area shape.)

Now define $t_{DA}$ as $0.000264\, kt/\phi\mu c_t A$. Substituting for $t$ from Eq. E.9 gives

$$t_{DA} = \frac{0.000264\, kt}{\phi\mu c_t A} = \frac{(0.000264)(kh)(24)(p_i - \bar{p})}{(5.615)(qB\mu)}$$

$$= \frac{0.00708\, kh\, (p_i - \bar{p})}{2\pi qB\mu},$$

or

$$2\pi t_{DA} = \frac{0.00708\,kh\,(p_i - \bar{p})}{qB\mu}. \quad \text{(E.10)}$$

Then, using Eqs. E.8 and E.10,

$$-\frac{0.00708\,kh\,(p_i - p^*)}{qB\mu} + \frac{0.00708\,kh\,(p_i - \bar{p})}{qB\mu} =$$

$$-p_D(t_D) + \frac{1}{2}\ln(2.246\,t_D) + 2\pi t_{DA}.$$

Simplifying,

$$\frac{(p^* - \bar{p})}{70.6\,qB\mu} = 4\pi t_{DA} + \ln(2.246\,t_D) - 2p_D(t_D)$$

$$= p_{DMBH}(t_{DA}). \quad \text{(E.11)}$$

Note that the left side of Eq. E.11 is the ordinate of the MBH plots (thus, we give it the name $p_{DMBH}$). Note also that $t_{DA} = 0.000264\,kt/\phi\mu c_t A$ is the abscissa of the MBH plots.

Eq. E.11 allows us to use the MBH charts to determine $p_D(t_D)$ for the drainage-area configurations considered in constructing the charts. To understand why this is so, note that Eq. E.11 can be put in the form

$$p_D(t_D) = 2\pi t_{DA} + \frac{1}{2}\ln(2.246\,t_D)$$

$$-\frac{1}{2}p_{DMBH}(t_{DA}). \quad \text{(E.12)}$$

For time, $t_D$, sufficiently small that no boundary effects have appeared (transient flow),

$$p_D(t_D) = \frac{1}{2}\ln(2.246\,t_D),\; t_D < t_{DB}.$$

For $t_D < t_{DB}$, then,

$$p_D(t_D) = \frac{1}{2}\ln(2.246\,t_D)$$

$$= 2\pi t_{DA} + \frac{1}{2}\ln(2.246\,t_D)$$

$$-\frac{1}{2}p_{DMBH}(t_{DA}),$$

or

$$p_{DMBH}(t_{DA}) = 4\pi t_{DA}$$

$$= \frac{(p^* - \bar{p})}{70.6\,qB\mu},\; t_D < t_{DB}. \quad \text{(E.13)}$$

Eq. E.13 shows that in transient flow (infinite-acting reservoir), $p_{DMBH}$ is a linear function of $t_{DA}$. Thus, in principle, a plot of $p_{DMBH}$ vs. $t_{DA}$ could be used to determine the time $t_{DA}$ up to which a reservoir of general shape is infinite acting (by observing the time at which a deviation from linearity occurs). In principle, then, this technique could have been used to determine the values in Table 1.1 in the column "Use Infinite System Solution With Less than 1% Error for $t_{DA}$ <," thus providing us with a means of determining the upper limit of applicability of the relationship $p_D(t_D) = 1/2\,(\ln t_D + 0.809)$ in reservoirs of *general drainage-area configuration*.

Now we turn our attention to developing a method for determining $p_D(t_D)$ after pseudosteady-state conditions have been established. Brons and Miller[5] pointed out that, at pseudosteady state,

$$p_{DMBH}(t_{DA}) = \ln(C_A t_{DA})$$

$$= \ln C_A + \ln t_{DA}, \quad \text{(E.14)}$$

where $C_A$ is a shape factor whose value depends on the specific drainage area configuration. This equation implies a linear relationship between $p_{DMBH}$ and $\ln t_{DA}$, with intercept $\ln C_A$ varying from case to case. The MBH charts (Fig. 2.12) show that this linear relationship does exist and that the intercept does depend on the specific drainage-area configuration. Further, the time at which $p_{DMBH}$ becomes a linear function of $\ln t_{DA}$ establishes the beginning of pseudosteady-state flow.

The task at hand is to show that Eq. E.14 leads to a method for determining $p_D(t_D)$ for general drainage-area shape for pseudosteady-state flow. If we substitute the value of $p_{DMBH}(t_{DA})$ from Eq. E.14 into Eq. E.12, the result is

$$p_D(t_D) = 2\pi t_{DA} + 1/2\ln(2.246\,t_D)$$

$$- 1/2\ln(C_A t_{DA})$$

$$= 2\pi t_{DA} + 1/2\ln\left(\frac{2.246\,t_D}{C_A t_{DA}}\right)$$

or

$$p_D(t_D) = 2\pi t_{DA} + 1/2\ln\left(\frac{2.246\,A}{C_A r_w^2}\right). \quad \text{(E.15)}$$

Thus, we have established a rule by which $p_D(t_D)$ can be established for general drainage shape in pseudosteady-state flow. Table 1.1 gives the times ("Exact for $t_{DA}$ <" and "Less Than 1% Error for $t_{DA}$ <") at which Eq. E.15 can be applied.

Finally, there is the problem of how to establish $p_D(t_D)$ for general drainage-area configuration when there is a gap between the upper limit of applicability of the transient solution and the lower limit of applicability of the pseudosteady-state solution. Eq. E.12, which applies at *all times*, can be used to fill this gap:

$$p_D(t_D) = 2\pi t_{DA} + 1/2\ln(2.246\,t_D)$$

$$- 1/2\,p_{DMBH}(t_{DA}).$$

To avoid both $t_{DA}$ and $t_D$ on the right side of the same working equation, we can rewrite this result as

$$p_D(t_D) = 2\pi t_{DA} + 1/2\ln(2.246\,t_{DA}A/r_w^2)$$

$$- 1/2\,p_{DMBH}(t_{DA}). \quad \text{(E.16)}$$

Thus, values of $p_{DMBH}$ could be read from their charts at a desired value of $t_{DA}$, and $p_D(t_D)$ could be calculated for use in subsequent reservoir analysis.

Earlougher provides values of $p_D(t_D)$ for general drainage-area shapes in Ref. 6 Appendix C. For applications, the reader should find the desired $p_D$ values in that reference.

In summary, this appendix has shown that well test analysis techniques are not limited to wells centered in cylindrical reservoirs or to infinite-acting reservoirs. One can derive or find in the literature dimensionless-pressure values for general drainage-area shapes; one can combine these values (using superposition) to take into account any arbitrary rate history before and during the test. Thus, the title of the appendix: A General Theory of Well Testing.

## References

1. Ramey, H.J. Jr. and Cobb, W.M.: "A General Pressure Buildup Theory for a Well in a Closed Drainage Area," *J. Pet. Tech.* (Dec. 1971) 1493-1505; *Trans.*, AIME, **251**.
2. Cobb, W.M. and Dowdle, W.L.: "A Simple Method for Determining Well Pressure in Closed Rectangular Reservoirs," *J. Pet. Tech.* (Nov. 1973) 1305-1306.
3. Dake, L.P.: *Fundamentals of Reservoir Engineering*, Elsevier Scientific Publishing Co., Amsterdam (1978).
4. Matthews, C.S., Brons, F., and Hazebroek, P.: "A Method for Determination of Average Pressure in a Bounded Reservoir," *Trans.*, AIME (1954) **201**, 182-191.
5. Brons, F. and Miller, W.C.: "A Simple Method for Correcting Spot Pressure Readings," *Trans.*, AIME (1961) **222**, 803-805.
6. Earlougher, R.C. Jr.: *Advances in Well Test Analysis*, Monograph Series, SPE, Dallas (1977) **5**.

# Appendix F
# Use of SI Units in Well-Testing Equations

This Appendix summarizes the changes required to solve the equations stated in the text by using International System (SI) metric units. To show the necessary changes and to allow the interested reader to apply the SI unit system to typical well-testing problems, this Appendix has four major parts: (1) conversion factors from "oilfield units" to SI units are tabulated in Table F-1 for the units used in the text; (2) a summary of preferred SI units for major variables is given in Table F-2; (3) major equations in the text are restated in Table F-3, with constants given in both oilfield and SI units; and (4) answers to all examples in the text are given in SI units. In addition, in the Nomenclature, preferred SI units are given (in parentheses) for each quantity used in the text.

TABLE F-1 – CONVERSION FACTORS

| To Convert From | To | Multiply By | Inverse |
|---|---|---|---|
| acres | $m^2$ | 4.047 E+03 | 2.471 E−04 |
| bbl | $m^3$ | 1.590 E−01 | 6.290 |
| cp | mPa·s | 1.0 | 1.0 |
| cp | μPa·s | 1.0 E+03 | 1.0 E−03 |
| cu ft | $m^3$ | 2.832 E−02 | 3.532 E+01 |
| ft | m | 3.048 E−01 | 3.281 |
| md | $m \cdot m^2$ | 9.869 E−01 | 1.013 |
| psi | kPa | 6.895 | 1.450 E−01 |
| °R | K | 5.555 E−01 | 1.80 |
| sq ft | $m^2$ | 9.290 E−02 | 1.076 E+01 |

A more complete table of conversion factors, emphasizing application to well-testing problems, is given by Earlougher.[1]

Table F-2 summarizes oilfield (customary) and preferred SI units (practical) for single variables and groups of variables of major importance in well test analysis.

TABLE F-2 – CUSTOMARY AND METRIC UNITS FOR MAJOR VARIABLES IN EQUATIONS

| | Customary Unit | Practical Metric Unit |
|---|---|---|
| Compressibility, $c_t$ | $psi^{-1}$ | $kPa^{-1}$ |
| Density, $\rho$ | lbm/cu ft | $kg/m^3$ |
| Gas flow rate, $q_g$ | Mscf/D | $m^3/d$ |
| Gas viscosity, $\mu_g$ | cp | μPa·s |
| Liquid flow rate, $q_o$ and $q_w$ | B/D | $m^3/d$ |
| Liquid viscosity, $\mu_{o,w}$ | cp | mPa·s |
| Permeability, $k$ | md | md |
| Pressure, $p$ | psi | kPa |
| Pseudopressure, $\psi(p)$ | $psi^2/cp$ | $MPa^2/Pa \cdot s$ |
| Radius, $r$ | ft | m |
| Slope, $m$ | psi/cycle | kPa/cycle |
| Temperature, $T$ | °R | K |
| Thickness, $h$ | ft | m |
| Time, $t$ | hr | h |
| Volume, $V$ | bbl | $m^3$ |
| Wellbore storage constant, $C_s$ | bbl/psi | $m^3/kPa$ |

TABLE F-3 – MAJOR EQUATIONS WITH CONSTANT VALUES IN CUSTOMARY AND SI METRIC

| Equation Number in Text | Equation | Numerical Value of Constants in Order of Appearance ($c_1, c_2, c_3 \ldots$) Customary | SI |
|---|---|---|---|
| 1.1 | $\frac{\partial^2 p}{\partial r^2} + \frac{1}{r}\frac{\partial p}{\partial r} = \frac{\phi\mu c}{c_1 k}\frac{\partial p}{\partial t}$ | 0.000264 | $3.557 \times 10^{-6}$ |
| 1.6 | $p_{wf} = p_i - c_1 \frac{qB\mu}{kh}\left\{\frac{2t_D}{r_{eD}^2} + \ln r_{eD} - \frac{3}{4} + 2\sum_{n=1}^{\infty}\frac{e^{-\alpha_n^2 t_D} J_1^2(\alpha_n r_{eD})}{\alpha_n^2 [J_1^2(\alpha_n r_{eD}) - J_1^2(\alpha_n)]}\right\}$ | 141.2 | $1.866 \times 10^3$ |
| 1.7 | $p = p_i + c_1 \frac{qB\mu}{kh} Ei\left(\frac{-c_2\phi\mu c_t r^2}{kt}\right)$ | 70.6<br>948 | $9.33 \times 10^2$<br>$7.036 \times 10^4$ |
| 1.9 | $\Delta p_s = c_1 \frac{qB\mu}{kh}\left(\frac{k}{k_s} - 1\right)\ln\left(\frac{r_s}{r_w}\right)$ | 141.2 | $1.866 \times 10^3$ |
| 1.11 | $p_i - p_{wf} = -c_1 \frac{qB\mu}{kh}\left\|\ln\left(\frac{c_2\phi\mu c_t r_w^2}{kt}\right) - 2s\right\|$ | 70.6<br>1,688 | $9.33 \times 10^2$<br>$1.253 \times 10^5$ |
| 1.12 | $p_{wf} = p_i - c_1 \frac{qB\mu}{kh}\left\|\frac{c_2 kt}{\phi\mu c_t r_e^2} + \ln\left(\frac{r_e}{r_w}\right) - \frac{3}{4}\right\|$ | 141.2<br>0.000527 | $1.866 \times 10^3$<br>$7.1 \times 10^{-6}$ |
| 1.13 | $\frac{\partial p_{wf}}{\partial t} = -\frac{c_1 qB}{c_t V_p}$ | 0.234 | $4.168 \times 10^{-2}$ |
| 1.16 | $\bar{p} - p_{wf} = c_1 \frac{qB\mu}{kh}\left\|\ln\left(\frac{r_e}{r_w}\right) - \frac{3}{4} + s\right\|$ | 141.2 | $1.866 \times 10^3$ |
| 1.17 | $p_i - p_{wf} = c_1 \frac{qB\mu}{kh}\left\|\frac{c_2 kt}{\phi\mu c_t r_e^2} + \ln\left(\frac{r_e}{r_w}\right) - \frac{3}{4} + s\right\|$ | 141.2<br>0.000527 | $1.866 \times 10^3$<br>$7.1 \times 10^{-6}$ |
| 1.19 | $J \equiv \frac{q}{\bar{p} - p_{wf}} = \frac{k_J h}{c_1 B\mu\left\|\ln\left(\frac{r_e}{r_w}\right) - \frac{3}{4}\right\|}$ | 141.2 | $1.866 \times 10^3$ |
| 1.20 | $\bar{p} - p_{wf} = c_1 \frac{qB\mu}{kh}\left\|\frac{1}{2}\ln\left(\frac{10.06\,A}{C_A r_w^2}\right) - \frac{3}{4} + s\right\|$ | 141.2 | $1.866 \times 10^3$ |
| 1.21 | $J = \frac{c_1\,kh}{B\mu\left[\frac{1}{2}\ln\left(\frac{10.06\,A}{C_A r_w^2}\right) - \frac{3}{4} + s\right]}$ | 0.00708 | $5.356 \times 10^{-4}$ |
| 1.26 | $C_s = c_1 \frac{A_{wb}}{\rho}\frac{g_c}{g}$ | 25.65 | 101.98 |
| 1.29 | $p_D = \frac{c_1 kh(p_i - p_w)}{q_i B\mu}$ | 0.00708 | $5.356 \times 10^{-4}$ |
| 1.46 | $p_i - p_{wf} = c_1 \frac{qB}{hL_f}\left(\frac{\mu t}{k\phi c_t}\right)^{1/2}$ | 4.064 | 6.236 |
| 1.47 | $r_i = \left(\frac{kt}{c_1\phi\mu c_t}\right)^{1/2}$ | 948 | $7.036 \times 10^4$ |

TABLE F-3 – MAJOR EQUATIONS WITH CONSTANT VALUES IN CUSTOMARY AND SI METRIC (Contd.)

| Equation Number in Text | Equation | Numerical Value of Constants in Order of Appearance ($c_1, c_2, c_3$ . . .) Customary | SI |
|---|---|---|---|
| 1.48 | $t_s = c_1 \phi\mu c_t r_e^2 / k$ | 948 | $7.036 \times 10^4$ |
| 1.52 | $t_p = c_1 N_p / q_{last}$ | 24 | 24 |
| 2.1 | $p_{ws} = p_i - c_1 \frac{qB\mu}{kh} \log\left[(t_p + \Delta t)/\Delta t\right]$ | 162.6 | $2.149 \times 10^3$ |
| 2.2 | $m = c_1 \frac{qB\mu}{kh}$ | 162.6 | $2.149 \times 10^3$ |
| 2.4 | $s = 1.151\left[\frac{(p_{1\,hr} - p_{wf})}{m} - \log\left(\frac{k}{\phi\mu c_t r_w^2}\right) + c_1\right]$ | +3.23 | 5.10 |
| 2.7 | $p_D = \frac{c_1 kh(p_{ws} - p_{wf})}{qB\mu}$ | 0.00708 | $5.356 \times 10^{-4}$ |
| 2.8 | $t_D = \frac{c_1 k \Delta t_e}{\phi\mu c_t r_w^2}$ | 0.000264 | $3.557 \times 10^{-6}$ |
| 2.9 | $C_{sD} = \frac{c_1 C_s}{\phi c_t h r_w^2}$ | 0.894 | 0.159 |
| 2.10 | $\Delta t_e = \Delta t/(1 + \Delta t/t_p)$ | – | – |
| 2.11 | $C_s = \frac{qB}{c_1}\frac{\Delta t}{\Delta p}$ | 24 | 24 |
| 2.12 | $t_D \cong 50\, C_{sD} e^{0.14 s}$ | – | – |
| 2.13 | $t_{wbs} \cong c_1 \frac{C_s e^{0.14 s}}{(kh/\mu)}$ | 170,000 | $2.247 \times 10^6$ |
| 2.14 | $r_{wa} = r_w e^{-s}$ | – | – |
| 2.15 | $L_f \cong 2\, r_{wa}$ | – | – |
| 2.16 | $(\Delta p)_s = 0.869\, m(s)$ | – | – |
| 2.17 | $E = \frac{J_{actual}}{J_{ideal}} = \frac{\bar{p} - p_{wf} - (\Delta p)_s}{\bar{p} - p_{wf}}$ | – | – |
| 2.18 | $E \simeq \frac{p^* - p_{wf} - (\Delta p)_s}{p^* - p_{wf}}$ | – | – |
| 2.19 | $s = \frac{h_t}{h_p} s_d + s_p$ | – | – |
| 2.20 | $s_p = \left(\frac{h_t}{h_p} - 1\right)\left[\ln\left(\frac{h_t}{r_w}\sqrt{\frac{k_H}{k_V}}\right) - 2\right]$ | – | – |
| 2.21 | $m_L = c_1 \frac{qB}{hL_f}\left(\frac{\mu}{k\phi c_t}\right)^{1/2}$ | 4.064 | 6.236 |
| 2.22 | $\log(L_f) = \frac{1}{2}\left[\left(\frac{p_{wf} - p_{1\,hr}}{m}\right) + \log\frac{k}{\phi\mu c_t} - c_1\right]$ | 2.63 | 4.497 |

TABLE F-3 – MAJOR EQUATIONS WITH CONSTANT VALUES IN CUSTOMARY AND SI METRIC (Contd.)

| Equation Number in Text | Equation | Numerical Value of Constants in Order of Appearance ($c_1, c_2, c_3 \ldots$) | |
|---|---|---|---|
| | | Customary | SI |
| 2.24 | $\log(\bar{p}-p_{ws})=\log\left(c_1\frac{qB\mu}{kh}\right)-\frac{c_2 k\Delta t}{\phi\mu c_t r_e^2}$ | 118.6<br>0.00168 | 227.37<br>−0.8385 |
| 2.25 | $p_i-p_{ws}=-c_1\frac{qB\mu}{kh}\left\{\ln\left[\frac{c_2\phi\mu c_t r_w^2}{k(t_p+\Delta t)}\right]-2s\right\}$ | 70.6 | $9.33\times10^2$ |
| | $-c_1(-q)\frac{B\mu}{kh}\left[\ln\left(\frac{c_2\phi\mu c_t r_w^2}{k\Delta t}\right)-2s\right]$ | 1,688 | $1.253\times10^5$ |
| | $-c_1\frac{qB\mu}{kh}Ei\left[\frac{-c_3\phi\mu c_t L^2}{k(t_p+\Delta t)}\right]$ | 3,792 | $2.814\times10^5$ |
| | $-c_1(-q)\frac{B\mu}{kh}Ei\left(\frac{-c_3\phi\mu c_t L^2}{k\Delta t}\right)$ | | |
| 2.28 | $\Delta p^*_{ws}=-c_1\frac{qB\mu}{kh}\left[-Ei\left(\frac{-c_2\phi\mu c_t L^2}{k\Delta t}\right)\right]$ | 70.6<br>3,792 | $9.33\times10^2$<br>$2.814\times10^5$ |
| 2.29 | $L=\sqrt{\frac{c_1 k\Delta t_x}{\phi\mu c_t}}$ | 0.000148 | $1.944\times10^{-6}$ |
| 2.30 | $V_R=\frac{(\Delta N_p)(B_o)}{(\bar{p}_2-\bar{p}_1)c_t\phi}$ | – | – |
| 2.31 | $p_{wf}=p_i+c_1\frac{q_g B_{gi}\mu_i}{kh}\left[\log\left(\frac{c_2\phi\mu c_{ti}}{kt_p}\right)-\frac{(s+Dq_g)}{1.151}\right]$ | 162.6<br>1,688 | 2.149<br>11.638 |
| 2.32 | $p_{wf}^2=p_i^2+\frac{c_1 q_g\mu_i z_i T}{kh}\left[\log\left(\frac{c_2\phi\mu c_{ti}}{kt_p}\right)-\frac{(s+Dq_g)}{1.151}\right]$ | 1,637<br>1,688 | 1.508<br>11.638 |
| 2.33 | $p_{ws}=p_i-c_1\frac{q_g B_{gi}\mu_i}{kh}\left[\log\left(\frac{t_p+\Delta t}{\Delta t}\right)\right]$ | 162.6 | 2.149 |
| 2.34 | $s'=s+Dq_g=1.151\left[\frac{(p_{1\,hr}-p_{wf})}{m}-\log\left(\frac{k}{\phi\mu_i c_{ti}r_w^2}\right)+c_1\right]$ | 3.23 | 2.10 |
| 2.35 | $p_{ws}^2=p_i^2-c_1\frac{q_g\mu_i z_i T}{kh}\log\left(\frac{t_p+\Delta t}{\Delta t}\right)$ | 1,637 | 1.508 |
| 2.36 | $s'=s+Dq_g=1.151\left[\frac{(p_{1\,hr}^2-p_{wf}^2)}{m''}-\log\left(\frac{k}{\phi\mu_i c_{ti}r_w^2}\right)+c_1\right]$ | 3.23 | 2.10 |
| 2.37 | $p_{wf}=p_i+c_1\frac{q_{Rt}}{\lambda_t h}\left[\log\left(\frac{c_2\phi c_t r_w^2}{\lambda_t t}\right)-\frac{s}{1.151}\right]$ | 162.6<br>1,688 | $2.149\times10^3$<br>$1.253\times10^5$ |
| 2.38 | $p_{ws}=p_i-c_1\frac{q_{Rt}}{\lambda_t h}\log\left(\frac{t_p+\Delta t}{\Delta t}\right)$ | 162.6 | $2.149\times10^3$ |
| 2.39 | $q_{Rt}=q_oB_o+\left(q_g-\frac{q_oR_s}{1{,}000}\right)B_g+q_wR_w$ | – | – |
| 2.40 | $\lambda_t=\frac{k_o}{\mu_o}+\frac{k_w}{\mu_w}+\frac{k_g}{\mu_g}$ | – | – |

TABLE F-3 – MAJOR EQUATIONS WITH CONSTANT VALUES IN CUSTOMARY AND SI METRIC (Contd.)

| Equation Number in Text | Equation | Numerical Value of Constants in Order of Appearance ($c_1, c_2, c_3 \ldots$) Customary | SI |
|---|---|---|---|
| 2.41 | $\lambda_t = c_1 \dfrac{q_{Rt}}{mh}$ | 162.6 | $2.149 \times 10^3$ |
| 2.42 | $k_o = c_1 \dfrac{q_o B_o \mu_o}{mh}$ | 162.6 | $2.149 \times 10^3$ |
| 2.43 | $k_g = c_1 \dfrac{\left(q_g - \dfrac{q_o R_s}{c_2}\right) B_g \mu_g}{mh}$ | 162.6<br>1,000 | 2.149<br>1.000 |
| 2.44 | $k_w = c_1 \dfrac{q_w B_w \mu_w}{mh}$ | 162.6 | $2.149 \times 10^3$ |
| 2.45 | $s = 1.151 \left[ \dfrac{p_{1\,hr} - p_{wf}}{m} - \log\left( \dfrac{\lambda_t}{\phi c_t r_w^2} \right) + c_1 \right]$ | 3.23 | 5.10 |
| 3.1 | $p_{wf} = p_i + c_1 \dfrac{qB\mu}{kh} \left[ \log\left( \dfrac{c_2 \phi \mu c_t r_w^2}{kt} \right) - 0.869s \right]$ | 162.6<br>1,688 | $2.149 \times 10^3$<br>$1.253 \times 10^5$ |
| 3.2 | $t_{wbs} = \dfrac{(c_1 + c_2 s) C_s}{kh/\mu}$ | 200,000<br>12,000 | $2.644 \times 10^6$<br>$1.586 \times 10^5$ |
| 3.4 | $k = c_1 \dfrac{qB\mu}{mh}$ | 162.6 | $2.149 \times 10^3$ |
| 3.5 | $s = 1.151 \left[ \dfrac{(p_i - p_{1\,hr})}{m} - \log\left( \dfrac{k}{\phi \mu c_t r_w^2} \right) + c_1 \right]$ | 3.23 | 5.10 |
| 3.7 | $t_{\ell t} \simeq \dfrac{c_1 \phi \mu c_t A t_{DA}}{k}$ | 3,800 | $2.82 \times 10^5$ |
| 3.8 | $\dfrac{p_i - p_{wf}}{q} = c_1 \dfrac{\mu B}{kh} \left[ \log\left( \dfrac{c_2 \phi \mu c_t r_w^2}{kt} \right) + 0.869s \right]$ | 162.6<br>1,688 | $2.149 \times 10^3$<br>$1.253 \times 10^5$ |
| 3.9 | $s = 1.151 \left[ \left( \dfrac{p_i - p_{wf}}{q} \right)_{1\,hr} \dfrac{1}{m'} - \log\left( \dfrac{k}{\phi \mu c_t r_w^2} \right) + c_1 \right]$ | 3.23 | 5.10 |
| 3.11 | $\dfrac{p_i - p_{wf}}{q_n} = c_1 \dfrac{\mu B}{kh} \left\{ \sum_{j=1}^{n} \dfrac{(q_j - q_{j-1})}{q_n} \log(t - t_{j-1}) \right.$ | 162.6 | $2.149 \times 10^3$ |
| | $\left. + \left[ \log\left( \dfrac{k}{\phi \mu c_t r_w^2} \right) - c_2 + 0.869s \right] \right\}$ | 3.23 | 5.10 |
| 3.12 | $p_i - p_{ws} = c_1 \dfrac{\mu B}{kh} \sum_{j=1}^{n} (q_j - q_{j-1}) \log(t - t_{j-1})$ | 162.6 | $2.149 \times 10^3$ |
| 3.15 | $p_i - p_{ws} = c_1 \dfrac{q_2 B\mu}{kh} \left[ \dfrac{q_1}{q_2} \log\left( \dfrac{t_{p1} + t_{p2} + \Delta t}{t_{p2} + \Delta t} \right) + \log\left( \dfrac{t_{p2} + \Delta t}{\Delta t} \right) \right]$ | 162.6 | $2.149 \times 10^3$ |
| 3.18 | $p_{wf} = p_i - c_1 \dfrac{q_2 B\mu}{kh} \left[ \log\left( \dfrac{k}{\phi \mu c_t r_w^2} \right) - c_2 + 0.869s \right]$ | 162.6 | $2.149 \times 10^3$ |
| | $- c_1 \dfrac{q_1 B\mu}{kh} \left[ \log\left( \dfrac{t_{p1} + \Delta t'}{\Delta t'} \right) + \dfrac{q_2}{q_1} \log(\Delta t') \right]$ | 3.23 | 5.10 |

TABLE F-3 – MAJOR EQUATIONS WITH CONSTANT VALUES IN CUSTOMARY AND SI METRIC (Contd.)

| Equation Number in Text | Equation | Numerical Value of Constants in Order of Appearance ($c_1, c_2, c_3 \ldots$) Customary | SI |
|---|---|---|---|
| 3.19 | $s=1.151\left[\frac{q_1}{(q_1-q_2)}\left(\frac{p_{1\,\mathrm{hr}}-p_{wf1}}{m}\right)-\log\left(\frac{k}{\phi\mu c_t r_w^2}\right)+c_1\right]$ | 3.23 | 5.10 |
| 3.20 | $p_i=p_{wf1}+m\left[\log\left(\frac{kt_{p1}}{\phi\mu c_t r_w^2}\right)-c_1+0.869s\right]$ | 3.23 | 5.10 |
| 3.21 | $s=1.151\left[\frac{b'}{m'}-\log\left(\frac{k}{\phi\mu c_t r_w^2}\right)+c_1\right]$ | 3.23 | 5.10 |
| 4.1 | $C_s=c_1\frac{A_{wb}}{\rho}$ | 25.65 | 101.98 |
| 4.2 | $C_s=c_{wb}V_{wb}$ | – | – |
| 4.3 | $C_{sD}=c_1C_s/\phi c_t h r_w^2$ | 0.894 | 0.159 |
| 4.4 | $k=c_1\frac{qB\mu}{h}\left(\frac{p_D}{p_i-p_{wf}}\right)_{MP}$ | 141.2 | $1.866\times10^3$ |
| 4.5 | $\phi c_t=\frac{c_1 k}{\mu r_w^2}\left(\frac{t}{t_D}\right)_{MP}$ | 0.000264 | $3.557\times10^{-6}$ |
| 4.6 | $t_D=\frac{c_1 kt}{\phi\mu_i c_{gi} r_w^2}$ | 0.000264 | $3.557\times10^{-3}$ |
| 4.7 | $\psi_D=\frac{kh\,T_{sc}[\psi(p_i)-\psi(p_{wf})]}{c_1 p_{sc} q_g T}$ | 50,300 | 3.733 |
| 4.8 | $s'=s+D\lvert q_g\rvert$ | – | – |
| 4.9 | $\psi(p)=2\int_{p_\beta}^{p}\frac{p}{\mu(p)z(p)}dp$ | – | – |
| 4.10 | $p_D=\frac{kh(p_i-p_{wf})}{c_1 q_g \mu_i B_{gi}}$ | 141.2 | 1.866 |
| 4.11 | $B_{gi}=c_1\frac{z_i T}{p_i}$ | 5.04 | 0.351 |
| 4.12 | $k=c_1\frac{q_g\mu_i B_{gi}}{h}\left(\frac{p_D}{p_i-p_{wf}}\right)_{MP}$ | 141.2 | 1.866 |
| 4.13 | $\phi c_{ti}=\frac{c_1 k}{\mu_i r_w^2}\left(\frac{t}{t_D}\right)_{MP}$ | 0.000264 | $3.557\times10^{-3}$ |
| 4.14 | $p_D=\frac{kh(p_i^2-p_{wf}^2)}{c_1 q_g \mu_i z_i T}$ | 1,422 | 1.309 |
| 4.15 | $k=c_1\frac{q_g\mu_i z_i T}{h}\left(\frac{p_D}{p_i^2-p_{wf}^2}\right)_{MP}$ | 1,422 | 1.309 |
| 4.16 | $\phi c_{ti}=\frac{c_1 k}{\mu_i r_w^2}\left(\frac{t}{t_D}\right)_{MP}$ | 0.000264 | $3.557\times10^{-3}$ |

TABLE F-3 – MAJOR EQUATIONS WITH CONSTANT VALUES IN CUSTOMARY AND SI METRIC (Contd.)

| Equation Number in Text | Equation | Numerical Value of Constants in Order of Appearance ($c_1, c_2, c_3 \ldots$) Customary | SI |
|---|---|---|---|
| 4.17 | $t_{DL_f} = \frac{c_1 kt}{\phi\mu c_t L_f^2}$ | 0.000264 | $3.557\times 10^{-6}$ |
| 4.18 | $L_f = \left[\frac{c_1 k}{\phi\mu c_t}\,\frac{t_{MP}}{(t_{DL_f})_{MP}}\right]^{1/2}$ | 0.000264 | $3.557\times 10^{-6}$ |
| 5.1 | $\psi(p_{wf}) = \psi(p_i) + c_1 \frac{p_{sc}}{T_{sc}}\frac{q_g T}{kh}\left[1.151\log\left(\frac{c_2\phi\mu_i c_{ti} r_w^2}{kt}\right) - (s + D\lvert q_g\rvert)\right]$ | 50,300<br>1,688 | 3.733<br>125.3 |
| 5.2 | $\psi(p) = 2\int_{p_\beta}^{p}\frac{p}{\mu z}\,dp$ | – | – |
| 5.3 | $\psi(p_{wf}) = \psi(\bar{p}) - c_1 \frac{p_{sc}}{T_{sc}}\frac{q_g T}{kh}\left[\ln\left(\frac{r_e}{r_w}\right) - 0.75 + s + D\lvert q_g\rvert\right]$ | 50,300 | 3.733 |
| 5.4 | $\psi(p_{wf}) = \psi(\bar{p}) + c_1 \frac{p_{sc}}{T_{sc}}\frac{q_g T}{kh}\left[1.151\log\left(\frac{c_2\phi\mu_{\bar{p}} c_{t\bar{p}} r_w^2}{kt}\right) - (s + D\lvert q_g\rvert)\right]$ | 50,300<br>1,688 | 3.733<br>125.3 |
| 5.5 | $p_{wf}^2 = \bar{p}^2 + c_1 \frac{q_g \mu_{\bar{p}} z_{\bar{p}g} T}{kh}\left[\log\left(\frac{c_2\phi\mu_{\bar{p}} c_{t\bar{p}}}{kt_p}\right) - \left(\frac{s + D\lvert q_g\rvert}{1.151}\right)\right]$ | 1,637<br>1,688 | 1.508<br>11.638 |
| 5.6 | $p_{wf}^2 = \bar{p}^2 - c_1 \frac{q_g \mu_{\bar{p}} z_{\bar{p}g} T}{kh}\left[\ln\left(\frac{r_e}{r_w}\right) - 0.75 + s + D\lvert q_g\rvert\right]$ | 1,422 | 1.309 |
| 5.7 | $\bar{p}^2 - p_{wf}^2 = aq_g + bq_g^2$ | – | – |
| 5.8 | $a = c_1 \frac{\mu_{\bar{p}} z_{\bar{p}g} T}{kh}\left[\ln\left(\frac{r_e}{r_w}\right) - 0.75 + s\right]$ | 1,422 | 1.309 |
| 5.9 | $b = c_1 \frac{\mu_{\bar{p}} z_{\bar{p}g} T}{kh} D$ | 1,422 | 1.309 |
| 5.10 | $q_g = C(\bar{p}^2 - p_{wf}^2)^n$ | – | – |
| 5.11 | $\bar{p}^2 - p_{wf}^2 = a_t q_g + bq_g^2$ | – | – |
| 5.12 | $a_t = c_1 \frac{\mu_{\bar{p}} z_{\bar{p}g} T}{kh}\left[\frac{1}{2}\ln\left(\frac{kt}{c_2\phi\mu_{\bar{p}} c_{t\bar{p}} r_w^2}\right) + s\right]$ | 1,422<br>1,688 | 1.309<br>125.3 |
| 6.1 | $p_i - p_r = -c_1 \frac{qB\mu}{kh} Ei\left(-\frac{c_2\phi\mu c_t r^2}{kt}\right)$ | 70.6<br>948 | $9.33\times 10^2$<br>$7.036\times 10^4$ |
| 6.2 | $p_D = -\frac{1}{2} Ei\left(\frac{-r_D^2}{4t_D}\right)$ | – | – |
| A.1 | $\frac{\partial(\rho u_x)}{\partial x} + \frac{\partial(\rho u_y)}{\partial y} + \frac{\partial(\rho u_z)}{\partial z} = -\frac{\partial}{\partial t}(\rho\phi)$ | – | – |

TABLE F-3 – MAJOR EQUATIONS WITH CONSTANT VALUES IN CUSTOMARY AND SI METRIC (Contd.)

| Equation Number in Text | Equation | Numerical Value of Constants in Order of Appearance ($c_1, c_2, c_3 \ldots$) Customary | SI |
|---|---|---|---|
| A.2 | $\frac{1}{r}\frac{\partial}{\partial r}(r\rho u_r) = -\frac{\partial}{\partial t}(\rho\phi)$ | – | – |
| A.4 | $\frac{\partial}{\partial x}\left(\frac{k_x\rho}{\mu}\frac{\partial p}{\partial x}\right)+\frac{\partial}{\partial y}\left(\frac{k_y\rho}{\mu}\frac{\partial p}{\partial y}\right)+\frac{\partial}{\partial z}\left[\frac{k_z\rho}{\mu}\left(\frac{\partial p}{\partial z}+0.00694\rho\right)\right] = \frac{1}{c_1}\frac{\partial}{\partial t}(\rho\phi)$ | 0.000264 | $3.553\times10^{-6}$ |
| A.5 | $\frac{1}{r}\frac{\partial}{\partial r}\left(\frac{r\rho k_r}{\mu}\frac{\partial p}{\partial r}\right)=\frac{1}{c_1}\frac{\partial}{\partial t}(\rho\phi)$ | 0.000264 | $3.553\times10^{-6}$ |
| A.8 | $\frac{\partial^2 p}{\partial x^2}+\frac{\partial^2 p}{\partial y^2}+\frac{\partial^2 p}{\partial z^2}=\frac{\phi\mu c}{c_1 k}\frac{\partial p}{\partial t}$ | 0.000264 | $3.553\times10^{-6}$ |
| A.9 | $\frac{1}{r}\frac{\partial}{\partial r}\left(r\frac{\partial p}{\partial r}\right)=\frac{\phi\mu c}{c_1 k}\frac{\partial p}{\partial t}$ | 0.000264 | $3.553\times10^{-6}$ |
| A.13 | $\frac{1}{r}\frac{\partial}{\partial r}\left(r\frac{\partial \psi}{\partial r}\right)=\frac{\phi\mu c_g}{c_1 k}\frac{\partial \psi}{\partial t}$ | 0.000264 | $3.553\times10^{-6}$ |
| A.14 | $\frac{1}{r}\frac{\partial}{\partial r}\left(r\frac{\partial p}{\partial r}\right)=\frac{\phi c_t}{c_1 \lambda_t}\frac{\partial p}{\partial t}$ | 0.000264 | $3.553\times10^{-6}$ |
| A.15 | $\lambda_t=\frac{k_o}{\mu_o}+\frac{k_g}{\mu_g}+\frac{k_w}{\mu_w}$ | – | – |
| A.16 | $c_t=S_o c_o+S_w c_w+S_g c_g+c_f$ | – | – |
| A.17 | $c_o=-\frac{1}{B_o}\frac{dB_o}{dp}+\frac{B_g}{B_o}\frac{dR_s}{dp}$ | – | – |
| A.18 | $c_w=-\frac{1}{B_w}\frac{dB_w}{dp}+\frac{B_g}{B_w}\frac{dR_{sw}}{dp}$ | – | – |
| D.5 | $B_g=c_1\frac{Tz}{p}$ | 0.00504 | 0.351 |
| E.1 | $\frac{c_1 kh(p_i-p_{wf})}{qB\mu}=\frac{1}{2}\left(\ln t_D+c_2\right)+s$ | 0.00708<br>0.809 | $5.356\times10^{-4}$<br>$-1.062$ |
| E.2 | $\frac{c_1 kh(p_i-p_{wf})}{qB\mu}=\left(\frac{2t_D}{r_{De}^2}+\ln r_{De}-0.75\right)+s = p_D+s,\ t_D>0.25\,r_{De}^2$ | 0.00708 | $5.356\times10^{-4}$ |
| E.3 | $\frac{c_1 kh(p_i-p_{wf})}{qB\mu}=p_D(t_D)+s$ | 0.00708 | $5.356\times10^{-4}$ |
| E.4 | $\frac{c_1 kh[p_i-p_{wf}(t_D)]}{B\mu}=\left[\sum_{j=1}^{n}\Delta q_j p_D\left(t_D-t_{Dj}\right)\right]+q_n s$ | 0.00708 | $5.356\times10^{-4}$ |
| E.5 | $\frac{c_1 kh(p_i-p_{ws})}{qB\mu}=p_D(t_D+\Delta t_D)-p_D(\Delta t_D)$ | 0.00708 | $5.356\times10^{-4}$ |

TABLE F-3 – MAJOR EQUATIONS WITH CONSTANT VALUES IN CUSTOMARY AND SI METRIC (Contd.)

| Equation Number in Text | Equation | Numerical Value of Constants in Order of Appearance ($c_1, c_2, c_3 \ldots$) Customary | SI |
|---|---|---|---|
| E.6 | $\frac{c_1 kh(p_i - p_{ws})}{qB\mu} = \frac{1}{2}\ln\left(\frac{t+\Delta t}{\Delta t}\right) - \frac{1}{2}\left[\ln(t_D + \Delta t_D) + 0.809\right] + p_D(t_D + \Delta t_D)$ | 0.00708 | $5.356 \times 10^{-4}$ |
| E.7 | $\frac{c_1 kh(p_i - p_{ws})}{qB\mu} = \frac{1}{2}\ln\left(\frac{t+\Delta t}{\Delta t}\right) + p_D(t_D) - \frac{1}{2}\left[\ln(2.246 t_D)\right] = \frac{1}{2}\ln\left(\frac{t+\Delta t}{\Delta t}\right) + \text{constant}$ | 0.00708 | $5.356 \times 10^{-4}$ |
| E.8 | $\frac{c_1 kh(p_i - p^*)}{qB\mu} = p_D(t_D) - \frac{1}{2}\ln(2.246\, t_D)$ | 0.00708 | $5.356 \times 10^{-4}$ |
| E.9 | $c_t V_p(p_i - \bar{p}) = c_1 qBt = c_t Ah\phi(p_i - \bar{p})$ | 0.2339 | $4.167 \times 10^{-2}$ |
| E.10 | $2\pi t_{DA} = \frac{c_1 kh(p_i - \bar{p})}{qB\mu}$ | 0.00708 | $3.975 \times 10^{-2}$ |
| E.11 | $\frac{(p^* - \bar{p})}{c_1 qB\mu} = 4\pi t_{DA} + \ln(2.246\, t_D) - 2p_D(t_D) = p_{DMBH}(t_{DA})$ | 70.6 | $9.33 \times 10^2$ |
| E.12 | $p_D(t_D) = 2\pi t_{DA} + 0.5\ln(2.246\, t_D) - 0.5\, p_{DMBH}(t_{DA})$ | – | – |
| E.13 | $p_{DMBH}(t_{DA}) = 4\pi t_{DA} = \frac{(p^* - \bar{p})}{c_1 qB\mu},\ t_D < t_{DB}$ | 70.6 | 41.25 |
| E.14 | $p_{DMBH}(t_{DA}) = \ln(C_A t_{DA}) = \ln C_A + \ln t_{DA} - c_3$ | – | – |
| E.15 | $p_D(t_D = 2\pi t_{DA} + 0.05\ln\left(\frac{2.246A}{C_A r_w^2}\right)$ | – | – |
| E.16 | $p_D(t_D) = 2\pi t_{DA} + 0.5\ln(2.246\, t_{DA} A/r_w^2) - 0.5\, p_{DMBH}(t_{DA})$ | – | – |

## Answers to Examples Expressed in SI Units

### Chap. 1

**Example 1.1.** $p_1' = 17\ 740$ kPa, $p_{10}' = 20\ 464$ kPa, $p_{100}' = 20\ 684$ kPa.

**Example 1.2.** $J = 4.612 \times 10^{-3}$ m$^3$/kPa·d, $k_J = 16$ md, $s = 16$.

**Example 1.3.**

| Geometry | Infinite Solution $t_{DA}$ | Infinite Solution $t$(hours) | Pseudosteadystate (Approximate) $t_{DA}$ | Pseudosteadystate (Approximate) $t$(hours) | Pseudosteadystate (Exact) $t_{DA}$ | Pseudosteadystate (Exact) $t$(hours) |
|---|---|---|---|---|---|---|
| Circular | 0.10 | 132 | 0.06 | 79.2 | 0.1 | 132 |
| Square-centered | 0.09 | 119 | 0.05 | 66.0 | 0.1 | 132 |
| Square-quadrant | 0.025 | 33 | 0.30 | 396.0 | 0.6 | 792 |

| Geometry | $C_A$ | $J$ (m$^3$/d·kPa) | $q$ (m$^3$/d) |
|---|---|---|---|
| Circular | 31.62 | $1.213 \times 10^{-2}$ | 41.81 |
| Square-centered | 30.88 | $1.213 \times 10^{-2}$ | 41.81 |
| Square-quadrant | 4.513 | $1.116 \times 10^{-2}$ | 38.47 |

$t = 30$ hours: $p_i - p_{wf}$

$$= -933\frac{qB\mu}{kh}\left[\ln\left(\frac{1.253 \times 10^5\, \phi\mu c_t r_w^{\,2}}{kt}\right) - 2s\right].$$

$t = 200$ hours: no simple equation can be written.

$t = 400$ hours: $\bar{p} - p_{wf}$

$$= 1.866 \times 10^3\frac{qB\mu}{kh}\left[\frac{1}{2}\ln\left(\frac{10.06A}{C_A r_w^{\,2}}\right) - \frac{3}{4} + s\right].$$

**Example 1.4.** $t = 75.8$ hours.

**Example 1.5.** $\Delta p = 112.73$ kPa.

**Example 1.6.** $t_p = 176$ hours.

**Chap. 2**

**Example 2.1.** $k = 48$ md, $p_i = 13\ 445$ kPa, $s = 1.43$.

**Example 2.2.** $\Delta t = 6$ hours, $\Delta t = 50$ hours.

**Example 2.3.** $k = 7.65$ md.

**Example 2.4.** $s = 6.37$, $r_{wa} = 1.036 \times 10^{-4}$ m, $(\Delta p)_s = 2668$ kPa, $E = 0.629$, $t_{wbs} = 7.42$ hours.

**Example 2.5.** $s = 12.3$, $s_p = 13.2$, $s_d = -0.18$.

**Example 2.6.** $\bar{p} = 30\ 413$ kPa.

**Example 2.7.** $\bar{p} = 30\ 420$ kPa.

**Example 2.8.** $L = 72.8$ m.

**Example 2.9.** $A_r = 5.707 \times 10^6$ m$^2$.

**Example 2.10.** $k = 9.96$ md, $s' = 4.84$; $k = 9.77$ md, $s' = 4.27$.

**Example 2.11.** $\lambda_t = 0.0457$ md/Pa·s, $k_o = 26.2$ md, $k_w = 1.49$ md, $k_g = 0.782$ md, $s = 1.50$.

**Chap. 3**

**Example 3.1.** $k = 7.65$ md, $s = 6.37$.

**Example 3.2.** $V_p = 4.992 \times 10^5$ res m$^3$.

**Example 3.3.** $k = 7.44$ md, $s = 6.02$.

**Example 3.4.** $k = 7.65$ md, $s = 6.32$, $p^* = 30\ 385$ kPa.

**Example 3.5.** $kh = 31.7$ md·m.

**Chap. 4**

**Example 4.1.** $k = 10.3$ md, $s = 5.0$, $C_s = 2.675 \times 10^{-4}$ m$^3$/kPa.

**Example 4.2.** $k_{wb} = 4.01$ md, $k_f = 8.03$ md, $E = 0.607$.

**Example 4.3.** $L_f = 18.2$ m, $k = 4.5$ md.

**Chap. 5**

**Example 5.1.** $q_g$ = AOF = $1.47 \times 10^6$ m$^3$/d.

**Example 5.2.** $q_g$ = AOF = $2.373 \times 10^5$ m$^3$/d.

**Example 5.3.** $q_g$ = AOF = $3.115 \times 10^5$ m$^3$/d.

**Example 5.4.** Results are tabulated in Table 5.10 and are plotted in Fig. 5.13.

**Example 5.5.** $k = 9.66$ md, $s' = -0.21$.

**Chap. 6**

**Example 6.1.** $k = 1433$ md, $c_t = 3.974 \times 10^{-4}$ kPa$^{-1}$.

**Example 6.2.** $k = 817$ md, $\phi c_t = 1.973 \times 10^{-7}$ kPa$^{-1}$, $c_t = 2.465 \times 10^{-6}$ kPa$^{-1}$.

**Appendix C**

**Example C.1.**

| $t$ (hour) | $p$ (kPa) |
|---|---|
| 0.001 | 98.46 |
| 0.01 | 94.04 |
| 0.1 | 65.71 |

**Example C.2.**

| $t$ (days) | $p_{wf}$ (kPa) |
|---|---|
| 0.1 | 18 857 |
| 1.0 | 18 506 |
| 100.0 | 18 375 |

**Example C.3.** $Q_p = 0.0370$ m$^3$.

**Example C.4.** $Q_p = 3.975 \times 10^4$ m$^3$.

**Appendix D**

**Example D.1.** $T_{pc} = 644$ K, $p_{pc} = 1965$ kPa.

**Example D.2.** $p_{bwb} = 13\ 307$ kPa.

**Example D.3.** GOR = 62.34 m$^3$/m$^3$.

**Example D.4.** $B_o = 1.22$ m$^3$/m$^3$.

**Example D.5.** $c_o = 5.09 \times 10^{-7}$ kPa$^{-1}$.

**Example D.6.** $c_o = 2 \times 10^{-5}$ kPa$^{-1}$.

**Example D.7.** $\mu_o = 1.21$ mPa·s.

**Example D.8.** $R_{sw} = 3.313$ m$^3$/m$^3$.

**Example D.9.** $B_w = 1.029$ m$^3$/m$^3$.

**Example D.10.** $c_w = 5.076 \times 10^{-7}$ kPa$^{-1}$.

**Example D.11.** $c_w = 0.957 \times 10^{-6}$ kPa$^{-1}$.

**Example D.12.** $\mu_w = 0.33$ mPa·s.

**Example D.13.** $T_{pc} = 216.6$ K, $p_{pc} = 4585$ kPa.

**Example D.14.** $B_g = 6.355 \times 10^{-3}$ m$^3$/m$^3$.

**Example D.15.** $c_g = 5.656 \times 10^{-4}$ kPa$^{-1}$.

**Example D.16.** $\mu_g = 17.8$ $\mu$Pa·s.

**Example D.17.** $c_f = 5.22 \times 10^{-7}$ kPa$^{-1}$.

## Reference

1. Earlougher, R.C. Jr.: *Advances in Well Test Analysis*, Monograph Series, SPE, Dallas (1977) 5.

# Appendix G
# Answers to Selected Exercises

## Chap. 1.

**Exercise 1.1.**

| $x$ | $Ei(x)$ | $\ln(1.781x)$ |
|---|---|---|
| 0.01 | −4.038 | −4.028 |
| 0.02 | −3.355 | −3.335 |
| 0.1 | −1.823 | −1.725 |
| 1.0 | −0.219 | 0.5772 |

**Exercise 1.2.**

| $r$ (ft) | $p$ (psi) |
|---|---|
| 0.333 | 2,812 |
| 1.0 | 2,837 |
| 10 | 2,888 |
| 100 | 2,940 |
| 1,000 | 2,988 |
| 3,160 | 3,000 |

$r_e = 1{,}989$ ft.

**Exercise 1.3.** $r_e > 1{,}989$ ft.

**Exercise 1.4.** (a) $t > 9.68$ seconds; (b) $t \geqslant 1.21$ seconds; (c) $t \geqslant 18$ weeks.

**Exercise 1.5.** $r_i = 1{,}989$ ft; $r = 1{,}490$ ft; $\Delta p = 2.45$ psi.

**Exercise 1.6.** $t_s = 25.2$ days.

**Exercise 1.7.** $r_i$ unchanged, double drawdown at each $r$.

**Exercise 1.8.** $(p_i - p_{wf})_{\text{total at A}}$

$$= -70.6\frac{q_A \mu B}{kh}\left\{\ln\left[\frac{1{,}688\phi\mu c_t r_{wa}^2}{k(t-t_A)}\right] - 2S_A\right\}$$

$$-70.6\frac{q_B \mu B}{kh} Ei\left[\frac{-948\phi\mu c_t r_{AB}^2}{k(t-t_B)}\right]$$

$$-70.6\frac{q_C \mu B}{kh} Ei\left[\frac{-948\phi\mu c_t r_{AC}^2}{k(t-t_C)}\right]$$

*Assumptions*: Sufficiently far from *each* well that *Ei*-function solution can be used for each well.

**Exercise 1.9.** (a) $p_{wf} = 2{,}680$ psi; (b) $p_{\text{shut-in}} = 2{,}961$ psi.

**Exercise 1.10.** $p = 2{,}486$ psi.

**Exercise 1.11.** No influence.
Ignore production before long shut-in in calculating $t_p$.

**Exercise 1.12.** (a) $t_p = 126$ hours; $t = 192$ hours (actual total producing time); (b)

| | $p$ (psi) | |
|---|---|---|
| $r$ (ft) | Super-position | Horner Approx-imation |
| 1 | 2,239 | 2,224 |
| 10 | 2,499 | 2,484 |
| 100 | 2,760 | 2,744 |
| 1,000 | 2,972 | 2,974 |
| 2,000 | 2,996 | 2,993 |

**Exercise 1.3.**

| Shape | Infinite-Acting (hours) | PSS (hours) | $J$ (STB/D/psi) | $q$ (STB/D) |
|---|---|---|---|---|
|  | 13.2 | 7.92 | 0.5262 | 263.1 |
| 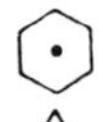 | 13.2 | 7.92 | 0.5262 | 263.1 |
|  | 11.9 | 9.24 | 0.5230 | 261.5 |
|  | 11.9 | 9.24 | 0.5225 | 261.3 |
| 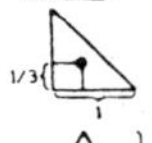 | 10.6 | 15.84 | 0.5177 | 258.9 |
|  | 1.98 | 79.2 | 0.4184 | 209.2 |
|  | 11.9 | 6.6 | 0.5256 | 262.8 |
| 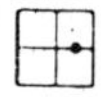 | 3.96 | 33.0 | 0.5061 | 253.0 |

## Chap. 2

**Exercise 2.1.** (a) error=0.00689; (b) no difference, $p_{ws}=1{,}150$ psi; (c) $r_e>1{,}233$ ft.

**Exercise 2.3.** $t_{wbs}\cong 7.85$ hours.

**Exercise 2.4.** MTR begins at $\approx 7.85$ hours; $k=24.5$ md.

**Exercise 2.5.** $s=0.064$; $(\Delta p)_s=32$ psi; $E=0.992$; $r_{wa}=0.469$ ft.

**Exercise 2.7.** $\bar{p}=4{,}308$ psi ($p^*$method); $\bar{p}\approx 4{,}405$ psi (modified Muskat method).

**Exercise 2.9.** $k=9.20$ md; $s+Dq_g\cong -0.952$.

**Exercise 2.10.** $k_o=32.5$ md; $k_w=3.48$ md; $k_g=1.18$ md; $\lambda_t=80.52$ md/cp; $s=-2.15$.

**Exercise 2.11.** $L=64$ ft.

**Exercise 2.12.** (a) in psi:

| | Δt (days) | | | |
|---|---|---|---|---|
| r (ft) | 0 | 0.1 | 1.0 | 10 |
| 1.0 | 2,837 | 2,948 | 2,973 | 2,992 |
| 10 | 2,862 | 2,948 | 2,973 | 2,992 |
| $10^2$ | 2,888 | 2,948 | 2,973 | 2,992 |
| $10^3$ | 2,914 | 2,949 | 2,973 | 2,992 |
| $10^4$ | 2,939 | 2,951 | 2,973 | 2,992 |
| $10^5$ | 2,965 | 2,965 | 2,976 | 2,993 |
| $10^6$ | 2,988 | 2,988 | 2,988 | 2,994 |
| $10^7$ | 3,000 | 3,000 | 3,000 | 3,000 |

(b)

| $\Delta t$ (days) | $r_i$ (ft) |
|---|---|
| 0 | 0 |
| 0.1 | 200 |
| 1.0 | 629 |
| 10 | 1,989 |

**Exercise 2.13.** $\bar{p}=4{,}418$ psi (using $t_{pss}$); $\bar{p}=4{,}411$ psi (using $t_p$).

**Exercise 2.14.** $t_{wbs}\approx 5$ hours; $k=48$ md; $r_i=371$ ft (at beginning), $r_i=813$ ft (at end); $s=10.93$, $\Delta p_s=950$ psi, $E=0.430$; $\bar{p}=4{,}325$ psi (MBH $p^*$ method); $\bar{p}=4{,}325$ psi (modified Muskat method).

## Chap. 3

**Exercise 3.1.** $k=9.55$ md; $s=4.45$; $A=67.9$ acres.

**Exercise 3.2.** $k=11.1$ md; $s=4.14$.

**Exercise 3.3.** $k=12.52$ md; $s=4.71$; $p^*\approx 4{,}380$ psi.

**Exercise 3.4.** (a)

| $t$ (hours) | Plotting Function |
|---|---|
| 0.5 | −0.301 |
| 1.5 | 0.428 |
| 2.5 | 1.331 |

(b) $p_{wf}=2{,}105$ psi.

## Chap. 4

**Exercise 4.1.** $k=9.68$ md; $s=4.59$; $E=0.636$; $t_{wbs}=7.3$ hours; $V_p=34.2\times 10^6$ cu ft; $r_e=986$ ft calculated; $r_i=226$ ft (beginning of MTR); $r_i=1{,}010$ ft (end of MTR).

**Exercise 4.2.** $s=5.0$ (for $C_{sD}\cong 10^3$); $t_{wbs}\approx 5$ hours; $k=10.3$ md; $C_s=0.0116$ RB/psi.

**Exercise 4.3.** $C_s=0.01$ RB/psi; $k_{wb}=4.01$ md; $k=8.03$ md; $E=0.606$.

**Exercise 4.4.** (1) $k=10.1$ md, $s=5.05$, $E=0.594$, $t_{wbs}=6.9$ hours; (2) $\bar{p}=2{,}854$ psia ($p^*$ method), $\bar{p}=2{,}864$ psia (modified Muskat method); (3) $V_R=3.01\times 10^7$ res bbl.

**Exercise 4.5.** $C_{SD}=10^3$; $s=5$; $k=9.92$ md; $V_R=3.01\times 10^7$ res bbl.

**Exercise 4.6.** $C_S=0.0103$; $k_{wb}=4.13$ md; $k=8.26$ md; $E=0.607$.

**Exercise 4.8.** Type-curve analysis: $L_f=200$ ft; $k=15.2$ md. Conventional analysis: $k=15.4$ md; $L_f=144$ ft. Square root analysis: $L_f=232$ ft.

## Chap. 5

**Exercise 5.1.** (a) AOF=107.0 MMscf/D; (b) AOF=100.0 MMscf/D.

**Exercise 5.2.** (a) AOF=6.6 MMscf/D (empirical method); (b) AOF=5.6 MMscf/D (theoretical method).

**Exercise 5.4.** $k=10.3$ md; $s'=0.533$.

## Chap. 6

**Exercise 6.1.** $t=851$ hours; $r_i=12{,}510$ ft; $\Delta p=68.6$ psi.

**Exercise 6.3.** $k=101$ md; $s=-2.10$; $E=1.46$; $r_i=290$ ft.

**Exercise 6.4.** $k=0.52$ md.

## Appendix A

**Exercise A.1.** $\dfrac{1}{r}\dfrac{\partial}{r}(r\rho u_r)=-\dfrac{\partial}{\partial t}(\rho\phi)$.

**Exercise A.2.** $\dfrac{1}{r}\dfrac{\partial}{\partial r}\left(\dfrac{r\rho k_r}{\mu}\dfrac{\partial p}{\partial r}\right)=\dfrac{1}{0.000264}\dfrac{\partial}{\partial t}(\rho\phi)$.

**Exercise A.3.** $\dfrac{1}{r}\dfrac{\partial}{\partial r}\left(r\dfrac{\partial p}{\partial r}\right)=\dfrac{\phi\mu c}{0.000264k}\dfrac{\partial p}{\partial t}$.

**Exercise A.4.** $\dfrac{1}{r}\dfrac{\partial}{\partial r}\left(r\dfrac{\partial\psi}{\partial r}\right)=\dfrac{\phi\mu c_g}{0.000264k}\dfrac{\partial\psi}{\partial t}$.

## Appendix B

**Exercise B.1.** Dimensional form:

$$\frac{1}{r}\frac{\partial}{\partial r}\left(r\frac{\partial \psi}{\partial r}\right)=\frac{\phi\mu c_g}{0.000264k}\frac{\partial \psi}{\partial t};$$

Dimensionless form:

$$\frac{1}{r_D}\frac{\partial}{\partial r_D}\left(r_D\frac{\partial \psi_D}{\partial r_D}\right)=\frac{\partial \psi_D}{\partial t_D}.$$

## Appendix C

**Exercise C.1.** $q_o=979$ STB/D.

$\Delta N_p=1{,}056$ STB.

**Exercise C.2.** $p_{wf}=3{,}150$ psia; $N_p=11{,}900$ STB.

**Exercise C.3.**

| $t$ (days) | Cumulative Water Influx (res bbl) |
|---|---|
| 100 | $3.278\times10^4$ |
| 200 | $5.377\times10^4$ |
| 400 | $8.959\times10^4$ |
| 800 | $14.68\times10^4$ |

**Exercise C.4.** Cumulative water influx $=2.342\times10^5$ res bbl.

## Appendix D

**Exercise D.1.** $q_{Rt}=145$ RB/D, $\lambda_t=35.0$ md/cp, $c_t=9.55\times10^{-6}$ psi$^{-1}$

**Exercise D.2.** $q_{Rt}=414$ RB/D, $\lambda_t=655$ md/cp, $c_t=1.74\times10^{-4}$ psi$^{-1}$.

# Nomenclature

$a = 1{,}422 \dfrac{\mu_{\bar{p}} z_{\bar{p}g} T}{kh} \left[ \ln\left(\dfrac{r_e}{r_w}\right) - 0.75 + s \right]$

$A$ = drainage area of well, sq ft ($m^2$)
$A_f$ = fracture area, sq ft ($m^2$)
$A_R$ = reservoir area, acres ($m^2$)
$A_{wb}$ = wellbore area, sq ft ($m^2$)

$b = 1{,}422 \dfrac{\mu_{\bar{p}} z_{\bar{p}g} TD}{kh}$

$b'$ = intercept of $(p_i - p_{wf})/q_n$ plot, psi/STB-D (kPa/$m^3$/d)
$B$ = formation volume factor, res vol/surface vol
$B_g$ = gas formation volume factor, RB/Mscf ($m^3/m^3$)
$B_{gi}$ = gas formation volume factor evaluated at $p_i$, RB/Mscf ($m^3/m^3$)
$B_o$ = oil formation volume factor, RB/STB ($m^3/m^3$)
$B_w$ = water formation volume factor, RB/STB ($m^3/m^3$)
$c$ = compressibility, $psi^{-1}$ ($kPa^{-1}$)
$c_f$ = formation compressibility, $psi^{-1}$ ($kPa^{-1}$)
$c_g$ = gas compressibility, $psi^{-1}$ ($kPa^{-1}$)
$c_{gi}$ = gas compressibility evaluated at original reservoir pressure, $psi^{-1}$ ($kPa^{-1}$)
$c_{gw}$ = compressibility of gas in wellbore, $psi^{-1}$ ($kPa^{-1}$)
$c_o$ = oil compressibility, $psi^{-1}$ ($kPa^{-1}$)
$c_{pr}$ = pseudoreduced compressibility
$c_t$ = $S_o c_o + S_w c_w + S_g c_g + c_f$ = total compressibility, $psi^{-1}$ ($kPa^{-1}$)
$c_{ti}$ = total compressibility evaluated at $p_i$, $psi^{-1}$ ($kPa^{-1}$)
$c_{t\bar{p}}$ = total compressibility evaluated at $\bar{p}$, $psi^{-1}$ ($kPa^{-1}$)
$c_w$ = water compressibility, $psi^{-1}$ ($kPa^{-1}$)
$c_{wb}$ = compressibility of liquid in wellbore, $psi^{-1}$ ($kPa^{-1}$)
$c_{wp}$ = compressibility of pure (gas-free) water, $psi^{-1}$ ($kPa^{-1}$)
$C$ = performance coefficient in gas-well deliverability equation
$C_A$ = shape constant or factor
$C_s$ = wellbore storage constant, bbl/psi ($m^3$/kPa)
$C_{sD}$ = $0.894\, C_s/\phi c_t h r_w^2$ = dimensionless wellbore storage constant
$D$ = non-Darcy flow constant, D/Mscf (d/$m^3$)
$E$ = flow efficiency, dimensionless

$Ei(-x) = -\displaystyle\int_x^\infty (e^{-u}/u)du$ = the exponential integral

$F'$ = $\Delta t_p/\Delta t_c$ = ratio of pulse length to cycle length
$g$ = acceleration of gravity, ft/$sec^2$ ($m/s^2$)
$g_c$ = gravitational units conversion factor, 32.17 (lbm/ft)/(lbf-$s^2$), dimensionless
$h$ = net formation thickness, ft (m)
$J$ = productivity index, STB/D-psi ($m^3$/d·kPa)
$J_{actual}$ = actual or observed well productivity index, STB/D-psi ($m^3$/d·kPa)
$J_{ideal}$ = productivity index with permeability unaltered to sandface, STB/D-psi ($m^3$/d·kPa)
$J_g$ = gas-well productivity index, Mcf/D-psi ($m^3$/d·kPa)
$J_1$ = Bessel function
$k$ = reservoir rock permeability, md
$k_f$ = formation permeability (McKinley method), md
$k_g$ = permeability to gas, md
$k_H$ = horizontal permeability, md
$k_J$ = reservoir rock permeability (based on PI test), md
$k_o$ = permeability to oil, md
$k_s$ = permeability of altered zone (skin effect), md
$k_V$ = vertical permeability, md
$k_w$ = permeability to water, md
$k_{wb}$ = near-well effective permeability (McKinley method), md
$L$ = distance from well to no-flow boundary, ft (m)
$L_f$ = length of one wing of vertical fracture, ft (m)
$m$ = $162.2\, qB\mu/kh$ = absolute value of slope of middle-time line, psi/cycle (kPa·cycle)
$m'$ = $162.6\, B\mu/kh$ = slope of drawdown curve with $(p_i - p_{wf})/q$ as abscissa, psi/STB/D-cycle (kPa/$m^3$/d·cycle)
$m''$ = slope of $p_{ws}^2$ or $p_{wf}^2$ plot for gas well, $psia^2$/cycle (kPa·cycle)
$m_L$ = slope of linear flow graph, psi/$hr^{1/2}$ (kPa·$h^{1/2}$)
$m_{max}$ = maximum slope on buildup curve of fractured well, psi/cycle (kPa·cycle)
$m_{true}$ = true slope on buildup curve uninfluenced by fracture, psi/cycle (kPa·cycle)
$M$ = molecular weight of gas
$n$ = inverse slope of empirical gas-well deliverability curve
$p$ = pressure, psi (kPa)
$\bar{p}$ = volumetric average or static drainage-area pressure, psi (kPa)
$p^*$ = MTR pressure trend extrapolated to infinite shut-in time, psi (kPa)
$p_D$ = $0.00708\, kh(p_i - p)/qB\mu$ = dimensionless pressure as defined for constant-rate problems

$p_{DMBH}$ = $2.303(p^* - \bar{p})/m$, dimensionless
$p_i$ = original reservoir pressure, psi (kPa)
$p_{MT}$ = pressure on extrapolated MTR, psi (kPa)
$p_o$ = arbitrary reference pressure, psia (kPa)
$p_{pc}$ = pseudocritical pressure, psia (kPa)
$p_{pr}$ = pseudoreduced pressure
$p_r$ = pressure at radius $r$, psi (kPa)
$p_{sc}$ = standard-condition pressure, psia (kPa) (frequently, 14.7 psia)
$p_{wf}$ = flowing BHP, psi (kPa)
$p_{ws}$ = shut-in BHP, psi (kPa)
$p_{1\,hr}$ = pressure at 1-hour shut-in (or flow) time on middle-time line (or its extrapolation), psi (kPa)
$q$ = flow rate, STB/D ($m^3/d$)
$q_D$ = dimensionless instantaneous flow rate at constant BHP
$q_g$ = gas flow rate, Mscf/D ($m^3/d$)
$q_{gt}$ = total gas flow rate from oil well, Mscf/D ($m^3/d$)
$Q_p$ = cumulative production at constant BHP, STB ($m^3$)

$$Q_{pD} = \frac{BQ_p}{1.119\,\phi c_t h r_w^2 (p_i - p_{wf})}$$

=dimensionless cumulative production
$R$ = universal gas constant
$R_s$ = dissolved GOR, scf gas/STB oil ($m^3/m^3$)
$R_{sw}$ = dissolved gas/water ratio, scf gas/STB water ($m^3/m^3$)
$R_{swp}$ = solubility of gas in pure (gas-free) water, scf gas/STB water ($m^3/m^3$)
$r$ = distance from center of wellbore, ft (m)
$r_{dt}$ = transient drainage radius, ft (m)
$r'_d$ = radius of drainage, ft (m)
$r_e$ = external drainage radius, ft (m)
$r_{eD}$ = $r_e/r_w$
$r_i$ = radius of investigation, ft (m)
$r_s$ = radius of altered zone (skin effect), ft (m)
$r_w$ = wellbore radius, ft (m)
$r_{wa}$ = effective wellbore radius, ft (m)
$s$ = skin factor, dimensionless
$s'$ = $s + Dq_g$ =apparent skin factor from gas-well buildup test, dimensionless
$s^*$ = $\log(k/\phi\mu c_t r_w^2) - 3.23 + 0.869s$

$$\bar{s} = \log\left(\frac{k}{\phi\mu c_t r_w^2}\right) - 3.23 + 0.869s$$

$S_g$ = gas saturation, fraction of pore volume
$S_o$ = oil saturation, fraction of pore volume
$S_w$ = water saturation, fraction of pore volume
$t$ = elapsed time, hours
$t_D$ = $0.000264\,kt/\phi\mu c_t r_w^2$ =dimensionless time
$t_{DA}$ = $0.000264\,kt/\phi\mu c_t A$ =dimensionless time based on drainage area, $A$
$t_{DL_f}$ = $0.000264\,kt/\phi\mu c_t L_f^2$ =dimensionless time based on fracture half-length
$t_{end}$ = end of MTR in drawdown test, hours
$t_{\ell t}$ = time at which late-time region begins, hours
= lag time in pulse test, hours
$t_p$ = cumulative production/most recent production rate=pseudoproducing time, hours
$t_{pss}$ = time required to achieve pseudosteady state, hours
$t_s$ = time for well to stabilize, hours
$t_{wbs}$ = wellbore storage duration, hours
$T$ = reservoir temperature, °R (°K)
$T_{pc}$ = pseudocritical temperature, °R (°K)
$T_{pr}$ = pseudoreduced temperature
$T_{sc}$ = standard condition temperature, °R (°K) (usually 520°R)
$u$ = flow rate per unit area (volumetric velocity), RB/D-sq ft ($m^3/d \cdot m^2$)
$V_p$ = reservoir pore volume, cu ft ($m^3$)
$V_R$ = reservoir volume, bbl ($m^3$)
$V_w$ = wellbore volume, bbl ($m^3$)
$x$ = distance coordinate used in linear flow analysis, ft (m)
$Y_1$ = Bessel function
$z$ = gas-law deviation factor, dimensionless
$z_i$ = gas-law deviation factor evaluated at pressure $p_i$, dimensionless
$z_{\bar{p}g}$ = gas-law deviation factor evaluated at $\bar{p}$, dimensionless
$\alpha_n$ = roots of equation $J_1(\alpha_n r_{eD})Y_1(\alpha_n) - J_1(\alpha_n)Y_1(\alpha_n r_{eD}) = 0$
$\gamma_g$ = gas gravity (air=1.0)
$\gamma_o$ = oil gravity (water=1.0)
$\Delta N_p$ = oil production during a time interval, STB ($m^3$)
$\Delta p^*$ = $p^* - p_w$, psi (kPa)
$(\Delta p)_d$ = pressure change at departure (McKinley method), psi (kPa)
$(\Delta p)_s$ = $141.2\,qB\mu(s)/kh = 0.869\,ms$ =additional pressure drop across altered zone, psi (kPa)
$\Delta p^*_{ws}$ = $p_{ws} - p_{MT}$ =difference between pressure on buildup curve and extrapolated MTR, psi (kPa)
$\Delta t$ = time elapsed since shut-in, hours
$\Delta t'$ = time elapsed since rate change in two-rate flow test, hours
$\Delta t_c$ = cycle length (flow plus shut-in) in pulse test, hours
$\Delta t_d$ = time at departure (McKinley method), hours
$\Delta t_{end}$ = time MTR ends, hours
$\Delta t_p$ = pulse-period length, hours
$\Delta t_x$ = time at which middle- and late-time straight lines intersect, hours
$\eta$ = $0.000264\,k/\phi\mu c$ =hydraulic diffusivity, sq ft/hr ($m^2/h$)

$\lambda_t$ = $(k_o/\mu_o + k_g/\mu_g + k_w/\mu_w)$
= total mobility, md/cp (md/Pa·s)

$\mu$ = viscosity, cp (Pa·s)

$\mu_g$ = gas viscosity, cp (Pa·s)

$\mu_i$ = gas viscosity evaluated at $p_i$, cp (Pa·s)

$\mu_o$ = oil viscosity, cp (Pa·s)

$\mu_{\bar{p}}$ = gas viscosity evaluated at $\bar{p}$, cp (Pa·s)

$\mu_w$ = water viscosity, cp (Pa·s)

$\rho$ = density of liquid in wellbore, lbm/cu ft (kg/m$^3$)

$\phi$ = porosity of reservoir rock, dimensionless

$$\psi(p) = 2\int_{p_o}^{p} \frac{p}{\mu z} dp$$

= gas pseudopressure, psia$^2$/cp (kPa$^2$/Pa·s)

# Bibliography


## A

Agarwal, R.G.: "A New Method To Account for Producing-Time Effects When Drawdown Type Curves Are Used To Analyze Pressure Buildup and Other Test Data," paper SPE 9289 presented at the SPE 55th Annual Technical Conference and Exhibition, Dallas, Sept. 21-24, 1980.

Agarwal, R.G., Al-Hussainy, R., and Ramey, H.J. Jr.: "An Investigation of Wellbore Storage and Skin Effect in Unsteady Liquid Flow – I. Analytical Treatment," *Soc. Pet. Eng. J.* (Sept. 1970) 279-290; *Trans.*, AIME, **249**.

Al-Hussainy, R., Ramey, H.J. Jr., and Crawford, P.B.: "The Flow of Real Gases Through Porous Media," *J. Pet. Tech.* (May 1966) 624-636; *Trans.*, AIME, **237**.

Amyx, J.W., Bass, D.M. Jr., and Whiting, R.L.: *Petroleum Reservoir Engineering: Physical Properties*, McGraw-Hill Book Co., Inc., New York City (1960).

## B

*Back Pressure Test for Natural Gas Wells*, Revised edition, Railroad Commission of Texas (1951).

Beal, Carlton: "The Viscosity of Air, Water, Natural Gas, Crude Oil and Its Associated Gases at Oil-Field Temperatures and Pressures," *Trans.*, AIME (1946) **165**, 94-115.

Brons, F. and Miller, W.C.: "A Simple Method for Correcting Spot Pressure Readings," *J. Pet. Tech.* (Aug. 1961) 803-805; *Trans.*, AIME, **222**.

Brown, George G., Katz, Donald L., Oberfell, George G., and Alden, Richard C.: *Natural Gasoline and the Volatile Hydrocarbons*, Natural Gas Assn. of America, Tulsa (1948).

## C

Carr, Norman L., Kobayashi, Riki, and Burrows, David B.: "Viscosity of Hydrocarbon Gases Under Pressure," *Trans.*, AIME (1954) **201**, 264-272.

Carslaw, H.S. and Jaeger, J.C.: *Conduction of Heat in Solids*, second edition, Oxford at the Clarendon Press (1959) 258.

Chatas, A.T.: "A Practical Treatment of Nonsteady-State Flow Problems in Reservoir Systems," *Pet. Eng.* (Aug. 1953) B-44 – B-56.

Chew, Ju-Nam and Connally, Carl A. Jr.: "A Viscosity Correlation for Gas-Saturated Crude Oils," *Trans.*, AIME (1959) **216**, 23-25.

Cobb, W.M. and Dowdle, W.L.: "A Simple Method for Determining Well Pressure in Closed Rectangular Reservoirs," *J. Pet. Tech.* (Nov. 1973) 1305-1306.

Cobb, W.M. and Smith, J.T.: "An Investigation of Pressure-Buildup Tests in Bounded Reservoirs," paper SPE 5133 presented at the SPE-AIME 49th Annual Meeting, Houston, Oct. 6-9, 1974. An abridged version appears in *J. Pet. Tech.* (Aug. 1975) 991-996; *Trans.*, AIME, **259**.

Craft, B.C. and Hawkins, M.F. Jr.: *Applied Petroleum Reservoir Engineering*, Prentice-Hall Book Co., Inc., Englewood Cliffs, NJ (1959).

Cullender, M.H.: "The Isochronal Performance Method of Determining the Flow Characteristics of Gas Wells," *Trans.*, AIME (1955) **204**, 137-142.

## D

Dake, L.P.: *Fundamentals of Reservoir Engineering*, Elsevier Scientific Publishing Co., Amsterdam (1978).

Dodson, C.R. and Standing, M.B.: "Pressure-Volume-Temperature and Solubility Relations for Natural-Gas-Water Mixtures," *Drill. and Prod. Prac.*, API (1944) 173-179.

## E

Earlougher, R.C. Jr.: *Advances in Well Test Analysis*, Monograph Series, SPE, Dallas (1977) **5**.

Edwards, A.G. and Winn, R.H.: "A Summary of Modern Tools and Techniques Used in Drillstem Testing," Pub. T-4069, Halliburton Co., Duncan, OK (Sept. 1973).

Edwardson, M.J. *et al.*: "Calculation of Formation Temperature Disturbances Caused by Mud Circulation," *J. Pet. Tech.* (April 1962) 416-426; *Trans.*, AIME, **225**.

## G

Gladfelter, R.E., Tracy, G.W., and Wilsey, L.E.: "Selecting Wells Which Will Respond to Production-Stimulation Treatment," *Drill. and Prod. Prac.*, API, Dallas (1955) 117-129.

Gray, K.E.: "Approximating Well-to-Fault Distance From Pressure Buildup Tests," *J. Pet. Tech.* (July 1965) 761-767.

Gringarten, A.C., Ramey, H.J. Jr., and Raghavan, R.: "Pressure Analysis for Fractured Wells," paper SPE 4051 presented at the SPE-AIME 47th Annual Meeting, San Antonio, Oct. 8-11, 1972.

Gringarten, A.C., Ramey, H.J. Jr., and Raghavan, R.: "Unsteady-State Pressure Distributions Created by a Well With a Single Infinite-Conductivity Vertical Fracture," *Soc. Pet. Eng. J.* (Aug. 1974) 347-360; *Trans.*, AIME, **257**.

## H

Hall, Howard N.: "Compressibility of Reservoir Rocks," *Trans.*, AIME (1953) **198**, 309-311.

Hawkins, M.F. Jr.: "A Note on the Skin Effect," *Trans.*, AIME (1956) **207**, 356-357.

Holditch, S.A. and Morse, R.A.: "The Effects of Non-Darcy Flow on the Behavior of Hydraulically Fractured Gas Wells," *J. Pet. Tech.* (Oct. 1976) 1169-1178.

Horner, D.R.: "Pressure Buildup in Wells," *Proc.*, Third World Pet. Cong., The Hague (1951) Sec. II, 503-523; also *Pressure Analysis Methods*, Reprint Series, SPE, Dallas (1967) **9**, 25-43.

## J

Jargon, J.R.: "Effect of Wellbore Storage and Wellbore Damage at the Active Well on Interference Test Analysis," *J. Pet. Tech.* (Aug. 1976) 851-858.

Johnson, C.R., Greenkorn, R.A., and Woods, E.G.: "Pulse-Testing: A New Method for Describing Reservoir Flow Properties Between Wells," *J. Pet. Tech.* (Dec. 1966) 1599-1604; *Trans.*, AIME, **237**.

## K

Kamal, M. and Brigham, W.E.: "Pulse-Testing Response for Unequal Pulse and Shut-In Periods," *Soc. Pet. Eng. J.* (Oct. 1975) 399-410; *Trans.*, AIME, **259**.

Katz, D.L. *et al.*: *Handbook of Natural Gas Engineering*, McGraw-Hill Book Co. Inc., New York (1959) 411.

## L

Larson, V.C.: "Understanding the Muskat Method of Analyzing Pressure Buildup Curves," *J. Cdn. Pet. Tech.* (Fall 1963) **2**, 136-141.

## M

Martin, J.C.: "Simplified Equations of Flow in Gas Drive Reservoirs and the Theoretical Foundation of Multiphase Pressure Buildup Analysis," *Trans.*, AIME (1959) **216**, 309-311.

Matthews, C.S., Brons, F., and Hazebroek, P.: "A Method for Determination of Average Pressure in a Bounded Reservoir," *Trans.*, AIME (1954) **201**, 182-191.

Matthews, C.S. and Russell, D.G.: *Pressure Buildup and Flow Tests in Wells*, Monograph Series, SPE, Dallas (1967) **1**.

McCain, W.D. Jr.: *The Properties of Petroleum Fluids*, Petroleum Publishing Co., Tulsa (1973).

McKinley, R.M.: "Estimating Flow Efficiency From Afterflow-Distorted Pressure Buildup Data," *J. Pet. Tech.* (June 1974) 696-697.

McKinley, R.M.: "Wellbore Transmissibility From Afterflow-Dominated Pressure Buildup Data," *J. Pet. Tech.* (July 1971) 863-872; *Trans.*, AIME, **251**.

Miller, C.C., Dyes, A.B., and Hutchinson, C.A. Jr.: "Estimation of Permeability and Reservoir Pressure From Bottom-Hole Pressure Build-Up Characteristics," *Trans.*, AIME (1950) **189**, 91-104.

## O

Odeh, A.S.: "Pseudosteady-State Flow Equation and Productivity Index for a Well With Noncircular Drainage Area," *J. Pet. Tech.* (Nov. 1978) 1630-1632.

Odeh, A.S. and Jones, L.G.: "Pressure Drawdown Analysis, Variable-Rate Case," *J. Pet. Tech.* (Aug. 1965) 960-964; *Trans.*, AIME, **234**.

## P

Perrine, R.L.: "Analysis of Pressure Buildup Curves," *Drill. and Prod. Prac.*, API, Dallas (1956) 482-509.

Pinson, A.E. Jr.: "Concerning the Value of Producing Time Used in Average Pressure Determinations From Pressure Buildup Analysis," *J. Pet. Tech.* (Nov. 1972) 1369-1370.

## R

Ramey, H.J. Jr.: "Non-Darcy Flow and Wellbore Storage Effects on Pressure Buildup and Drawdown of Gas Wells," *J. Pet. Tech.* (Feb. 1965) 223-233; *Trans.*, AIME, **234**.

Ramey, H.J. Jr.: "Practical Use of Modern Well Test Analysis," paper SPE 5878 presented at the SPE-AIME 51st Annual Technical Conference and Exhibition, New Orleans, Oct. 3-6, 1976.

Ramey, H.J. Jr.: "Rapid Methods for Estimating Reservoir Compressibilities," *J. Pet. Tech.* (April 1964) 447-454; *Trans.*, AIME, **231**.

Ramey, H.J. Jr.: "Short-Time Well Test Data Interpretation in the Presence of Skin Effect and Wellbore Storage," *J. Pet. Tech.* (Jan. 1970) 97-104; *Trans.*, AIME, **249**.

Ramey, H.J. Jr. and Cobb, W.M.: "A General Pressure Buildup Theory for a Well in a Closed Circular Drainage Area," *J. Pet. Tech.* (Dec. 1971) 1493-1505; *Trans.*, AIME, **251**.

"Review of Basic Formation Evaluation," Form J-328, Johnston-Schlumberger, Houston (1976).

Russell, D.G.: "Determination of Formation Characteristics From Two-Rate Flow Tests," *J. Pet. Tech.* (Dec. 1963) 1347-1355; *Trans.*, AIME, **228**.

Russell, D.G. and Truitt, N.E.: "Transient Pressure Behavior in Vertically Fractured Reservoirs," *J. Pet. Tech.* (Oct. 1964) 1159-1170; *Trans.*, AIME, **231**.

## S

Saidikowski, R.M.: "Numerical Simulations of the Combined Effects of Wellbore Damage and Partial Penetration," paper SPE 8204 presented at the SPE-AIME 54th Annual Technical Conference and Exhibition, Las Vegas, Sept. 23-26, 1979.

Schultz, A.L., Bell, W.T., and Urbanosky, H.J.: "Advancements in Uncased-Hole, Wireline Formation-Tester Techniques," *J. Pet. Tech.* (Nov. 1975) 1331-1336.

Slider, H.C.: "A Simplified Method of Pressure Buildup Analysis for a Stabilized Well," *Trans.*, AIME (1971) **271**, 1155-1160.

Slider, H.C.: *Practical Petroleum Reservoir Engineering Methods*, Petroleum Publishing Co., Tulsa (1976) 70.

Smolen, J.J. and Litsey, L.R.: "Formation Evaluation Using Wireline Formation Tester Pressure Data," *J. Pet. Tech.* (Jan. 1979) 25-32.

Standing, M.B.: *Volumetric and Phase Behavior of Oil Field Hydrocarbon Systems*, Reinhold Publishing Corp., New York (1952).

Standing, Marshall B. and Katz, Donald L.: "Density of Natural Gases," *Trans.*, AIME (1942) **146**, 140-149.

Stegemeier, G.L. and Matthews, C.S.: "A Study of Anomalous Pressure Buildup Behavior," *Trans.*, AIME (1958) **213**, 44-50.

## T

*Theory and Practice of the Testing of Gas Wells*, third edition, Pub. ECRB-75-34, Energy Resources and Conservation Board, Calgary (1975).

Trube, Albert S.: "Compressibility of Natural Gases," *Trans.*, AIME (1957) **210**, 355-357.

Trube, Albert S.: "Compressibility of Undersaturated Hydrocarbon Reservoir Fluids," *Trans.*, AIME (1957) **210**, 341-344.

## V

van Everdingen, A.F. and Hurst, W.F.: "The Application of the Laplace Transformation to Flow Problems in Reservoirs," *Trans.*, AIME (1949) **186**, 305-324.

Vela, S. and McKinley, R.M.: "How Areal Heterogeneities Affect Pulse-Test Results," *Soc. Pet. Eng. J.* (June 1970) 181-191; *Trans.*, AIME, **249**.

## W

Wattenbarger, R.A. and Ramey, H.J. Jr.: "An Investigation of Wellbore Storage and Skin Effect in Unsteady Liquid Flow – II. Finite-Difference Treatment," *Soc. Pet. Eng. J.*(Sept. 1970) 291-297; *Trans.*, AIME, **249**.

Wattenbarger, R.A. and Ramey, H.J. Jr.: "Gas Well Testing With Turbulence, Damage, and Wellbore Storage," *J. Pet. Tech.* (Aug. 1968) 877-887; *Trans.*, AIME, **243**.

Winestock, A.G. and Colpitts, G.P.: "Advances in Estimating Gas Well Deliverability," *J. Cdn. Pet. Tech.* (July-Sept. 1965) 111-119. Also, *Gas Technology*, Reprint Series, SPE, Dallas (1977) **13**, 122-130.

# Author Index

# Subject Index